The crux of the UFO problem is that humans can't always be trusted.

If you seek the core truths behind this enigma, it's best to keep that in mind.

I became a seeker in early childhood, by fate, by choice and by design. This is what I've uncovered, the narrative of my life-long treks intertwined with my discoveries along the way. Although it is written in the format of a novel, this book is non-fiction.

Every word of it is true.

--Wade Vernon

FORCE FIELDS

Alien Visitations To A Planet
Living In The Dark

Wade Vernon

FORCE FIELDS

Alien Visitations To A Planet
Living In The Dark

Copyright©2014 by Wade Vernon

Excerpt from "The Presentation of Self in Everyday Life" by Bill Zavatsky is from *Theories of Rain and Other Poems* (New York: SUN, 1975). Copyright (c) 1975 by Bill Zavatsky. Used with the author's permission.

Note: Every effort has been made to locate the copyright owner of material reproduced in this book. Omissions brought to the author's attention will be corrected in subsequent editions.

Front cover art: Ryan Fugate / Ryanfugatefineart.com
Front & Back Cover Graphics Layout: Wade Vernon
Rights to cover art ©2013 Wade Vernon
Book *InDesign* Text Formatting: Cristi Peterson
Author photo: Tom Campion/2013
Contact the author at:
ForceFields13@mail.com
or via letter at:
P.O. BOX 101
Austin, TX 78767

Library Of Congress Cataloging-in-Publication Data
Vernon, Stephen Wade
Force Fields: Alien Visitations To A Planet Living In The Dark
Copyright Registration #: TXU001903762
Registration Date: February 3, 2014
ISBN 978-0-9914335-1-3
Printed in the United States Of America
IngramSpark / Expanded Paperback Edition : July 2014

For Kimberly, whose unfailing visions steadfastly
pointed exactly where to look, and
for Ellen, who gently, patiently assisted the boy
struggling with his sneakers, itching for another
hike way out into the fields, and
for Monah, who was there for me always,
through thick, thin and thinner still. . . .

PROLOGUE

Mysteries have always drawn me in like a strong magnetic field. Puzzling over them, they have forged my mind, and who I have become. As a young boy, I had an almost punchdrunk sense of wonder in the face of the world. I could stare transfixed for hours at the night time sky in our small East Texas town, at the moon and the stars and the planets as they made their way across the heavens. The wonders I beheld up there drove me to question and probe tirelessly beneath the surface of things. What are we doing here, what is this mystery of being, of existence, and what is man's place in this vast universe? And in this realm of unanswered questions, as I grew to adolescence and early manhood, nothing captured the focus of my attention as much as UFOs, flying saucers and the very real possibility that an alien intelligence was presenting itself to us.

In my forty plus years of research, including a couple of close UFO encounters of my own, I have come to realize that we are not alone, and indeed we haven't been alone for a very long time. The veracity of this statement is as clear to me as is my amazement at the degree to which so many millions of people willfully choose to keep themselves blinded to it, and its ramifications. UFO and alien visitation deniers are doing a bang up job of guarding their place in the dark. Alas, the perennial joke is on the jokers--as in, partaking in the biggest ongoing joke of all time. Epic. Cosmic.

Some have said that some of the aliens call our beautiful world Sol-3 (since our sun is known as Sol, this makes perfect sense, but you are free to accept or reject this notion). For my fellow humans and me, it's Earth. But really, you can call us the Hypnosians, the proud residents of the planet Hypnos, a warrior race generally unhappy with peace and all its attendant boredom. Legions of us are hypnotized by the magnifying and distorting forces of our 24/7 sensory deluged, news-entertainment-and-visual information-overloaded hypno-tronic culture of smart phones, flat screen HDTV's with 100's of cable channels, and monitors palm sized to gigantic streaming millions of web pages, always at the ready for hungry force feedings. Sensory overload is now our default setting.

Bathed in the fierce glow, enthralled, it's difficult not to surrender to the relentless barrage of *watch this, eat this, look like this, need this, buy this, watch this. . . .* Our poor numb, whimpering psyches. If we weren't so dumbed down, and our neurons beaten half senseless by our relentless sensory sledgehammers, we might have risen up by now to tell our masters what to do with their condescending lies— or rather, The Big Lie—the one that was long past the half century mark at the turn of the new millennium, the lie from which all other lies emerge and are made more palatable, acceptable, believable, workable. The ongoing alien presence on Earth: arguably the biggest story of all time, but it's kept locked away by the keepers of the keys just as securely as back in the 1940's when it all began. Isn't it time the government and military-industrial machine came clean? More than three score years beyond Roswell, and we're still kept in the dark, the material deemed so frightening and damaging to National Security by the Brookings Institution and myriad other think tanks and top notch advisers that it's classified higher than the H-bomb material, Ultra compartmentalized, Need To Know basis. Need to know? How about, Deserve To Know? How about, Already Paid The Admission Fee? The folly of it all: as tax payers we're bankrolling our own cover-up. Shouldn't we demand to see what we're paying for, and share in the abundant benefits behind the carefully hidden free energy and field propulsion technology?

Roy, my favorite *Blade Runner* Nexus 6 replicant said, "I've seen things you people wouldn't believe." I believe I can say the same, particularly in view of some of the startled reactions I've gotten from a few well connected UFO community pundits when I've shared with them some of the off-the-record revelations high level insiders have imparted to me. Perhaps my book is a wake up call, from me to Sol-3. Are you ready to open your eyes and see?

I bring you the spoils of my cross country treks: investigations, hypnosis sessions, interviews with top tier sources, including famous abductees, highly respected UFO researchers, a naval admiral and a NASA aerospace engineer. Please indulge me if I jump around, different places, different times, tracking my search for answers to the big questions of this global enigma--inspired by my own UFO experiences at the heart of it all, driving my quest. I've been a running boy, and the boy has grown into a man who is tired of fleeing from the truth of his past. As for my occasionally unpredictable narrative leaps, perhaps they mirror the visitors' space-time jumps as they prowl our planet, day and night, dropping by for a rest stop, replenishing fuel, taking joy rides, collecting research material, more lab specimens (human or bovine), maybe drawing some juice out of the high tension lines of our electrical power grid–the m.o.'s. vary, depending upon where they're from, their mission, the particulars of their unique cultural mores, or lack thereof. (Footnote to the starry-eyed New Agers: possession of a stardrive technology, the nuts and bolts hardware enabling their sleight of hand zigzags, their atmospheric acrobatics, their interstellar leaps, does not insure that they're all benevolent angels visiting us from on high. As Arthur C. Clarke said so long ago, *a sufficiently advanced technology would appear indistinguishable from magic.* Point well made--and I'm sold that not all of the traveling magicians are as nice and charming as Houdini, Copperfield or Criss Angel. Gene Roddenberry was a brilliant prophet, and *Star Trek* nailed it in the 1960's: there are good guys and bad guys, from here, and from myriad planets of star systems scattered throughout our galaxy, and from other galactic island universes too, and multi-dimensional spaces and times as well, the multiverse. The cosmos is flooded with life, and alas, not everybody has good manners or wants to join the United Federation of Planets. That said, it's clear to me that they're not all demons or dark beings, either. The fundamentalists and evangelicals are simply wrong about that. A wide range of visiting life forces is represented here, with a full spectrum of intentions and agendas. More to come about all of that, and then some.)

Further entangling this complex web of wonder and deceit is the fact that not all of the flying saucers (the UFOs which are bona fide spacecraft) come from beyond our world. Some of them are ours, built (and/or reverse engineered from crash retrievals--some even say a few of them may have been gifted to us as operational learning tools by aliens), tested and flown right here at top secret facilities, under such restrictive and compartmentalized security that sometimes even standing presidents have been denied a viewing. (Hint: antigravity propulsion research was a multi-billion dollar black project, even in the 1950's. That's a lot of R & D time.)

Not everybody who reads this book will believe all of my words. That's a given. I'm prepared for some skepticism and derision. I'll be the first to admit that amid the thousands of people sharing solid, true UFO encounters, this field hosts its share of crackpots, cultists, mental cases and hoaxers just out for a buck. Maybe the most I can hope for is to inspire some people to open their eyes and see with greater clarity the reality of the big picture. The truth really is out there. Honest to God, it is. But sometimes it has to be disentangled from lies. Filtered from the sensory overloaded blur we live in.

UFO books can at times be dry, but there is humor in this story, and a cast of real life characters with their own quirks, personalities and foibles. There is also whimsicality here, just as the UFO phenomenon itself so often seems to be poking fun at us, unpredictably darting about, teasing us, taunting us, oftentimes scaring us, sometimes scaring the hell out of us, then vanishing into thin air. I refer to them as discs when their quoted sources do, and as disks, as they go by this name too, the fluctuating moniker perhaps ideally suited to things so intrinsically difficult to pin down and definitively define. Some insiders say there are scores of different alien species piloting them. The Universe seems to favor diversity. A lesson for us all?

The humor found here also reflects my perception that there's humor in the cosmos itself--the cosmos being where some of these vehicles and their occupants come from. The Universe can be a really funny place. Although our astoundingly brilliant physicist Einstein once said that God does not play dice with the Universe, I do think the Creator was having quite a blast, big bang style, when He tossed gigantic buckets of profusely iridescent plasma and proto-star matter and energy against the black expanses of space-time, rendering mega-light years wide, psychedelic pop art canvases sprawled out in nearly every direction the Hubble space telescope looks. Such spontaneous joy behind those explosive artworks. Such wonder in His astonishingly scaled art installations. The intergalactic curators of these virtually infinite cosmic art galleries must surely be laughing at how small minded we earthlings can be.

Everything in this book is the truth as best as I know it; there are no exaggerations or misrepresentations, but plenty of documentation and ruminations. This is a mystery story, and although I believe I've correctly answered a lot of the questions one poses when trying to get to the core truths this enigmatic and complex global UFO phenomenon presents to us, I do not profess to have all the answers. Some mysteries remain. My search goes on, and I keep questioning. Perhaps that's best. Mysteries, like humor, can add such vivid color to our lives. How bland and boring Sol-3 would be without them.

So. On we go. Before we saddle up and take off, the question I pose to you now is: how much truth can your reality handle?

BOOK I.
BEGINNINGS

It's clear to me now that my journeys have described a huge sign of the cross, like a giant hand blessed the roads I traveled. Tyler, Corpus Christi, Houston, New Orleans, Austin, L.A., Miami, Santa Fe and back again, with countless stops along the way. The transept of the cross pierced through Texas, deep in the heart of it, just like Texas pierced through me. North, south, east, west, Father, Son, Holy Spirit–one of those towns was bound to be the right one for me. Through Christ Our Lord, amen. Forgive me, I'm a recovering Catho-holic.

Freedom's always been high on my list, all the way back to my earliest remembered childhood days in Tyler, Texas. Breaking out of my crib when I was one and a half is one of my first clear memories. The wooden bars squeaked and groaned, straining under the back and forth motions of four determined little hands. My sister Natalie, all fired up in her cowgirl hat and boots, grinned in at me. I'm not sure who started the breakout, but I was grateful she understood my need to escape, to explore. This toddler was born to roam.

Natalie's face lit up as she rocked back and forth on the heels of her boots, and we laughed and giggled as the bar cracked and gave way with a crisp thwack, opening nearly enough space for my little body to get through. We stopped to see if our parents had heard, but Hank Williams blaring from the kitchen radio camouflaged the operation. Natalie nodded and we kept at it.

Natalie was three. Despite my tender age, I got the message when she said, "Just one more, Wade. Just one more to go." In minutes the second bar splintered, cracked and gave way, and she was helping me climb out to the free world, though early American furniture and wall to wall carpet were the extent of the adventures awaiting us out there. Triumphantly, we marched down the hall to the den brandishing the broken bars like batons, our parents momentarily speechless. Monah glanced up from her ironing and cackled, "Clay, honey, put down the paper a minute and look what the cat dragged in. You kids just slay me." We all had a good laugh. And I slept that night in a reinforced crib. But by then, the pattern was set for life.

Tyler, Texas in 1960 was a Norman Rockwell painting come to life, possessing an innocence and small town goodness found only in books, movies and memories anymore. Looking back today, it's hard to believe such a place ever existed, but it did. Lost in time, like the dusty vacuum tubes powering our crackling radios, resting on sensible Formica countertops. Funny how the advent of the integrated circuit heralded the gradual disintegration of our nuclear families.

Back then, strangers on the street would say, "Howdy" and "Good morning," and neighbors, unannounced, brought over freshly baked dishes just to be friendly, like Mrs. Korkelson and her aromatic stuffed grape leaves. Crime was practically unheard of. We left our screen doors unlatched all day, and rarely locked our doors at night. Perched on the southeast edge of town near the end of Barbara Street, wide open spaces taught me to love nature: the joy of the wind in my face and the sound of it rustling the leaves of the oaks and sycamores; the warmth of the sun on my bare skin on a breezy, blue sky day; the drunken feeling of running wild and free through miles of open fields of grass and wildflowers without any cares, sometimes with friends, sometimes alone.

In the other direction, Barbara Street joined Troup Highway. Just to the east of the intersection was a drive-in Frost Top with a giant mug of root beer on the roof that rotated all day long. It gave our part of town a year round holiday atmosphere. It looked so real to me that I swore someday I'd climb up there to swim in it, especially on blistering summer days when the street tar melted in little blotches and you had to watch how you walked, extra carefully if your feet were barefoot like mine so often were.

At night the frosty mug with its foamy head was all lit up, a wet beacon drawing carloads of thirsty summer children begging their parents for root beer floats and ice cream sundaes. Wild teenagers pulled up in convertibles blasting Elvis and the Everly Brothers, flirting and kissing and causing a commotion, playing games us kids hadn't learned yet. But I knew someday that would be me. Growing up might not be so bad after all. The Brothers wove their musical magic through those days, their mantra *dream* thrice repeated, conjuring scenes of early days and forgotten innocence.

We lived at the quiet end of Barbara--quiet so long as our rowdy gang of young hooligans wasn't in trouble. I'll never forget playing Indians on the warpath and tying up Cathy Horton to her carport pole. She was a settler and we were angry natives. We never heard a girl scream so loud as when we set fire to the toilet paper swaddled around her feet (we stomped it out pretty quickly). That was her brother Charlie's idea, like feeding their little brother Joey mud pies. Well, maybe I had a hand in both those capers, too. Wild kids, running free.

From the beginning, I tasted the heady excitement of life on the edge; just a few doors down from our sturdy red brick and white clapboard home, the Smiths had the last house on the block. Beyond it, Barbara dutifully ceded her paved street and sidewalks to the wilderness, huge open fields of grass and berries and wildflowers leading to a pine forest that seemed to go on forever. Miles and miles of pines against a wide open sky. God, those trees stood tall; they had so much life in them, like scores of powerful rockets clustered on a launch pad stretching out to the horizon, boldly aimed towards the heavens, fueled up with pine sap to the bursting point. The viscous trails of amber sap streaming down their proud trunks scented the air with the promise of vitality, excitement, adventure. The forest beckoned to me, come, little boy, and explore.

In my baggy shorts and Keds, I tore down those well worn trails day after day, drawn to the woods as though some magnetic force in the heart of the pines drew me in, pulling me deeper into their fold, and I gleefully surrendered. I loved those woods because they were full of mystery and the promise of new adventures. Sometimes I drifted alone for hours among the towering pines. Looking up, I imitated their sun streaked limbs and extended my arms, gliding around their massive trunks on the swirling winds, entranced by the wispy clouds high overhead, spirited away on invisible air currents. The wind whispered to me that my life was without boundaries, full of unlimited possibilities. With all that space around me, it's no wonder freedom became such a cherished thing. The endless fields of pines and the wide open skies were mine, and I was theirs. They made me a wild animal. They made me feel immortal.

I was always driven by a sense of insatiable curiosity. That eventually got me into trouble out in those fields. I strayed a little too far that day–the right place, the wrong time, the classic ambush scenario. One door slammed shut forever with a loud bang, while countless others flew open, ripped free from their hinges, carried off by the mighty whirlwinds. Nature at her wildest, the supernatural, my one true love. My only love. Always and forever. Soul mates to the end.

Until I uncovered and released the blocked memories of what happened out there (thirty years after the fact), I always reminisced so many happy times in Tyler. The biggest challenge was knowing what game to play, and what flavor of popsicle to get when the ice cream truck beckoned us kids with its sing song melodies. Talk about amnesia. I thought that only happened in grade-B movies or on soap operas. I was wrong about that. Sometimes it happens in real life, too.

Some terrifying nightmares kept coming back for awhile, recurring without warning. Sometimes I did look back as a man and recall them; their foggy details would momentarily surface, then silently retreat back to the murky depths. Sleepwalking episodes too, and waking up screaming loudly, with Mom rushing in to comfort me. And I had deep seated fears about being trapped or held hostage. They popped up when I was four and watched *The Wizard Of Oz* on TV for the first time. I bolted from the den when the witch's flying monkeys snatched up Dorothy and locked her in the tower. There was something uncomfortably familiar about that scene. I couldn't bear to see someone kept against their will. I shouted out to Natalie from the living room to tell me when Dorothy escaped; dutifully, she called out when the coast was clear, so I could safely return and watch it to the end.

With a lot of protesting from Natalie and me–Aaron was just an infant then–we moved away from Tyler when I was six. It was hard to say goodbye to Charlie and Cathy and the other kids. It was just as tough to say goodbye to my fields. As young as I was, I still knew I'd never find anything like them again. From then on I was on the run, following Dad wherever his company sent us. Do it enough and it gets in your blood. It becomes a way of life. My way is the highway.

Once I was on my own, I ran to escape the boredom of stillness, of staying put too long. Escape from bondage, real or imagined, has been my way for years. It's taken me from one city and relationship and bed to another, and another, and another. One of my shrinks said I'm afraid of people getting to know me too well (I've been to a few, like 50,000 mile check-ups). Maybe he was right. Intimacy–sure, it can offer pleasure from the closeness and touching. And pleasure can make you want to linger. But everything passes, and when the newness fades and the pleasure goes, pain moves right on in. And pain makes the open road look mighty inviting. I lick my index finger and hold it in the wind. The whirlwinds haven't failed me yet.

Eventually I gave both coasts a try, even though my Texas pals said I would hate them–Miami was too Latino and crazy, L.A. too mean and smoggy. I didn't pay them any mind. I knew I needed to be free and explore. My gut instincts told me that important lessons awaited me out on the edge, at land's end. Besides, friends just envied my freewheeling ways. Not everybody gets to run wild and free and follow the yellow brick road.

I always knew I was a seeker, not just running to be running, or running to convince myself I was free. I was searching for the missing puzzle pieces that could make me whole, never dreaming I would find them out in those Tyler fields. Fierce lessons were forced on me in those fields. Innocence was taken, replaced by a knowingness of things that could shatter a full grown man. After all the years, and all those thousands of miles, the pieces, memory shards of the experiences were still there, waiting for me. They all fit into the puzzle, completing it without gaps, top to bottom, edge to edge. What a picture they made, once assembled. Very clear, real, and too frightening for some to cope with. Like a friend, a pretty Linguistics Ph.D. candidate at FIU in Miami told me in 1991, after listening spellbound to my hypnotic regression tapes, "I don't question the reality, or the validity of your experiences." Her words came slowly, solemnly. "But I'm not comfortable with their implications. It would mean that there are others, with greater powers, and that sometimes they treat us horribly. Like we mean nothing to them. And I can't handle that. So I choose not to believe the picture you're showing me. It's too unsettling."

The big picture, once assembled . . . Jesus. I mean, I've always known that I was different. But not that different. I've had to learn to make peace with that. Live with it. Live.

* * * * * * *

My father Clay was a research scientist for a large multi-national corporation. A University Of Virginia baseball scholarship had opened the doors of academia to him, and it was there he discovered his innate gift for science and research. He was truly at home in a lab.

9

Mom (Monah) was a free-spirited homemaker with a wacky, Lucille Ball sense of humor who dabbled in art—mostly oil on canvas—and she had a passion for reading poetry and great English literature, with an occasional pulp bestseller thrown into the mix. She plowed through books and novels, expanding our home library as we grew up, and read to us until we could read on our own.

The summer after first grade my parents unplugged our black and white Zenith TV. Library visits ensued, igniting our minds, and Natalie and I brought home stacks of books. The Corpus Christi neighborhood kids poked fun at our wacky family—a summertime den with a dark television screen, its power plug draped across the top for months? Natalie and I caught the literary bug, a terminal dose for me. Starter volumes soon lead to Ray Bradbury's *Dandelion Wine* and *The Martian Chronicles*, opening me up to other lives, other ways, other worlds.

Clay's company transferred us to Houston in June of 1965; by then my voracious reading habits were set in stone. The UFO Wave the country was experiencing in late 1965 and early 1966 captured my interest via magazines, newspapers and books as we settled into a suburban ranch style house in the deeply wooded locale of west Houston known as the Memorial Area.

Memorial Drive lazily snakes and meanders along Buffalo Bayou all the way from downtown Houston on out to the far western edges of the city, and if you've never driven it before, words can't capture fully the majestic beauty of the pine trees that tower over all the neighborhoods. A soothing, reassuring sense pervades, as though the thicket of trees are protecting you from all harm, like the welcoming arms of Mother Earth. The seemingly endless vistas of tall pines spreading the scent of pine sap lend an enchanted forest atmosphere to the subdivisions, including ours on Pebblebrook Drive. Not the wilderness refuge I had in Tyler, but a densely wooded locale nonetheless. Life in the forest, an ideal setting for a mystery to unfold.

Today when I Google Earth my old West Houston neighborhood, sliding the magnifier toggle provides a bird's eye view of a controlled descent. From a few miles up it's mostly vast patches of green. Then, closer in, higher resolution reveals a semblance of structure and order beneath the blanket of forest. Roads, buildings and power grids appear, a random yet orderly jigsaw puzzle replicated thousands of times across the country, signifying the relentless industriousness of the human colony at work and play. Continue descending, and streets and houses emerge; Memorial Plaza reveals itself, safely tucked into the protective canopy of the woods. A peaceful, sleepy village on the edge of Space City USA, the perfect stage for Act Two of my introduction to the UFO visitation experience.

Therein lies the dramatic irony and incongruity of my time line: as an amnesiac childhood experiencer, my later childhood's burgeoning interest in flying saucers and the UFO Wave hitting the U.S. in the mid-to-late 1960's (in fact, the UFO flap spanned the whole globe then) would culminate in a close up visual encounter on Pebblebrook, but it wasn't my first. It was to be a reunion of sorts, a reintroduction of the mysterious ones into my life. But in the Mobius strip of human time, experience and memory, the details of my later sighting as an eleven year old Houstonian–which then seemed with certainty to be my first close up UFO experience–eventually led me back to my primary encounter at age three. The hidden, deeply suppressed subconscious memories I harbored from Tyler served to add more fuel to the fire of my flying saucer curiosity igniting at age nine, leading up to the Pebblebrook close encounter at age eleven.

By 1966, my reading tastes leaned towards nuclear physics and science fiction. In the sci-fi realm, I'd ventured from Bradbury's poetic musings, on to Isaac Asimov, Robert Heinlein, Theodore Sturgeon and Arthur C. Clarke, and others, like Poul Anderson and the cult sci-fi writer Philip K Dick. This dovetailed nicely with the television shows of the day: *Voyage To The Bottom Of The Sea*, *Lost In Space*, *Time Tunnel*, *The Invaders* and *Star Trek* all fed into my adolescent sci-fi fantasies, replete with visions of jetting out of the ocean in a flying sub, hurtling through space in a flying saucer, talking to smart computers, traveling back and forth through time, warping across the galaxy faster than light and firing a phaser pistol at hostile aliens.

The year 1966 was also a turning point for the direction of my readings. The media flood of UFO sightings and reports coming in all over the U.S. occurred with such relentlessness it seemed a dam had broken somewhere, my introduction to the concept of a UFO Wave. *The Reader's Digest* article, *Outer Space Ghost Story* by John G. Fuller (May 1966), was a turning point for me. Its title stood out on the *Reader's Digest* cover, and I tore through the story at the breakfast table before school one morning. It related the Sep-

tember 3, 1965 close up UFO sighting by 18 year old Norman Mus-
carello, and a police officer he'd coaxed from the local Exeter police
station to come back out to where he'd originally seen the object
around 2:00am. It was described as 80 - 90 feet wide, silent, and
with brilliant red pulsating lights about its rim, gently falling like a
leaf, hovering, wobbling and yawing. Sure enough, once back out in
the countryside near a local farm house, the huge and mysterious,
elongated round-shaped object reappeared for the two of them. It
bathed everything around it in a brilliant red light, turning a white
clapboard farm house blood red. Horses kicked at their corrals, and
dogs howled. The pervasive sense in this tale was that this was for
real; it would be hard to imagine a respected patrol officer almost
drawing his revolver in fright to shoot at a mirage in the night sky
coming down towards him. Something profoundly mysterious and
exciting was going on here. I was captivated, my hunger for more
information stoked. I checked out from the library Mr. Fuller's excellent book, *Incident At Exeter*, soon discovering that this Exeter,
N.H. sighting I'd just read about in the *Digest* was to become a UFO
pattern repeated over and over during that wave. Almost countless
reports of often similarly described unknown aerial objects from utterly credible witnesses in towns all over New Hampshire and surrounding states emerged, often given by police officers, scientists,
clergy and other respected officials, in the late Fall of 1965.[1]

Fuller's investigation yielded over 60 eyewitnesses. Most reported luminous disk-shaped objects, predominantly white or orange,
or changing colors. Many saw the red pulsating perimeter lights.
Most operated silently, while some had high-frequency hums. In
all, 73 reports mentioned objects being near or over high-power
transmission lines. Pondering his research, Fuller opined, "Con-
fidential comments made to me by Coast Guardsmen and military
men in the area support the laymen's testimony. Collusion, hoax or
mistaken identity by so many people seems improbable. The continued official reticence surrounding the subject of UFOs seems as
mysterious as the Exeter story itself."[2]

Incident At Exeter left me with more questions than answers.
And like any good scientist, my questions needed answers. Living in
a home with parents who revered reading and the perpetual enrich-
ment of the mind, we had subscriptions to newspapers and most of
the mainstream weekly news magazines, so I tracked the progression of this strange phenomenon in *The Houston Post, The Hous-*

ton Chronicle, Time, Newsweek, LOOK, U.S. News & World Report and *Reader's Digest.* All of them carried more and more UFO stories, from all over the U.S. and around the world, each article in turn leading me to other readings.

This repetitive cycle of published sightings triggered a whole series of "yes" and "on" buttons in my psyche, and my intensely inquisitive mind dutifully kicked into overdrive, imprinting a salvo of question marks onto every UFO story I could get my hands on. A research scientist was born. My science nerd buddies and I idolized *Star Trek's* Mr. Spock, the green blooded, coldly logical thinking machine from Vulcan, even sometimes going so far as to emulate his emotionless, dispassionate assessment of the facts in our conversations. Lost in the rarified air of our intellectual pursuits, we gamely accepted our outsider status in school. Skinny as a rail, I got roughed up sometimes. Ohh, to see those bullies now. . . .

True to my role, I went to work, using the scientific method before we'd even learned about it in school. (I was, after all, the boy genius who was designing experimental fusion reactors in my spare time in fifth grade, which utilized computer controlled diminishing radius magnetic fields and pulse laser beams to initiate and contain controlled fusion reactions; and who plowed through *The Neutrino: Ghost Particle Of The Atom*–penned by my sci-fi pal Isaac Asimov– as a sixth grader, loving every page of it.) The scientist's first step: make observations and gather data. Since my mom was a book-a-holic herself, she was happy to shuttle me back and forth to my favorite haunt, the Elizabeth L. Ring Library in Spring Branch.

Even after only a few months of reading and research, certain absolutes seemed starkly clear to me: 1) a significant percentage of the UFO reports came from intelligent, sober minded individuals, people with highly developed observational skills, like scientists, pilots, police officers, and military men. 2) The objects or craft being reported exhibited aerodynamic capabilities far in advance of anything we had, oftentimes performing maneuvers defying our known laws of physics, abrupt 90-degree turns at thousands of miles an hour, for example. In short, I was hooked.

By September of 1968, the beginning of my 6th grade school year, I'd already consumed a small library's worth of UFO material. Though information would emerge indicating that sightings and encounters had occurred in our country as early as the turn of the 20th century, and on up into and through the 1940's, the first standout report was private pilot Kenneth Arnold's sighting on June 24, 1947, of nine silvery, shiny pie-plate shaped objects that zipped along in formation at up to 1,200 miles per hour as viewed through the window of his plane while flying in the vicinity of Mt. Rainier. Later, he stated that the objects moved "like a saucer would if you skipped it across the water." *East Oregonian* newspaper reporter Bill Bequette paraphrased Arnold's account, and "saucer-like" became "flying saucers."[3]

Quickly on the heels of this: America's first UFO Wave; the formation of Project Sign (instigated by Lt. General Nathan F. Twining), a government project to study them, soon to dissolve in 1948; Brig. Gen George Schulgen, of the Army Air Force's air intelligence division, after completing a preliminary review of the many UFO reports--then called "flying disks" by military authorities--concluded by the end of July 1947 that the mystery objects were real craft; Schulgen requested that Twining and his command, including the intelligence and engineering divisions located at Wright-Patterson Air Force Base, conduct a more exhaustive review of the data; the now famous Twining Memo of September 23, 1947, essentially his answer to Schulgen, stated that the phenomena being reported were real and not visionary or fictitious, seemed as large as man-made aircraft, though disc shaped; further, that their high speed evasive maneuvers lent credibility to the concept that some of the objects were controlled manually, automatically or remotely; and it was deemed possible that some foreign nation had a form of propulsion, possibly nuclear, which was outside of our domestic knowledge.[4]

My first inkling that there might well be a cover up in place: the case of Captain Thomas F. Mantell of the Kentucky Air National Guard. On January 7th, 1948, Mantell was piloting his WWII era P-51 Mustang accompanied by three others on a routine training flight when he was dispatched by Godman Tower (Godman Army Air Field, KY) to pursue a UFO that had been reported by numerous locals in nearby towns as well as viewed by the tower. He radioed the tower, "The object is directly ahead of and above me now . . . it appears to be a metallic object . . . and it is of tremendous size . . . I'm still climbing . . . I'm trying to close in for a better look." Mantell steadily climbed to 22,000 feet--too high for a WWII fighter plane without oxygen. His companion planes turned back. By all accounts Mantell may have passed out from lack of oxygen at about 30,000 feet; at least his plane leveled off at that height. His doomed aircraft then began to plunge back toward earth. He crashed a few moments later on a farm near Franklin, Kentucky. Mantell's watch had stopped at 3:16p.m., and his body was still strapped in his plane. By 3:50 PM, the giant object was not visible from Godman, but reports continued as the UFO progressed southward into Tennessee.

The reports of the incident spread like wildfire. Theory and speculation spanned radio shows, television, and newspapers. *The New York Times's* story began with the headline, "Flier Dies Chasing A Flying Saucer," and another story headlined with, "Plane Exploded Over Kentucky as That and Near States Report Strange Object." Common speculation that Mantell was chasing a UFO was countered by the Air Force, which initially concluded that Mantell and his fellow pilots were chasing the planet Venus.

Mantell was a courageous WWII pilot who had been awarded a Distinguished Flying Cross as well as an Air Medal with 3 Oak Leaf Clusters for Heroism. And he died chasing the planet Venus in broad daylight (described by both Godman Tower personnel viewing through field glasses as well as multiple town's people in the vicinity, for hours, as a very large disc or round object slowly cruising by overhead). Got it.

I saw this pattern of idiotic rhetoric from our government leaders repeated over and over again, from cases I read about from the 1950's on through the late 1960's--and in fact the script still seems to be in play today, well into the second decade of the 21st century, though the motivating factors behind the rubber-stamped lies may have shifted. In early 1948, right about the time of the inception of the Air Force's newly minted Project Sign, the term UFO, for "unidentified flying object" insinuated itself into the international vocabulary. Edward J. Ruppelt, who played an important role in the Air Force investigations, claimed to be the originator of this acronym. Ruppelt and his fellow investigators operated out of the Air Materiel Command at Wright Patterson Air Force Base in Dayton, Ohio.

There was an almost nonstop UFO research project active in the U.S. Air Force from January 1948 to December 1969, but their support varied, and the names changed through the years. David M. Jacobs's book *The UFO Controversy in America* covers these and other developments up to 1974. It was often a half-hearted endeavor, and generally had a threadbare staff, often of flunkies. The classification level wasn't very high, generally 2A. The existence of the projects were generally in the public eye, not concealed. In December 1948 Project Sign was terminated, replaced by Project Grudge, with its objective being a "Detailed Study of Flying Discs." The Project Grudge (1949) final report came out in August, and in December 1949 the Air Force provided the news media a summary. Their primary conclusion: "There is no evidence that objects report- ed upon are . . . an advanced scientific foreign development; and therefore, they constitute no direct threat to the national security . . . all evidence and analysis indicated that UFOs were . . . the misinterpretation of . . . conventional objects [or] a mild form of mass hysteria and war nerves" or hoaxes crafted by publicity seekers and "psychopathological persons."[5] Mass hysteria, war nerves, and pathological crazies: See: CIA: Disinformation.

Captain Edward J. Ruppelt, appointed to head Grudge, in turn appointed the astronomer Professor J. Allen Hynek to be his chief scientific consultant. In March 1952, the Air Force once again changed the code name, this time to Project Blue Book, giving it the formal title Aerial Phenomena Group. The Air Force also initiated a new reporting protocol whereby Blue Book immediately received information of new sightings. According to Jacobs, "By June 1952 Project Blue Book was a dynamic, ongoing organization."[6] (Note: It is referred to as Project Blue Book as well as Project Bluebook, depending upon the source).

Compelling photographic evidence emerged during this era. On August 25th, 1951, a case now known as The Lubbock Lights oc- curred. While relaxing in a backyard in Lubbock, Texas, three col- lege professors witnessed a formation of about twenty to thirty soft- ly glowing bluish-green lights silently passing overhead, followed by a similar grouping a few hours later. They saw at least ten more over the next two months, in addition to over 100 other people in and around Lubbock. On the night of August 30th, Carl Hart took five photographs of the strange objects from his backyard. Captain Ruppelt was sent to investigate. In part, he concluded that, "In each photograph, the individual lights in the formation shifted position according to a definite pattern." The objects never made a sound, and their flight paths were almost always north to south. Photos can be found on the web, at **ufoevidence.org**.[7,8]

True to form, the increasingly aggressive and dogmatic UFO debunker Dr. Donald Menzel of Harvard sought to discredit the sightings, both in *Time* and *LOOK* magazines and in his book *Fly-*

ing Saucers (1953) ; he claimed the objects were merely refracted city lights. The Lubbock Lights were eventually ascribed to light reflecting from a flock of birds (plover) by ATIC. Any reasonably intelligent person casually perusing the photographs would find this explanation ludicrous. Plovers in flight morphed into a procession of large bluish-green glowing orbs flying in a tight formation?

In Richard M. Dolan's incomparable "UFO Bible-Volume I," *UFOs and the National Security State, Chronology of a Cover-up 1941-1973* (2002), he reveals Ruppelt stated that access to the information about the Lubbock Lights was restricted to two groups: (1) ATIC's UFO investigative team and (2) another, unnamed group of people who, "due to their associations with the government, had complete access to our files." Dolan quotes Ruppelt, "These people were scientists–rocket experts, nuclear physicists, and intelligence experts. They had banded together to study our UFO reports because they were convinced that some of the UFOs that were being reported were interplanetary spaceships, and the Lubbock series was one of these reports."[8] (This "unnamed group" eventually gets a name.)

Between 1952 and 1956, two civilian UFO groups came into existence: APRO (Aerial Phenomena Research Organization), led by James and Coral Lorenzen, and NICAP (National Investigations Committee On Aerial Phenomena), led during its formative years by Donald E. Keyhoe, a retired Marine Corps major. These organizations, especially NICAP, lobbied Congress to hold hearings on the UFO question.

I discovered that Maj. Donald E. Keyhoe had written a few UFO books, and bought an old copy of *Flying Saucers From Outer Space* (1953). Upon first reading it in 1966, I found the book fascinating and enlightening, seeing as how this was evidently a sober minded ex-military man coming forward and flat out telling the public that the Air Force and numerous government agencies were lying about UFOs, and keeping some heavy duty secrets from us. Today, Keyhoe's seminal work firmly stands as a sad commentary on our compartmentalized information society: sixty years on, and the lies, derision and obfuscation of the truth on the part of our military and government (and now, high tech industries) haven't really changed. We're still kept out of the loop. Old wine, new bottle. Sad but true.

Perhaps the most dramatic and attention grabbing UFO sightings of this era, featuring extremely high speed maneuvers, high speed right angle turns and radar tracking coordinated with visual confirmations, were those encountered in July, 1952, culminating in two nights during which discs flew, singly and in formation, over Washington, D.C. As though building up to this big event, during the first two weeks of July, flying singly, in pairs, or in group formations, the strange machines were seen all over the world. From the 8th to the 12th of July, our Midwest states were flooded with reports; since most of them came from Air Force pilots, Intelligence could keep them secret. Sighting reports poured into Dayton and Washington from Colorado, Michigan, Indiana, Missouri, Florida and Virginia, many of them from either commercial or military pilots. Most of the objects were reported as disks, seen at night, often simultaneously tracked on radar, and two of them tracked at speeds of 1,700 miles per hour and 3,000 miles per hour. One of the sightings, on a Saturday night over Indianapolis, was witnessed by thousands of people.[9]

Just past midnight on July 20th, 1952, eight air traffic experts, headed by Senior Controller Harry G. Barnes, began their watch at the Air Traffic Control center at Washington National Airport. The Center utilized long-range surveillance radar, able to track planes 100 miles away. It was a clear night with light traffic.

At exactly 12:40am, seven sharp blips suddenly appeared on the radar scope. The strange objects practically materialized out of thin air. Senior controller Barnes contacted the tower Operator Howard Cocklin, who confirmed both the radar blips on his scope and a visual sighting. "I can see one of the things. It's got a bright orange light—I can't tell what's behind it," Cocklin reported. Barnes immediately contacted Air Defense Command, then resumed monitoring the main scope. By now, the mysterious objects had split up; two of them cavorted over the White House, while a third one was near the Capitol—both areas classified as restricted air space. While minding his scope, Barnes called Andrews Field, across the Potomac in Maryland. Concerned controllers there confirmed radar tracking as well.[10]

Jet interceptors were dispatched, while Barnes and his controllers tracked the objects. At one point they watched a new blip track a sharp, high speed 90-degree turn—a maneuver clearly beyond our aircraft's capabilities.

The real show stopper of the night: Tower operator Joe Zacko had been monitoring his ASR scope—one built for tracking high speeds—

when a UFO suddenly materialized on the glass; one look told him it was moving incredibly fast. He watched its blips flash across the screen as the object zoomed over Andrews Field toward Riverdale. When the trail suddenly vanished, Zacko called Cocklin. Together, they determined the UFO had been blazing across Washington skies at an unheard of two miles per second—7,200 miles per hour. It was clear from the trail that the UFO had descended vertically into the ASR beam, and then leveled off momentarily; then, climbing at an unbelievable rate, it had streaked out of the beam again.

The objects circled Washington for almost two hours, always in view of the anxious controllers. The simultaneous, solid radar returns coordinated with visual sightings likely meant only one thing: up there in the night sky, some type of superior flying machines were engaging a reconnaissance mission over Washington. From their precise, often extraordinarily rapid maneuvers, it seemed clear they were under intelligent control—perhaps even operated by highly intelligent beings onboard. If landing was their intent, our nation's capital would have been a logical point for contact. Whether they were friendly or hostile was unclear.[11]

It wasn't until almost 3:00am when the Air Force jets reached Washington. The UFOs vanished just before their arrival. Presumably they had either spotted the distant fighters or monitored their call to the Center. Then, five minutes after the jets gave up their pursuit and departed, the UFOs reappeared, once again swarming the skies over Washington, unchallenged. The air show demonstrating their indisputably superior technology and flight mastery resumed.

As dawn began to break, the mysterious objects ended their five hour aerial blitz over our capital. E. W. Chambers, a radio engineer, was leaving the WRC transmitter station around 5:30am when he witnessed five huge discs overhead circling in a loose formation. As he watched, awestruck, the discs angled upwards and climbed steeply into the early morning sky.[12]

The UFOs were gone before most people awoke. Public interest, verging on hysteria, grew quickly once the story broke. At first the Air Force did their best to minimize the significance of the sightings over Washington. Initially, officers denied that Andrews Field radar operators had tracked the objects, and one spokesman maintained that a defective Control Center radar scope was to blame. Another officer outright denied that any fighters had been dispatched to the area. But their snow job fashioned to quell public fears fizzled. Long distance calls, telegrams and thousands of letters from concerned citizens swamped the Pentagon. The public wanted answers. Beholden to their constituents, congressmen demanded action. Radio commentators and newspapers insisted on a press conference.[13]

The Air Force hoped the public's interest would dwindle once the memory of the Washington sightings faded; Major General John A. Samford, Director of Air Force Intelligence, stubbornly held out and refused to meet with the press. But whatever hopes he and the Air Force harbored for this phenomenon to dutifully retreat were dashed when another wave of UFOs invaded Washington's air space on the night of July 26, 1952, shortly after 9:08pm. This time, the objects remained too high to be seen by most of the capital's residents; but like before, anxious controllers at the Center tracked them on radar. Again, Andrews Field and Washington Airport tower men witnessed the UFOs' maneuvers, simultaneously pinpointing them at locations where airline pilots saw mysterious lights.

Air Force jets were again dispatched; this time, when the first fighters arrived, some of the objects were still lurking. Lieutenant William L. Patterson tried to chase the nearest craft, flying at a top speed of over 600 m.p.h., but it outpaced him quickly.

Air Force Intelligence officer Major Dewey Fournet, Jr., the Pentagon's top investigator, dashed to the center, accompanied by Albert M. Chop and and a radar specialist. Barnes and his men were quizzed for two hours as they observed the UFO blips, while the radar specialist made sure the scope was always functioning properly. When Fournet and the others emerged, they declined to comment to newsmen, but confirmed Patterson's failed jet intercept.

Accounts of the second Washington UFO blitz broke in newspapers all over the country. Within two days newspaper editors from coast to coast were firing off their opinions and demands for an explanation from the Air Force.[14] See UFOs buzzing the Capitol at:

ufoevidence.org/photographs/section/1950s/Photo394.htm

General Samford finally caved under the fierce barrage of public pressure and editorials, and scheduled a press conference for July 29th (Keyhoe's sources indicated it was under direct orders from higher up, namely, Lieutenant General Nathan Twining--soon to become Air Force chief of staff). The confrontational Washington UFOs of July forced officials into a corner, resulting in an historical precedent and turning point: Samford and his cohorts resorted to lies and debunking, a template for all subsequent Air Force and scientific community responses for decades. The truth would not be told. The die was cast. Take it or leave it, America.

   Dr. Donald Menzel, the Harvard astronomer--and ultimately one of the premier UFO debunkers--proffered a blanket explanation that was to become one of the standby, when-all-else-fails, thinly veiled lies used to explain away particularly intrusive and persistent sightings that revealed themselves both visually and with strong returns and active blips on radar: *temperature inversion.* General Samford must have felt a sense of shame when he played out the silly script for the reporters who had gathered, seeking the truth for their anxiety plagued readers. The question and answer session lasted over eighty minutes, during which time Samford and his comrades adroitly danced around the issue of the multiple solid radar returns coordinated with both ground visual and pilot visual confirmations. The temperature inversion theory (briefly, that atmospheric hot air layers sandwiched with cold air layers can reflect and refract both ground and aerial light sources, creating somewhat of a light show, like a desert mirage) was used throughout, even when skeptical reporters scoffed at using atmospheric aberrations to explain these sightings. Ultimately, Samford stooped so low as to suggest that some of the solid radar returns might have come from birds, dumbfounding some of the well informed reporters with such a notion. (Donald Keyhoe learned later that radar experts had flatly denied the idea that any of the returns on those nights were from birds.)[15]

   Mike Wallace interviewed Donald Keyhoe years later, the black and white broadcast originally televised on March 8, 1958; video clips online reveal an occasionally testy Wallace grilling Keyhoe, sometimes almost mercilessly raking him over the coals. Particularly telling is when Wallace seems to say, how dare you question the final authority of Dr. Donald Menzel (apparently one of the crowned kings of the scientific community then). Menzel's gradually emerging secret persona in this UFO rite of passage makes these clips significant, and not to be missed. See the first of them at:

www.youtube.com/watch?v=PxkdCciZMhk

   Sometimes when I took a breather from devouring UFO books, I'd work in what I called "my lab," a converted corner of the laundry room that my parents indulged me with: bracketed wall shelves for my books, chemistry set and cloud chamber snagged via the *Edmunds Scientific Catalog*, and a long, narrow table under them used for chemistry experiments and studies under a swing arm lamp. My mind was forever tinkering, either in books or by building and then taking things apart. I built scores of models, like a lot of boys then: from airplanes to submarines, and cars to movie monsters. Then came launching self crafted model rockets from the *Estes* rocket catalog. I definitely kept the local hobby store staff busy.

One day when I was disassembling the simple rotor of a battery powered car model--a choice red Ferrari--I toyed with the engine rotor, the core of whirling electromagnets. As I spun the rotor faster and faster, something told me that this was one of the keys to a flying saucer's propulsion system: a whirling electromagnetic field, amped up to a higher and higher voltage and rpm, amped up to the point that the gravitational field distorts, just like it does around our sun, even to the point that the field generated yanks the craft out of our visible space and catapults it somewhere else. Wink of an eye.

Although I rarely spoke about my flying saucer readings with him, my father Clay did share my interest in UFOs (though not with the same 24/7 obsession, he had a full time job as a scientist). He had read John G. Fuller's excerpt from *Incident At Exeter* in the *Reader's Digest* right after I had. He was leaning towards a guarded scientific, somewhat skeptical viewpoint, while the more I read and researched, the more my gut sense said that some of these things were highly advanced airborne vehicles that we hadn't created. Too many solid reports from unimpeachable sources like pilots, military men and police officers sold me on that idea. It was that simple.

One Saturday night in 1966, the two of us watched the late movie, the local *Houston Chronicle* TV Section having simply billed it by the title, *UFO*. Made in 1956, it turned out to be a unique film, a part-documentary, part-drama which examined UFO sightings in the U.S., focusing on the era of 1947 to the early 1950's. It opens with a somewhat melodramatic, big orchestra overture, with the title and subtitle emblazoned on the black and white screen, *UFO: Unidentified Flying Objects / The True Story Of Flying Saucers*. The film covered the high points of a lot of what I'd been reading, including the Arnold sighting, the Mantell tragedy, and in a fascinating segment, a dramatic reenactment of the 2 big UFO nights in July of 1952 over Washington, D.C. Punching up the effect of the narrative is footage showing the April 7, 1952 issue of *LIFE* magazine with President Harry Truman on the cover; over his head in the upper right corner, the bold all-caps banner says: THERE IS A CASE FOR INTERPLANETARY SAUCERS. Readers thumbing through the pages come upon the article with its title boldly querying, "Have We Visitors From Space?," with a highlighted quote on one page stating, "The Air Force is now ready to concede that many saucer and fireball sightings still defy explanation."

Even more eye-catching are the two UFO film clips presented in full at the end of the movie. One, lensed by Nick Mariana of Great Falls, Montana on August 15, 1950, shows what appears to be two high altitude glowing disk-shaped objects, streaking across the sky and remaining in view for about ten seconds. The extensive government film analysis concluded that the objects moved almost directly against the wind, and weren't balloons; a magnified, frame by frame analysis revealed a steady, horizontal motion, and that they weren't freefalling. Finally, they weren't meteors, birds, nor any type of known aircraft. Multiple viewings sold me on all of that.

The second film clip was shot by Delbert Newhouse, an active Navy officer and USN Chief Photographer with 21 years experience, on July 2, 1952. Newhouse, his wife and two kids were vacationing and driving cross country from Washington, D.C. to Portland, Oregon. About 7 miles past Tremonton, Utah, his wife noticed a strange group of objects in the sky. They pulled over to have a better look, and Newhouse quickly went through his luggage to get his 16mm Bell & Howell movie camera and load the film. He described them as about 12 objects in rough formation, appearing like 2 saucers, one inverted over another, and he estimated them to be about the size of a B-29 at 10,000 feet. Newhouse stated they were unlike anything he'd seen before, having at that time logged over 2,000 flight hours. Both clips are currently available on YouTube, easily accessed via Googling the UFO observer's name and sighting date, or location, like "Tremonton UFO film;" additionally, the *UFO* docudrama film is currently in Netflix's online streaming library.

I've viewed these film clips numerous times. I'm always struck by the magnified-image clips of the Tremonton film, especially when I freeze frame them. I see about a dozen shiny disks engaged in a slowly dancing flight pattern that at times resembles a DNA double helix inscribed onto the sky. Their choreographed aerial dance tells me that their saucers are in conference mode, twelve captains in a conference call, plotting surveillance missions, strategies and flight sorties. A little over a fortnight later came the first of the D.C. saucer invasions. An almost gleeful skydance conference call, in advance of the Washington air show. And yet, looking as organized as an industrious ant colony in motion. Like a hive mind at work.

This fascinating *UFO* movie (in many ways offering even more disclosure than we're treated to today), closes with a narrator's voiceover as his words appear on the screen, testifying that everything was authentic, and backed up by documentation and eye-witness accounts supported by affidavits and official government reports. Further, that the evidence was presented "with integrity and objectivity to establish the fact that Unidentified Flying Objects, commonly known as "flying saucers," do exist." The real kicker for me was the final words, "Some kind of flying objects have been photographed in the sky. If they cannot be identified as objects known to man . . . What are they? If they are not man-made . . . Who made them? If they are not from this planet . . . Where are they from?"[16]

This film was ten years old when I viewed it that night. At the time, it seemed to me like the Air Force and government had been toying with the notion of full disclosure back then. It brought to mind a statement that Donald Keyhoe had made in *Flying Saucers From Outer Space*, that during the year following the July 1952 "UFO invasion" of Washington, he'd had a lot of behind the scenes access at the Pentagon, and witnessed Air Force Intelligence struggling with the issue, What shall the public be told about the flying saucers? He'd seen more than one skeptic, after closed door sessions with Air Force Intelligence men, emerge from the private meeting badly shaken by what had been revealed to them.[17]

And yet, only a few months before the night Clay and I saw this film, a well documented UFO flap in Michigan occurred, and the Air Force was still playing the silly debunk and disinformation game.

In March of 1966, a series of sightings in Michigan made headlines across the country. Both *Time* and *Newsweek* carried the story, in their April 1st and April 4th issues, respectively. I was completely fascinated that both magazines featured an identical sketch of one of the primary objects reported, a metallic disc the shape of a somewhat flattened football with a round cupola on top, lights on the perimeter and a surface cross-hatched like a quilt. Even more striking: the same sketch made the local Houston newspapers carrying the reports. It amazed me that it was so out in the open.

Newsweek opened with, "For more than a week . . . in the countryside around Ann Arbor, Mich., the reports piled in." The article elaborated that in the village of Dexter, a shimmering, strangely lighted vehicle began making night time appearances, followed by similar reports in nearby locales witnessed by sheriff's deputies in three counties, officers who said they might not have believed it if they hadn't seen it with their own eyes. They concurred that the aerial objects moved at incredible speeds, made very sharp turns, dove down at them and hovered, and were extremely maneuverable.[18,19]

Frank Mannor, 47, a truck driver who lived with his family in a farmhouse in the countryside outside of Dexter, Michigan, witnessed an object with pulsating lights seemingly land in, and then rise up above and hover over a field near his house around 7:30pm. He and his wife and kids watched it for fifteen minutes from about a half mile away, and then he and his son Ronnie hiked out to about 500 yards from it. Mannor said it had "a blue-green light on the right hand side and on the left a white light . . . The body was like a yellowish coral rock . . . you couldn't see it too good, because it was surrounded with heat waves, like you see on the desert. The white light turned to a blood red as we got close to it. . . ."[20]

Mrs. Mannor phoned the Dexter police. When Patrolman Robert Hunawill arrived at the scene, he was astounded to see a strange, lighted object hover over his police cruiser, then shoot up to join three other "objects" crossing the swamp. Hunawill wasn't the only witness to the Mannor's sighting; it was shared by a total of fifty-two independent witnesses, including a dozen police officers.[21]

Calls to the police, the Air Force and the local university came pouring in. The night after the Mannor incident, on March 20, 1966, eighty-seven Hillsdale college coeds and some faculty members, together with local police and the county civil defense director, William Van Horn, witnessed another strange display of lights near the college campus;[22] they described a glowing, football shaped object hovering over a swampy area a few hundred yards from the women's dormitory. The witnesses watched the object for four hours.[23]

The Air Force dispatched an investigative team, led by Dr. J. Allen Hynek, their long standing scientific consultant for Project Blue Book. As the *Time* article stated, "Through its Project Blue Book, the Air Force had looked into 10,147 other Unidentified Flying Objects since flying saucers entered American mythology in 1947. Because of inadequate sighting data, 646 elude technical explanation."[24]

Pressured by the Air Force to deal with the Michigan UFO situation, Hynek held a press conference on March 25th at the Detroit Press Club. In front of a packed house, he proposed "swamp gas" as a possible explanation for the Dexter sightings. In a five-page statement, Hynek said he felt certain that the two strongest sightings—those at the Mannor farm and at Hillsdale College—were caused by marsh gas. "Rotting vegetation produces the marsh gas, which can be trapped by the ice and winter conditions. When a spring thaw occurs, the gas may be released in some quantity. The flame is a form of chemical luminescence and its low temperature is one of its peculiar features."[25]

Hynek's final explanation for the hovering, darting, zigzagging vehicles with clearly defined shapes, witnessed by scores of people (many of them police officers and myriad other reputable observers) was *swamp gas*. He and the Air Force paid hell for it.

The press wasted no time in latching onto the swamp gas theory. Cartoonists lampooned the idiotic, luminous gas notion at Hynek's expense. Numerous newspapers proffered the half baked hypothesis to prove their allegations that the Air Force and the scientific community had adopted an official policy of ridicule and deception. Needless to say, the Air Force's credibility floundered, Hynek's along with it.[26]

In the *Time* coverage, Blue Book's stats revealed that, "646 cases elude technical explanation because of inadequate sighting data," whereas in *U.S. News & World Report*, in their April 4, 1966 article entitled "'Flying Saucers'—Illusions or Reality?," 646 sightings were "unexplained," but another 1,248 were incomplete investigations due to insufficient evidence. So therefore, we could in fact say that potentially there were 1,894 cases representing genuinely unknown phenomena, i.e., UFOs that remained unidentified. That's nearly 20% of the cases as of that article's date. But then, it seemed increasingly more clear that Project Blue Book was never really about disseminating the truth, so why fret over the twelve-hundred or eighteen-hundred some odd reports?[27, 28]

I found particularly entertaining the Blue Book pronouncement that no UFOs have ever given any indication of a threat to our national security. I guess that holds true, provided our government doesn't mind hypersonic aircraft of unknown origin buzzing and blitzing the restricted air corridors over the White House and the Capitol, unfettered. No problem! We don't really care all that much for our crooked politicians and dirty-money lobbyists anyway. Have at 'em! As for the assertion that none of the objects manifested performance capabilities beyond the range of present day scientific knowledge: do we have, even in the second decade of the 21st century, any commercial jet aircraft that can cruise handily at over 7,000 miles per hour, in excess of Mach 9, in the atmosphere? By all means, book me a flight right now to Paris, window seat please.

The August 22nd, 1966 issue of *U.S. News & World Report* featured yet another UFO article, this one sporting a spunky advertiser's jingle title, " 'UFO's' : They're Back In New Sizes, Shapes, Colors." (You could almost feel the Zeitgeist stretching out its welcoming arms to make room for the Visitors in the Collective Unconscious, sandwiched between Ultra-Brite toothpaste and Dodge Charger commercials.) It opened with the zest of a used car lot commercial voice over: "The mystery of "unidentified flying objects" is heating up again this year. UFO reports are increasing. Latest models have new features." It then referenced the latest Gallup Poll that had tallied 5 million Americans as being certain they'd seen flying saucers.

The playful advertising verbiage continued with, "Strange objects in the sky are being reported now at a near-record rate–with the 1966 models more varied and colorful than ever . . . lights of all kinds abound on the 1966 models. . . ." Factoid bullets followed, describing most UFOs as disk shaped, 2 to 100 feet in diameter. Some were football shaped, some brightly glowing, often reddish, some with large varicolored lights on top or bottom, some with smaller perimeter lights, most lights pulsating or changing colors. Most UFOs were silent, but some emanated buzzing, whirring or high pitched whines. In daylight sightings, most appeared as light colored metal, sometimes aluminum or dull chrome. More sightings were reported at night than day. More reports had been coming from the Northeastern U.S. They were still exhibiting erratic, almost playful flight characteristics, with a tendency to zigzag, bounce up and down, oscillate and wobble, then dart off so fast as to seem to dematerialize.[29]

There was such a fascinating dichotomy playing out in the media; the really intense, close up, well documented sightings like those in Exeter, New Hampshire as chronicled by John G. Fuller got dismissed as "stars and planets" combined with the distorted lights from temperature inversions, but when the phenomenon was addressed as a generality, it was spoken of as though referring to real, solid flying objects. The *U.S. News & World Report* article also revealed that the Air Force's receipt of 646 UFO sighting reports in the first seven months of 1966 reflected nearly double the number reported in the same period in 1965, and showed an annual rate that was exceeded only once before--in 1952. The article then rattled on about how most of the sightings turned out to be natural phenomena (stars, planets, balloons, etc., take your pick). But for me this was confirmation that this UFO Wave of 1966 was similar in scope to Keyhoe's wave of 1952. They weren't going away. And the Air Force propaganda of mental pablum was still the same, fourteen years later: that UFO's had given no sign of being a threat, nor provided any conclusive evidence that they were extraterrestrial vehicles. But wasn't that a given, seeing as how all the unexplained ones were probably swamp gas, temperature inversions, Venus (on the move in a really big way), or stray balloons?

* * * * * *

When the UFO docudrama film ended around midnight that Saturday night in 1966, my father and I went out to the back patio to ponder what we'd seen, entertaining the exciting possibility that we were being visited. I think he enjoyed my enthusiasm for the ETH (Extra-Terrestrial Hypothesis) thrust of the film. The starlit sky over our dark suburban lot was cloudless, and the glowing band of the Milky Way looked vivid and close enough to touch down on Earth. It harkened back to warm summer nights when I would lay on a blanket in the backyard, my father's binoculars glued to my eyes as I slowly, steadily tracked across our galaxy's shimmering, overwhelming ocean of stars and nebulae. So many varicolored stars out there, spread out like a mind-blowing patchwork of illuminated, glittering gems tossed against the blackest swath of velvet. Each one of those, a thermonuclear stellar furnace unto itself, providing heat and light to its own planets like our friend Sol, 93 million miles away, did for us day in, day out.

Always, half drunk from the awesome beauty and order of it all, I sensed the presence of a greater intelligence lurking behind the grand design, invisible threads of a boundless consciousness underlying the fabric of space-time. So many other worlds circling around those distant lights. Worlds harboring intelligent life, some of them more advanced than ours--I knew it in my gut. You're in the news nearly every day, so show yourselves to me, I would silently command to my as yet unmet visitors. With every UFO book and article I read, or movie footage I viewed, my mind would shout it out, *Prove to me you're here!*

And prove to me they did, after years of yearning, early in my fall school semester in Houston, 1968.

I loved our 6th Grade specialized classrooms at Memorial Junior High, with students migrating through multiple wings and buildings for different courses, just like college. Coach Wilson, my Science and Social Studies teacher, quickly latched onto me for class discussions when he discovered my passion for physics, astronomy and cosmology. When it came time for research projects, I grabbed at the chance to do mine on the Milky Way Galaxy.

Normally I was shy and nervous giving oral reports in front of a class. But science was my trusty armor, so once I got to the chalk board and launched my discussion on our galaxy, I relaxed and switched to autopilot. I started with simple concepts about our Sun, our solar system and its relative position in our galaxy. Then I progressed to what a light year is (the speed of light clocks in at 186,282 miles per second, x 60 for light minutes, x 60 for light hours, x 24 for light days, x 365 for light years, or 5 trillion, 874 billion, 589 million miles; sometimes rounded up to 6 trillion miles for conversation) the size of our galaxy in light years (about 100,000 light years in diameter), how our galaxy harbored between 250 billion to 400 billion stars, what a parsec was (3.258 light years), and how the vast cosmic distances between stars would really require something like *Star Trek's* warp drive engines for practical navigation among them. It helped that I was reading *The Making Of Star Trek* (Whitfield & Roddenberry, 1968) at the time, so I knew that Warp 8 was 500 times the speed of light; as Spock would have said, *fascinating*.

Finally, I ventured to the concepts of the size and age of the Universe, how many galaxies there might be (one or more of them out there for each star in our galaxy), the Big Bang Theory of cosmic expansion, and how it was more popular than the Steady State Theory. All of the material I'd dug up was second nature to me, most of it familiar from fourth and fifth grade readings at home, and I thought I was being clear, but by the time I got to the formula for the red shift for starlight (basically a variation of the doppler shift), VM=38r, explaining how we can determine a star or galaxy's receding velocity by its distance from us, I think I'd lost some of my fellow students. (The most current mathematical model is now: $v = H_0 D$, and the Universe is believed to span about 13.8 billion light years from Earth to the farthest galaxies detected, though this number, like the value of any good real estate, is mutable, so it pays to check in with pundits regularly for updates, in the event of a bullish market; additionally, this cosmological model sidesteps the currently very popular multiverse theory, whose explosively larger mega-sprawl of available galactic space-time properties truly represents a real estate broker's dream of avarice beyond measure.)

Closing my presentation, I asked if there were any questions. I looked out at a room of silent smiles and blank faces. Coach Wilson chuckled and grinned and said, "Well . . . anybody have any questions?" Volleys of laughter and shaking heads.

Returning to my desk, Patti Ryland, a racy blonde girl who ran with the fast crowd, scowled at me and muttered, "Great work, brainiac, nobody understood a damn word of that!"

But a few got it, like Greg Polasek, a fellow bookworm and Memorial outcast who introduced himself after class. He was a tall and serious looking guy, and with the close cropped black hair reminded me of Mr. Spock. He even wore the pullover tops, chinos and scuffed boots of a Starfleet Academy officer. En route to our next class, he gave me a quick rundown of his favorite intellectual pursuits, and when he said the words *flying saucer*, I knew I'd made a pal.

Greg was completely in tune like I was with the UFO Waves in America, and we swapped titles of books we'd read, including our favorite sci-fi authors. Breaking the ice led to lunches together, and discussions about our shared intellectual interests–and the tyranny and folly of the school's in-crowd. Weekend sleep-over's were an inevitability. And so it was fitting that I would have a preamble of sorts to my close up UFO encounter at home on Pebblebrook, via a group UFO sighting with Greg and his family one Friday night in their backyard in late September of 1968.

The Polasek home was a rambling ranch house nestled in a quiet hamlet of homes south of Memorial Drive to the west of West Belt, not far from where I lived. Their homestead also enjoyed the protective canopy of the dense forest, the towering pines and oaks of Memorial, the soothing maternal essence of the deep woods that seemed to hug the homes on their generous plots of land, a gentle and reassuring framework beneath the sheltering sky.

The Polasek property boasted a huge backyard, a vast sloping green lawn dotted with pine trees leading down to the wild forest along Buffalo Bayou. On that Friday afternoon, Greg and I took a hike into their woods. The air fairly bristled with the crisp freshness of the Indian Summer, and inhaling the pine scented oxygen invigorated the body and the senses. Sunlight filtering down through the trees dappled the ground in light and shadows; autumn colors of brown, red, rust, yellow and gold seemed more vivid than usual, and the trek of steady feet over the twigs, broken branches and pine cones on our path sounded in clear snaps and crackles.

Around twilight, still with a bit of light for hiking, we saw the evening star, Venus, her pale yellow presence a steady beacon low in the western sky. Presently one, then two stars winked on. And then, a new light appeared, as bright as Venus, perhaps a bit brighter, about 40 degrees above the horizon. Seconds later, it began to dart erratically. Then it made a series of jumps, playing hop scotch in the sky. Greg stopped abruptly on the trail, breaking our silence with, "Look at what that light is doing!"

The seemingly distant light, a bright white dot, like a small, solid circle, launched into even more bizarre aerial acrobatics. Starting about twenty degrees higher than the treetops, it slowly reduced altitude, spiraling downwards in a corkscrew pattern until it was below treetop level, then stopped abruptly, then rose back up in a similar fashion. Then, once at its highest altitude, it engaged the hop scotch maneuver again, jumping across the darkening sky, as though mounted on an invisible pogo stick.

We stared silently, engrossed. "Wow. Amazing," I said. "What do you make of it?"

"I don't know. Let's make sure my family sees it too," Greg said. He briskly bounded up the hillside and through the stand of trees leading back to his house.

Shortly, we'd gathered his father, mother and older brother, Rick, a tall basketball jock at Memorial Senior High, in their backyard to watch the strange light as it continued to trace odd patterns in the western sky. The sun was setting, and its lingering afterglow steadily faded from twilight into full darkness. His father, a taciturn and thoughtful electrical engineer, puffed on his pipe as he studied the zigzagging object. His mother, a kind faced woman, folded her arms, smiled, and shook her head bemusedly at the aerial show being put on for us in the clear, dark sky above Buffalo Bayou.

The show was just starting. Continuing its methodical, up and down corkscrews, the bright light stopped dead in the mid-heaven, about midway from the horizon to the zenith; reaching its apex, it plunged into another downward spiral, flaring up with an intense light burst like a firefly, and then vanished—winked out—only to re-appear a second later near the horizon, returning to its original lu-minosity. A stellar firefly dancing in the sky, beautiful in its mystery.

"Wow, did you see that!" Rick shouted. "Can you figure out how fast that jump was, Dad?" Greg's father was a math wizard, so cal-culating trigonometric functions on the fly was a breeze for him.

Pensively puffing up small clouds of smoke, Mr. Polasek slowly pulled his pipe from his mouth and remarked, "Hmm. It just now made that jump at nearly 2,000 miles an hour, and then stopped dead in its tracks. Interesting." No sooner had he uttered those words, then the object reversed its maneuver. Hovering near the horizon amidst clusters of trees, it flared up to a brighter intensi-ty, fluttered and wobbled upwards, winked out . . . and then reap-peared high up in the mid-heaven. Lights off. Lights on. Through-out, it remained many miles away, but retained its luminosity, still the brightest light in the heavens. An hour passed, and still the object defied physics and dazzled us with its aerobatic tricks. Mrs. Polasek brought out sandwiches and sodas, and we all sat down on lawn chairs to eat and watch the show. Wolfing down the last of his pastrami on rye, Rick suggested that Greg and I accompany him in his MG westbound on the Katy Freeway and try to track it—maybe get closer, get a better look at it. Since the object was steadily re-ceding west, fading into the visible horizon, that seemed like a great idea. The Polaseks wished us luck.

It was my first time in a convertible sports car, my introduction to the joy of hauling ass. Greg insisted I sit shotgun up front, and he wedged himself into the jump seat. Rick tore down I-10 like a rocket, and we kept the object in view the entire time, entranced by its ongoing bizarre repertoire of unorthodox aerial maneuvers. In minutes we'd passed the Dairy Ashford exit, heading out beyond the Houston city limits, our topless speedster affording us a wide open view of the night sky. The steady growl of the MG's engine was matched by the roar of the wind in my ears. I looked back at Greg, our hair flopping around wildly, and we both laughed, caught up in the moment and the excitement, the thrill of a high speed chase after an unknown thing receding into the dark sky. What could be more exciting to two young trekkies than chasing a UFO?

Approaching the town of Katy, it looked like we might be gradually gaining on it; but just when we thought we were closing in, the object retreated west again. It lost some luminosity, but was still clearly visible as a darting and falling light against the background of stationary stars. It seemed to be playing tag with us. Barely twenty minutes into our pursuit, Rick resignedly shouted out that it was pointless. We made a U-turn at the next exit. Mission aborted.

The next morning I got home from Greg's in time for breakfast, anxious to share my adventure. Clay listened attentively as I recounted over scrambled eggs and bacon the details of our group sighting and the high speed UFO chase. He pensively gazed at me somewhat glassy eyed when I told him about the 2,000 mile per hour speed bursts Mr. Polasek had calculated. His reaction didn't surprise me; as a research scientist he naturally respected the word of an electrical engineer, and the implications of the object's performance profile must have hit home. It wasn't the first clear sign I had that my dad wasn't altogether comfortable with the idea that aliens might already be here. In the late summer of 1967 I'd bought a paperback copy of *The Interrupted Journey*, John G. Fuller's engrossing non-fiction book about the now famous Barney and Betty Hill encounter with a UFO in September of 1961 in the White Mountains of New Hampshire, describing how aliens abducted them and took them onboard their craft for medical examinations. When I'd shared the gist of the storyline with Clay, he'd loudly guffawed, "Good GAWD, Wade, do you believe everything you read?!" He dismissed the Hill story and book outright, without reading a word of it, saying its bestseller status was proof positive that "there's a sucker born every minute." I accepted his hardcore scientific viewpoint and inherent skepticism, but I didn't share his derisive opinion about the Hills and their encounter and Dr. Simon's hypnosis sessions. Fuller's presentation of the whole story rang true for me.

I found it to be a page-turner of a book, and the further I read, the more I felt an inexplicable sense of familiarity with the Hill's tale, a deep, gut level awareness that this was a true story, and that the Hill's bizarre encounter was real–skeptics be damned. By the time I reached the passages taken from the tape recordings of their time regression hypnosis sessions with their psychiatrist, Dr. Benjamin Simon, I got chills up and down my arms. Prompted by Dr. Simon during a hypnotic trance session, Barney Hill roughly sketched with pad and pencil a drawing of the alien saucer, as well as the leader's face; his eyes, large, dark, almond shaped, and slanted upwards, are strikingly similar to those repetitively described in the ensuing years of UFO/alien encounter reports. America's first gray, as it were.

Another striking feature of the Hill's encounter was the element known as *missing time.* The full scope of what had happened to them on that lonely stretch of road near Indian Head takes the better part of the narrative to unfold, but their first inkling of it popped up when they realized they'd arrived back home in Portsmouth over two hours late, with no apparent reason for the time lag, and with more peculiarities to follow.

The full story of the Hills' experience is quite intense. *The Interrupted Journey* haunted me for a long time. It represented a turning point in my UFO research, being the first time I'd found a purported alien encounter that seemed believable, with two credible witnesses, much more so than the fairytale-like saucer stories peddled by George Adamski, which I believed to be bunk. Whereas Adamski's ufonauts had claimed to come from Venus, Mars and Saturn, Betty Hill's Q & A session with the alien "leader" onboard their spacecraft culminated in her being shown a star chart indicating the alien's home port, another star system in our galaxy (which would also become part of UFO folklore via the Marjorie Fish star pattern 3-D model, which seemed to pinpoint their visitor's home star system as Zeta One Reticuli, 39.17 light years away, in the southern constellation of Reticulum; *Zeta One Reticuli* is by now so culturally suffused it could be a TV game show question).

My magnetic attraction to UFOs–coinciding with reading the Fuller material–now revved up in intensity. The group sighting with the Polaseks only served to whet my appetite. My next UFO encounter followed just a few weeks later. They'd shown themselves from a distance. Now it was time for their close up.

Friday, October 11, 1968. Almost exactly four hundred seventy six years ago to the day, Christopher Columbus saw a light, and the *Pinta* reported sighting land. Perhaps it was synchronistically apropos that on this night, I would see a light too, but not from this land. Not from anywhere around here, I'm guessing. It was to be a time of great discovery, just the same.

It had been an uneventful evening; Clay and Monah's habitual Friday-nighter celebrations (what I used to refer to as *The Graduate* parties, as the drinks, dancing and fun often revolved around the

Simon & Garfunkel movie soundtrack LP blasting for hours from the living room stereo--hey, hey, hey Mrs. Robinson and all that) had dwindled to only once or twice a month, and this was one of their quiet, off nights. My parents, Natalie and Aaron had all retired before me, so ever the night owl, I stayed up, watched *Star Trek*, read Arthur C. Clarke's *Childhood's End* for the third time, and watched *The Tonight Show*. Johnny Carson had been in rare form that night, always a fun way to launch the weekend. When it was over at midnight, I switched off the TV, killed the den lights, and went to the bathroom to brush my teeth and prepare for bed.

Everyone else was asleep. I burrowed under my bed covers and listened to my transistor radio, to relax and drift away. Spirit's spooky song *Mechanical World* took me on a fantasy journey to a future society run by invasive machines–at least that's the mental image I created every time I heard it play; for me, the refrain was a mournful, *someone's gonna take my mind away* . . . though apparently those aren't the lyrics. The haunting chorus for me was an elegiac lament about a mind struggling against a dystopian world overtaken by advanced AI, an ideal accompaniment to my life's path then, determined as I was to stuff my head with facts about particle accelerators, laser beams, artificial intelligence, receding galaxies, space warping flying saucers and the quantum mechanics of nuclear fusion: *someone's gonna take my mind away. . . .*

As the song faded, I was stirred from my reverie by radio static jumbled with a vibrating hum, as though I wasn't tuned into the station properly. Amid the static cracklings, the volume steadily decreased. I was losing the signal to some sort of interference, so I spun the dial and tuned in elsewhere. Curiously, a quick scan revealed that every station on the AM band was fading out in a strange mish-mash of vibrating voices, crackling static and distorted songs. No matter how high I turned the volume knob, the signal strength steadily declined. Soon I lost the entire bandwidth. Funny, I'd gotten a brand new battery barely a week before. But the radio abruptly fell silent, and went dead, so I set it on my night table with the power button still on. Strange.

Still too alert to fall asleep, I knelt up on my bed and parted the curtains of my large bedroom window. The wide glass panes revealed a clear, moonless, starry night. All was tranquil and silent, not even an occasional wind gust stirred through the towering pine trees. I considered getting one of my telescopes and setting it up to stargaze through my window, but decided to look for constellations instead. None of the star patterns looked familiar, but their beauty captivated me. For several minutes I soaked up the silent splendor and wonder of the starry sky. It was around half past midnight, maybe twelve forty a.m.

I suddenly had an inexplicable urge to look up overhead, to the zenith. I pressed the top of my head against the glass, resting the back of my head on the window sill, and gazed up over the eaves of the house. Enjoying this unusual perspective, I wondered to myself what had compelled me to do it. It was as if someone had softly said, *look straight up.* I silently laughed at myself for bothering with it.

A few seconds later, it appeared: a mysterious, unknowable thing plunging downwards from the heavens like a flat stone sinking into dense water, wobbling on its way down. It glowed, lit up with its own luminescence, a brilliant ball of white light—brighter than the full moon—but with a definite, solid outline, lacking a halo. Startled, I resumed upright posture. Dumbfounded, I watched as it darted, bobbed, zigzagged, and leveled off its descent a few hundred feet above me, swelling in size as it rapidly approached. Then it glided towards the far end of our neighbor's rooftop, towards the east, with a swirling and zigzag motion as graceful and silent as a swan.

It moved with a jerking, darting motion whenever it changed directions, then smoothed into a graceful glide after each maneuver. A ring of red lights around its middle flashed periodically. It traced circles and squares against the dark, starlit sky, like a celestial pinball banking off of the bumpers and paddles of an invisible pinball machine. Somewhere in the back of my mind was a tape silently playing, this shouldn't be here, this can't be here . . . this can't be here, but it is here, it can't be here, but it is here, it can't be, but it is, it can't be, but it is . . . is this what I think it is?

Abruptly, it stopped dead in the air, directly over the far end of our neighbor's roof. It was roughly twenty feet wide, about the size of a small plane, lit up with a fierce white light. I had the eerie feeling that I'd been seen, spotted. I knew it in my gut. The realization washed over me like a chilly fog. That, and the sense that the hunter had tagged its game. Ahh, got him. Not a comfortable feeling. A very uncomfortable one. There was a definite sense of a highly advanced intelligence present, a powerful presence, almost like a living machine. That awareness scared me, because nothing about this thing, not its appearance, nor its silence, nor its flight characteristics gave me the sense that it was from here. And nothing about it felt warm and friendly, either. Not overtly hostile, but not friendly either. An unfriendly, uninvited visitor. Ahh, but you did invite them here, Wade. You did invite them. Be careful what you wish for.

Fear washed over me in cold waves, and it scared me even more when I realized that my fright was increasing more and more with each second that it hovered there. And the silence—how could it be so silent? There wasn't the slightest hint of air current around it, no propulsive thrust, nothing. I suddenly got it how Barney Hill had felt like a rabbit about to be trapped. Got it. You nailed it, Barney. My sense of solitary helplessness served to ratchet up my fear levels ever higher, the elevating fright compounded by the realization that I'd never felt a fear like this before, one that didn't seem to have any upper limits. A steadily increasing sense of helpless paralysis washed over me. The mercury in my inner fear thermometer seemed sure to burst through the top. The hairs at the nape of my neck bristled as though statically charged. The silence of the night highlighted sounds and sensations normally ignored—the pulsing throb of the blood coursing through the artery in my neck, the liquid sounds of my nervous swallowing, the feel of my fingertips as my hands firmly clutched the hardwood window sill for balance.

This thing, this unknown object, had spotted me, and now it watched me, intently. I was being scrutinized and studied, assessed, hypnotized by the red flashing lights. Fear, a word I hadn't associated with the night for years, tugged and pulled at me, just as surely as this enigmatic, silent, brilliant thing hovering nearby drew my attention in. Illuminated with its own unearthly light, it blazed against the blackness of the night sky.

Multiple realizations came flooding in. My mind swiftly exhausted an array of potential name tags, failing to identify it; because of i.d. failure, the word *alien* popped in; with this word, my fear intensified, leading to a deep sense of dread. The object offered no familiar reference points to reassure me. Although my little brother Aaron snored peacefully nearby in his bed, and Natalie and my parents slept in nearby rooms, I felt completely alone. You were right, Barney. I am a rabbit. And there is no escape.

A new row of pulsating lights suddenly came to life around its perimeter: red, silver and white flashing lights, rotating in a counter-clockwise, then clockwise direction. They flared in sequence almost in time with my quickening pulse, ba-dum-bump, ba-dum-bump, ba-dum-bump, igniting with an intensity that neutralized my mind and nervous system. It was like being blasted with lasers. I could feel the part of me that I called my self draw in upon itself, as if retreating further within the shell of my body would protect me. As if anything or anyone could now.

And then, how could this be, the object dematerialized and instantly rematerialized a good fifty feet lower in altitude, with no apparent motion carrying it from up there to down here. No linear vectoring going on, no motion "from" and motion "to," no "shortest distance between points A and points B is a line" business. For this vehicle, the shortest distance was not being at A, and then being at B. Wink of an eye. The outside chance that this aircraft was one of ours seemed dashed in an instant. Nothing we had could wink in and out like that, dematerializing and then rematerializing in a dead stop, with utter silence. And the absolute, cold finality of that chilled me to the bone. Right about now in Barney Hill's encounter, the leader in the saucer's window telepathically told him, *don't be afraid.* But no consoling words were offered here. No window, no leader, no reassurances. The feeling here was more like, *we don't care if you're scared.*

As the object gently descended and came into a more stationary hover about 30 feet above the neighbor's roof, it revealed the form of a disc shaped craft. It had only appeared spherical from below. The flashing lights around the disc's perimeter continued to blast me with their intensity, the furiously oscillating rhythm unabated, ba-dum-bump, ba-dum-bump, ba-dum-bump. I was under its spell, drawn into a hypnotic trance. As it swirled closer, something on its surface shifted, a mechanical reconfiguration, perhaps for a different flight mode—who knows? With the sensory overload and my intensified fears, it's a wonder I could function well enough simply to observe. The disc traced squared-off circles as it slowly, silently descended. If the perimeter lights had made equally intense sounds, the sonic blast would have awakened the whole neighborhood. Why was it coming so close? What did it want?

The brilliant, blazing thing eased down. I felt myself fading out, surrendering to its shocking light. My willpower and motor abilities vanished like water droplets impacting a hot frying pan. I wanted to call out, to shout to my family, but my vocal chords were taut, my throat dry and drawn tight. If I could break away I could wake up my family, but that required thought and motion. I was completely paralyzed; my hands, frozen in a death grip, fiercely grasped the windowsill. The dawning awareness of my total helplessness terrified me, but by now the shock of my utter incapacitation distanced me from the thing outside. I had split apart and become an impartial observer standing next to the terrified eleven year old boy kneeling up on the bed. I hadn't yet learned the term disassociation, but I was doing it. I had disassociated, split apart, left the scene. There physically, but in all other respects, gone. Very far gone. Deep inside my head, and beyond. Fading out from me, and fading into it.

The impossibly brilliant saucer descended, closer and closer, its red, silver and white perimeter lights furiously blasting my central nervous system like salvos of daggers. I could feel my consciousness ebbing as I faded into it, into the intense white light. Through the years, I had desperately pined for them to prove to me they were physically real. They were making good on that request in a big way. The white light overtook everything. It blotted out the blackness of the sky and the surrounding trees, it made the neighbor's roof and house disappear, it became the only reality, just the blinding, all encompassing white light and its multicolored photon fusillade. Everything was emptying into the ocean of white light. I was blacking out with my eyes wide open.

The fierce barrage of lights shut me down, shut me off. And there was nothing I could do about it. It was going to absorb me completely, like the ocean swallows a raindrop. The white light would absorb me . . . it was absorbing me . . . it did absorb me. Total whiteout. Zap. Pffft. All gone now. Power down. Off. *Someone's gonna take my mind away. . . .*

Fade out.

Fade in. . . . An indeterminate period of time later, my puzzled eyes struggled to regain their vision. Awareness of the here and now returned; awareness of having sore limbs and knees, awareness that feeling was returning to my arms and legs. Still on the bed, still looking up, hands still clutched to the windowsill. Realizing I'd survived whatever had transpired, but not really knowing for sure what had happened after the blackout. Remembering the shock and overpowering fear. Not feeling the shock anymore, but still feeling afraid. Still feeling fear. Because the disc was still here.

Like a computer operating system rebooting, my numbed out body/mind/spirit started functioning again. My eyes came into focus on the saucer. It hovered over the Landry's rooftop. Stationary. Vigilant. Silent. The perimeter lights were gone. Only the white glow remained. Perhaps there was no need for the laser cannons now–they'd done their work and the mission was complete. Abruptly, the disc silently darted upwards and over to the right, towards the southern edge of the Landry's backyard; it swung out with a smooth swinging motion like an operating room's high intensity lamp being tilted up and away from the anesthetized patient upon completion of the operation. The patient would experience a temporary sense of numbness. But a full recovery was expected.

The glowing saucer, now roughly fifty feet high again, morphed back into a glowing ball of light. The ball of light smoothly glided over to the right, where it parked itself behind the upper branches of a pine tree, like a sailboat in its final cruise back into the docking slip. It stopped dead in the air, and waited.

The absurdity of its presence there struck me. A real, solid, physical thing of substantial size, mass and volume, glowing incandescently, had silently anchored itself fifty feet up in the air. Floating like a weightless helium balloon. Hanging out and doing . . . what? Reviewing the notes from its scan of my central nervous system? Filing a new flight plan? Waiting to see if I'd venture outside and meet with them? What to do. I wasn't sure. Curiosity told me one thing, common sense another.

Self preservation instincts won the debate. I unclenched my hands and pushed myself away from the window sill. I collapsed onto my bed and dove under the covers, pulling them up to my neck. I lay there in the dark with my head on my pillow, partly regretting that my fear was too powerful for me to venture outside. I couldn't risk being snatched up and taken away. I knew at that moment, burrowed into the safe cocoon of my bed, that one day I might regret not running outside–or not trying to wake up my family and have them go with me. But the fear of being taken was a powerful force. I kept as still and quiet as possible.

A few minutes later, the red power light on my transistor radio lit up, and the speaker sputtered back to life. I'd forgotten that I'd left the power button switched on when it went dead. I held it close to my ear, noticing that the crackling static and strange vibrating noises from before persisted. Having, even at age eleven, a fairly comprehensive knowledge of UFO sightings and case files, it wasn't lost on me that many close encounter reports included details about power outages and electrical system failures, as well as car engines, televisions and radios experiencing interference and going dead with UFOs in close proximity. So my radio's lingering bursts of static, my UFO detector, kept me still for awhile. After perhaps twenty minutes, a dj announced, "It's three a.m." The incongruity and significance of the time didn't register with me for a long time, not until years later in fact. Shock and fear have a way of blocking out things.

At some point during the night, I heard Natalie get up and use the bathroom down the hall. I bolted out of bed, and when I heard the water running, I gently tapped on the door and said, "Natalie! I just saw a UFO out my window!"

She started, then laughed and said, dismissively, "Oh Wade!" When she opened the door, leaving the light on, she strode past me in her bathrobe, grinned condescendingly and shook her head. "I'm telling you," I said, "it was the real thing. It came down right over the Landry's rooftop!" All she could offer was to mumble "oh jeez" and wave me off. She retreated to her bedroom and closed the door.

The next morning, when I told Clay all about what I'd seen, I wasn't prepared for his utter derision; nor was I expecting that he and Natalie would do a tag team about it, turning the whole thing into a huge joke, passing the laughs and ridicule back and forth. "And did a little green man with antennae on his head look out the window of it and wave to you?" Clay chortled, putting his index fingers to the sides of his head, making scary faces and humming silly sci-fi movie music, with Natalie chiming in. I kept telling them, "You're wrong. You're so wrong! I hope they scare the daylights out of you someday!" but that only elicited even more roars of laughter and derision, so I let it go.

It was disheartening to have a scientist I'd looked up to for so long behave that way. Even more so that he was my father, and had expressed such pride over the years in my scientific pursuits and obsessions, occasionally star gazing and planet watching with me and my telescopes. His reaction clearly revealed his uneasiness with a subject matter that he probably knew was far more serious than he wanted to admit to himself. He definitely wasn't naive when it came to good science–like the speed calculations Mr. Polasek had made. In the coming years I realized that my father and sister shared a lot of traits–and one of them was an almost primal fear that the ETH explanation might be the real deal. Like many intellectuals, they were bright enough to have known, at least in the back of their minds, that some of these sightings were likely indicative of alien vehicles in our atmosphere, but they weren't prepared to accept the implications of such a staggering reality. So there was no way they'd ever accept the ETH as truth. It would have shattered their world view and their alpha status. It reminded me of Carl Sagan's supercilious persona, arrogantly dismissing the notion that more intelligent and technologically superior aliens might be out there--and visiting here.

A few weeks later, my parents hosted some neighbors, the Farrell's, for drinks on a Saturday night. Tim Farrell was a hot shot attorney, and Eileen a freelance artist. They were friendly, fun and attractive and I liked them a lot. I stopped by the living room to say hello, and Eileen and I got into a discussion about a controversial movie that had come out recently, *2001: A Space Odyssey.* I'd thoroughly enjoyed it as a successful translation from book to screen. Eileen, realizing I completely understood what author Arthur C. Clarke and director Stanley Kubrick were saying, told me she didn't "get it" and insisted I explain it to her. It made sense to her when I said that the primary theme was the evolution of man, from cave dweller to modern man, and from there via an evolutionary leap to a Star Child. And how the monolith opened the Star Gate at the psychedelic climax, transforming Bowman's pod into a hyper-light shuttle to the Visitor's home star on the other side of the galaxy.

I loved talking with Eileen; she always spoke to me as an equal, despite our age difference. Like a few of our neighbors, she knew I was a sharp kid and a junior mad scientist. Our talk that night ventured to the possibility of other intelligent life in the universe, which led to me telling her about my recent UFO sighting. She pulled me away from the others, and implored me to spill the full story for her. She became so engrossed with all of the details that she had me take her to my bedroom so I could show her how the object maneuvered over the Landry's roof outside my window. I had her undivided attention. As I wrapped up the recounting of my flying saucer experience, Clay strode briskly in with a flashlight, a man on a mission. He told us, "Here. This is what you saw." He switched on the flashlight, held it to the window and played the beam back and forth across the night sky outside. When the beam hit a cloud bank, he said, "There. See that. See how the beam looks like a circle of light when it hits that cloud? Boom. There. And boom, there and there.

See? What you saw that night was a searchlight beam playing over a cloud, back and forth. Made it look like a disc. Boom. There. And boom. There. See that?"

Eileen shot me a wide-eyed grin at the sound of this. I told him, "Dad. I have two telescopes. I started reading college physics books when I was in fourth grade. You know what a good observer I am. This was not a searchlight beam bouncing off of clouds. It was a solid object. And like I said, there weren't any clouds that night. It was clear." He wandered off shaking his head and scoffing, asserting that my sighting was a case of mistaken identity. Eileen's eyes followed him out. Then she said, "Well, I know what a good amateur astronomer you are, and all the physics you study and understand, and how bright you are. And I believe that you saw a real UFO." She smiled at me knowingly and shook her head. "That is so funny about Clay! I just don't get it that your dad can't accept what you saw. How funny."

BOOK II.
INTERLUDE

In June of 1969, Clay's company transferred us to New Orleans, variously known as The Big Easy, the Crescent City or The City That Care Forgot. Uprooting proved tougher this time. Accelerated learning classes at Memorial Junior High got downgraded to a Louisiana school system so dusty and backwards that by the 8th grade, even in a pricy private Catholic school, we still hadn't covered the basic geometric theorems I'd learned in 5th grade Math class in Houston. Fifth grade. I missed my Houston pals too, so my task was clear: find a new set of bookworm outsiders, and try to ignore the feeling of neurons fizzling out and expiring in somnambulistic Heart Of Dixie classrooms. Bless me, St. Andrew, for I have sinned.

Relocating also meant I was abandoning the site of my shocking UFO encounter. So close up. Its nearest approach, the point I called *the white out*, had indelibly burned itself in my memory. It was reminiscent of Theodore Sturgeon's short story, *A Saucer Of Loneliness*: the protagonist also blacked out in the near presence of a saucer, and it left its indelible mark on her life, too. I was already on a path of one who is different. My extremely close up visitation amped it up exponentially. Blacking out in a saucer's blinding light. Like being branded with an alien marker. A permanent, invisible set of dog-tags for life; no one else can see or hear them clanging around your neck, but for you they're ever present, as permanently attached and deeply fixated as a conscious subroutine.

Whatever happened that night on Pebblebrook was with me always now. It rolled around in my head, the glowing, bouncing orb as steadfast as the sun. Sometimes, its final approach replayed in my mind's eye several times a week. So many questions remained: where was it from, who or what was operating it, how long had it lingered that night, what did they want from me? The mystery was as much a part of me as any of my vital organs. As unreal as it was, the event that night had become one of the most real things that had ever happened to me. I wanted them to come back so badly. I didn't fit in down here with my fellow humanoids at all. I'd even let them take me away next time, no guarantee of return required.

As the mad flush of hormones kicked in with the onset of puberty, the colorful, non-stop carnival atmosphere of the French Quarter became my backdrop. Testosterone and brio fueled nights on the town with other rebellious boys, older high schooler's with cars, became a regular weekend staple, sometimes even on school nights. Our fake photo ID's transformed the Quarter and Uptown bars into a nocturnal playground for young teens, every nightclub, disco or hole in the wall pub overseen by owners and managers who didn't give a damn about underage kids. If you resembled your mug shot and the card said you were of age, you were in. My extended bar hopping interlude could have been scored to the tune of The Who's classic *Baba O'Riley*. Teenage wasteland indeed.

Adolescence set fire to my blood, my body, my loins, my mind. The steadily growing realization that I liked guys more than girls added more fuel to the fire, already stoked by the brutal regimentation and guilt laden dogma of my Catholic school stalag brainwashing. Weekly force-feedings of Catholicism's twisted dogma became dietary staples: same-sex sex was forbidden and evil, punishable by damnation and eternal hellfire--except for priests, who could rape parish boys with joyful abandon, knowing the fold protected them. Having an emotionally distant scientist father, an ex-star college baseball jock who didn't really warm up to me unless I was competing on a swim team and winning breast stroke blue ribbons like I had in Houston, ratcheted up the adolescent angst a few more notches. Bottled up frustrations got vented with some of my wild local boys. When I wasn't escaping by acting out, indulging in neighborhood vandalism and occasional truancy, I retreated into the protective inner sanctum of my UFO readings and research, astronomy, nuclear physics, flying saucers, and science fiction. I'd even won a 2nd Place ribbon at the St. Andrew's 8th Grade Science Fair with my elaborate flying saucer triptych of photos, facts and illustrations, classifying UFOs as a sociological phenomenon, the fruit of months of late night labors. The science nerd made good. It was deemed eligible for entry into the regional science fair at Tulane University.

After a few years of relative calm in the realm of UFO reports, October 11, 1973 brought another landmark case into the headlines and on television. Two shipyard workers from Pascagoula, Mississippi claimed to have been taken onboard a UFO, examined, and then released. It was the first solidly documented abduction case since the Hill's New Hampshire encounter in 1961. Being so close by to the Mississippi Gulf Coast, the New Orleans media covered the story in depth. I can still vividly recall my excitement while reading the UPI report in the New Orleans *Times-Picayune* on October 12th:

PASCAGOULA, MISS. (UPI) Two shipyard workers who claimed they were hauled aboard a UFO and examined by silvery skinned creatures with big eyes and pointed ears were taken to a military hospital Friday to be checked for radiation.

Officials said Charles Hickson, 42, and Calvin Parker, 19, would make no further public statements concerning their weird tale until they had talked further with Federal authorities. Both work at Walker Shipyards, where Hickson is a foreman. Neither man suffered any apparent injuries but as a precautionary measure were taken to nearby Keesler Air Force Base Hospital to be checked for radiation exposure, officers said.

Jackson County Chief Deputy Barney Mathis said the men told him they were fishing from an old pier on the west bank of the Pascagoula River about 7p.m. Thursday when they noticed a strange craft about two miles away emitting a bluish haze. They said it moved closer and then appeared to hover about three or four feet above the water, then "three what-ever-they-were came out, either floating or walking, and carried us into the ship," officers quoted Hickson as saying. "The things had big eyes. They kept us about 20 minutes, photographed us and then took us back to the pier. The only sound they made was a buzzing, humming sound." They left in a flash.

The sheriff's office said it received several other calls during the night from residents of the area about sighting a strange blue light in the sky. Numerous UFO sightings also have been reported in many parts of the state during the past couple of weeks.

Captain Glen Ryder of the sheriff's department, who questioned both men Thursday night, said he thought at first "they were pulling my leg. We did everything we knew to break their stories," Ryder said, but both stories fit. "If they were lying to me, they should be in Hollywood."

Mathis said Hickson appeared to be a reasonable man and was not a heavy drinker, according to his wife and employers. Authorities said both men said they were not drinking when the incident occurred but admitted "they went to have a drink or two after it was over." They had to have something to settle their nerves, said

Mathis. He quoted Hickson as saying: "I was so damned scared I didn't know what it was."

Officers said Parker reported he passed out when the three creatures--purportedly with pointed ears and noses and a pale skin type covering--emerged from the craft. He said he didn't regain consciousness until he'd been released back on the pier.

Deputies took statements from both men and left them together in a room with a hidden tape recorder in an effort to check out the story. Mathis said there was nothing on the tape to indicate a hoax.

Hickson estimated he and Parker were inside the unidentified craft for 15 or 20 minutes. He told officers he was placed on some kind of table and examined from head to foot by what he described as something like an electronic eye.[1]

The full story received heavy media coverage. Both Hickson and Parker came across as salt of the earth types, making it harder to imagine that their tale had been fabricated. Within days, the Pascagoula encounter became major international news. The Aerial Phenomena Research Organization (APRO) sent University of California professor James Harder to investigate (he'd received his Ph.D. in Fluid Mechanics at UC Berkeley in 1957, and was APRO director of research from 1969-1982)[2] ; Dr. J. Allen Hynek, formerly with the US Air Force's defunct Project Blue Book (and who would go on to found the Center For UFO Studies, CUFOS, that year), also arrived to look into the story. Harder and Hynek interviewed Hickson and Parker together. Harder hypnotized Hickson, but he became so frightened that the session had to be aborted.[3]

In addition to the taped interview with local officials, the transcript of their deeply emotional private conversation on a hidden tape recorder greatly bolstered the probable veracity of their account.

The interview was conducted in tandem by Sheriff Fred Diamond and Captain Glen Ryder at approximately 11:00pm Thursday evening. The facts derived here: they were two friends fishing on the shores of the Pascagoula River around 7:00pm. A blue light appeared in the sky, circled, and in seconds the glowing object hovered about 25 yards away. It made a quiet buzzing sound, and finally came to a rest in a silent hover a few feet above the ground, effectively trapping the two men, making the river the only possible escape route for them. A doorway appeared, and out floated 3 robot-like figures, about 5 feet tall each; two of them "lifted and float-ed" Hickson off the ground with claw-like pincers. He couldn't feel a thing. Parker passed out on the spot.

Hickson was taken onboard, and was kept suspended in the air with no visible sign of support and no bonds, presumably in some sort of zero-gravity field that kept him floating stationary in a state of near paralysis; all he could move was his eyes. He watched as a very large eye-like device slowly moved and scanned all over his body, up and down. When he tried speaking to one of the robots (in a later interview, Hickson and Parker made up their minds that they were robotic machines programmed to carry out a mission with mil-itary precision), the only response he got was a buzzing noise; the other two robots were silent. They each had a slit for a mouth, with metallic conical protuberances in the position of the nose and ears. Hickson never noticed if they had eyes. The grayish skin appeared wrinkled, though this was possibly a body covering. Estimated time onboard: 20 - 30 minutes. The craft was described as about 8 feet tall, oblong in shape, and very brightly illuminated within, though Hickson never noticed a specific light source. In a later, casual in-terview with Ralph Blum, researcher and freelance writer of *Beyond Earth: Man's Contact With UFOs* (Blum & Blum, 1974), Hickson related how as a professional welder--a burner and a fitter--he was accustomed to bright light, but that his eyes bothered him for two or three days after the abduction. He also stated that he and Calvin had come to the conclusion that Hickson was the only one taken onboard, that Parker was left outside in a paralyzed state of shock. When the robots floated Hickson back to the waterfront outside, he found his young friend standing with his arms extended, an almost indescribable look of terror on his face. The object quietly buzzed and departed, disappearing from view in seconds.

Charlie told the interviewers that he'd tried to call Keesler Air Force Base, and they'd told him to call the sheriff. After a few more questions, the interview was over.[4]

Sheriff Diamond asked Charlie to come back in the morning to make a complete statement. Hickson said he didn't want any pub-licity, or to get his family upset. Then Diamond and Captain Ryder left Hickson and Parker alone in the room with the tape recorder still running, unbeknownst to them. The compelling content of this recording made on the sly lends an undeniable stamp of authen-ticity to Hickson and Parker's UFO encounter, making it difficult for even a hardened skeptic to denounce outright its veracity. From *Beyond Earth: Man's Contact With UFOs*:

CHARLIE: (voice shaky) I can't take much more of that.

CALVIN: (frantic) I got to get home and get to bed or get some nerve pills or see the doctor or something. I can't stand it. I'm about to go half crazy.

CHARLIE: I tell you, when we through, I'll get you something to settle you down so you can get some damn sleep.

CALVIN: I can't sleep yet like it is. I'm just damn near crazy.

CHARLIE: Well, Calvin, when they brought you out--when they brought me out of that thing, goddamn it I like to never in hell got you straightened out.

CALVIN: (voice rising) My damn arms, my arms, I remember they just froze up and I couldn't move. Just like I stepped on a damn rattlesnake.

CHARLIE: (sighing) They didn't do me that way. (Now both men were talking as if to themselves.)

CALVIN: I passed out. I expect I never passed out in my whole life.

CHARLIE: I've never seen nothin' like that before in my life. You can't make people believe.

CALVIN: I don't want to keep sittin' here. I want to see a doctor.

CHARLIE: They better wake up and start believin' . . . they better start believin'.

CALVIN: You see how that damn door come right up?

CHARLIE: I don't know how it opened, son. I don't know.

CALVIN: It just laid up and just like that those son' bitches--just like that they come out.

CHARLIE: I know. You can't believe it. You can't make people believe it--

CALVIN: I paralyzed right then. I couldn't move--

CHARLIE: They won't believe it. They gonna believe it one of these days. Might be too late. I knew all along they was people from other worlds up there. I knew all along. I never thought it would happen to me.

CALVIN: You know yourself I don't drink.

CHARLIE: I know that, son. When I get to the house I'm gonna get me another drink, make me sleep. Look, what we sittin' around for. I gotta go tell Blanche . . . what we waitin' for?

CALVIN: (panicky) I gotta go to the house. I'm gettin' sick. I gotta get out of here. (Charlie gets up and leaves the room, and Calvin is alone.) It's hard to believe . . . Oh, God, it's awful . . . I know there's a God up there. . . . (As Calvin prays, his words trail off and become inaudible).[5]

Upon contacting the sheriff's office, the first thing Hickson and Parker wanted to do was to take a lie-detector test, which they both passed. Blum had access to Hynek and Harder immediately following their hypnosis sessions and interviews with the two abductees; he was privy to their recorded press conference before it hit the evening news. Hynek's statement:

> There's simply no question in my mind that these men have had a very real, frightening experience, the physical nature of which I am not certain about--and I don't think we have any answers to that. But I think we should very definitely point out that under no circumstances should these men be ridiculed. They are absolutely honest. They have had a fantastic experience and also I believe it should be taken in context with experiences that others have had elsewhere in this country and in the world.[6]

Harder followed with:

The many reports made over the past twenty, thirty years point to an objective reality that is not terrestrial. When you've eliminated all the probable explanations, and you still have something that you know is real, you're left with the less probable explanations, and I've been left with the conclusion--reduced, perhaps, to the conclusion--that we're dealing with an extraterrestrial phenomena. I can say so beyond any reasonable doubt.[7]

In a follow up interview, Harder allowed that trying to ascertain where the craft came from or why they were here were matters for speculation, but that he felt confident Hickson and Parker's experience was a real one. He based his self-assured statement on the fact that a very strong feeling of terror was virtually impossible to fake while in a hypnotic trance state. When asked if he thought the visitors posed any threat, or should we fear them, he commented that when a mouse is picked up in a laboratory environment, it's most definitely a frightening thing for the mouse; but that didn't mean that the lab techs meant the mouse any harm.

Worthy of mentioning: when Hickson and Parker were shuttled to Keesler Air Force Base for their physicals (wanting to be checked for radiation, which the local hospital wasn't equipped for), they were driven onto the base with a full military escort in tow--two cars full of air police followed their unmarked car. According to Tom Huntley, who'd driven them, they were subjected to full physicals by doctors, "who looked like space creatures--all wrapped in white and masked and gloved." After the exams, Hickson and Parker were taken to a conference room, where they were met by a large group of military brass, including colonels, majors and four air police officers. The head of base intelligence interrogated them, seeming very cool, as though he'd heard it all before. They were asked if there was room in the UFO to stand up, what sort of clothes the creatures wore. When Charlie mentioned the claw hands, two colonels exchanged looks over that.[8]

How fascinating that Keesler Air Force Base officers possessed such a keen interest in the Pascagoula abduction as to ask specific questions about the interior of the UFO, considering the USAF had declared the very craft themselves non-existent in December of 1969, when Project Blue Book folded (questioning Hickson and Parker intently as if to say, we know it's all a fantasy, guys, but we'd like to get all the details anyway, just for the hell of it). Later on, when Blum spoke with the public information officer at the base, he said there was nothing to tell. The case was closed. Keesler had promised to send a copy of their report to Sheriff Diamond, but the sheriff never received one from them.[9]

Following the Pascagoula UFO encounter, Calvin Parker was eventually hospitalized in Laurel, Mississippi, for an emotional breakdown. In an interview several years after the UFO event, Hickson speculated that Parker fared worse after the encounter because he had never previously experienced a profoundly frightening ordeal (whereas Hickson had seen combat in the Korean War, and thus had a good bit of familiarity with terrifying experiences).[10,11]

In January, 1974, Charles Hickson appeared on *The Dick Cavett Show*. Among the guests were Gemini 4 and Apollo 9 astronaut Jim McDivitt (full title, Brigadier General James A. McDivitt), who'd sighted a UFO in space that had remained unidentified, describing it as a white cylindrical object with an extension like an antenna, which he'd observed and photographed during the Gemini 4 mission[12], Air National Guard pilot Captain Larry Coyne who'd had a close call with a UFO while piloting his helicopter, Dr. J. Allen Hynek, and the ever supercilious Dr. Carl Sagan, then professor of astronomy at Cornell. It was repulsive to watch Sagan haughtily dismiss the panelist's UFO claims, mocking each man's testimony. It was thoroughly enjoyable watching McDivitt tear into Sagan, point-counterpoint style. He defended Charlie Hickson, saying, "You know I think the experience Mr. Hickson had can't be overlooked. You're trying . . . to overcomplicate this thing from a scientific standpoint. I'd be the last one in the world to argue with the scientific approach, but I think that personal experiences like some of these people have had certainly can't be dismissed.

CAVETT: You don't dismiss Mr. Hickson?

MC DIVITT: No, I don't. Not one single bit!

SAGAN: I don't dismiss him, but I do note that there are large numbers of people who have a wide range of other kinds of experiences that we don't ordinarily believe--like those people who believe that they've made contact with deities, or people who are themselves deities. . . ."[13]

Sagan's proud persona, posing and hissing like the king cobra of scientism, starkly clashed with Cavett's more evenhanded guests.

He utilized the tired argument against interstellar travel: since we don't have the technology to cross the vast distances between stars to get to them, then of course they can't get here. Particularly entertaining was his reference to our crude attempts at space travel, "it's tough to travel between the stars, because the stars are *extremely* far apart. For example Pioneer 10 is the fastest spacecraft that we've ever launched, and it will be our first interstellar spacecraft to leave the solar system--it's traveling so fast that it will travel the distance to the nearest star in only eighty thousand years. That's our fastest spacecraft. Our slower ones take longer, of course . . . ."[14] (This statement was in odd conflict with a proclamation Sagan had made before the American Astronautical Society in 1966, that Earth may have been visited by various galactic civilizations on the order of 10,000 times during geological time.)[15]

When Ralph Blum asked physicist and UFO researcher/author/lecturer Stanton Friedman about Pioneer 10 and Carl Sagan's 80,000-years figure, Friedman pointed out that Pioneer 10 was never intended for interstellar travel and had no form of propulsion once it had left earth's atmosphere. "To describe Pioneer 10 as a means of interstellar travel," Friedman said, "is like tossing a bottle into the Atlantic and calling it an ocean liner."[16]

My gut sense about Sagan was that he revered his elevated standing in the scientific community far more than he valued the whole truth--especially when said truth might encompass the presence of intellects beyond Earth with vastly superior technologies that might make ours seem child-like in comparison. Not an easy thing to swallow when one is accustomed to a self-ordained position at the top of the galactic totem pole. Accepting and acknowledging the presence of superior intelligences elsewhere would necessitate he surrender his throne. Openly admitting he was not the crown of creation: not a workable hypothesis for the good doctor.

Hynek, the voice of reason, summed up the Cavett show discussion by saying, "Well, if these are intelligences, then they know something about the physical world that we don't know, and they also know something about the psychic world that we don't know--and they're using it all."[17]

As the Pascagoula encounter began to fade from the spotlight, it could at least be said that Dr. James Harder and Dr. J. Allen Hynek, both highly respected in their professions, believed Hickson's and Parker's story. In one of his final comments to the media about it,

Hynek stated, "There was definitely something here that was not terrestrial."[18]

When I read these summary words from such a respected source, I thought the Hickson/Parker abduction would solidify the ETH, transforming the hypothesis into an accepted reality. The Pascagoula incident would break down the barrier of denials and secrecy that the U.S. government and its many departments had foisted upon us, denying the reality of UFOs and alien visitations.

I snatched up a copy of *Beyond Earth* when it appeared in 1974, and found it to be a superb source for material on the Pascagoula experience, as well as a wide range of other UFO cases. Certainly one of my favorites--and one that merited more attention--was the 1967 abduction encounter of Nebraska police officer Herbert Schirmer. For a young sci-fi addicted kid, the aliens the officer encountered were a real treat, straight out of *The Time Tunnel*--skin tight silver flight suits that came up and wrapped over their heads like a helmet, with an antenna protruding from the head's left side, and a serpent-like emblem emblazoned on their chests. Onboard, the control room was all about high-tech controls, with a vision screen displaying a demonstration of how they draw electricity from power lines, and another showing a group of saucers flying in outer space that were supposedly "warships." Schirmer is also told by the frigidly dispassionate leader that, while they were communicating, information in the form of data was being transmitted into his mind. The Blums attest that Schirmer had been thoroughly checked out medically and psychologically, and that he had an impeccable history regarding his family, his health and his background.[19]

Regarding the possibility that the Pascagoula UFO case would become a springboard for disclosure, Ralph and Judy Blum's hopeful beliefs and assertions were wrong, as were mine. Opposite their book's Title page, under the heading, ABOUT THE AUTHORS, the Blums optimistically opined, "After a year of studying the evidence and listening to the people who have had close contact with them, it is impossible to say that UFOs don't exist. We predict that by 1975 the government will release definite proof that extraterrestrials are watching us."[20] Alas, Ralph and Judy, we're a full forty years beyond your prediction, and there's no sign of that magic word, disclosure.

We're well into the second decade of the 21st century, and still the government and military cover up remains intact.

BOOK III.
CONTACT: BEHIND THE GRAY WALL

"Do you consider us so boring or so repulsive that of all the millions of beings, imaginary or otherwise, who are prowling around in space looking for a little company, that there is not one who might possibly enjoy spending a moment with us? On the contrary, my dear–my house is full of guests. . . ."

–Jean Giradoux
The Madwoman Of Chaillot

The summer before my senior year, Clay's company ricocheted us back to Houston. Having finally forged friendships with some neighborhood boys, and my French Quarter/Uptown bar hopping pals, that was rough. I had enjoyed the Crescent City and its colorful diversions; the Big Easy is an exotic locale like no other. But oil company brats, like Army brats, are an adaptable lot (we have to be), and my siblings and I managed to cope with another uprooting.

After graduating high school in Houston, I attended UT Austin, which soon became my favorite city in Texas. I loved the beautiful campus, the superb professors and departments, the warm weather, the amazing prevalence of movie-star looking girls and guys, the liberal minded outlook and the myriad outdoor watering holes for lap swimming, like Barton Springs Pool and Deep Eddy Pool. The healthy atmosphere and health conscious locals suited me, a wandering lost soul in search of healing and recovery from a turbulent adolescence he'd never quite overcome. In time, I would see Austin as the jewel in the belt buckle of the Sun Belt. Alas, today about 70,000 newcomers flocking here every year think so too.

My UFO obsessions remained, as did the curiosity about my Houston close encounter. More than ten years on, and flashbacks could still distract me during college classes, so I vowed to uncover the mysteries of that night in October of 1968. Opting to try time regression hypnosis like the Hills had, I found a psychologist in the Yellow Pages (Summer 1979, pre-web & Google) who was skilled at clinical hypnosis. I started with the "A's" and skimmed down until a name and accompanying display-ad caught my eye. Thus providence–and a lot of dumb luck–led me to Dr. Edmond M. Bazerghi.

Dr. Bazerghi listened attentively as I recounted the details of my

Houston sighting. He asked about my family dynamics, my current personal life, how school was going, who I was dating. In my second session he administered an MMPI, the Minnesota Multiphasic Personality Inventory. I found it fascinating that the results looked ok (whatever "ok" indicates in a psych test), except for the scoring of what he referred to as the Subtle Fake Good Scale. When I asked him to translate, he said, "Your scores on this scale mean you're glossing over some things. I'm inclined to believe you've had a much tougher and more problematic relationship with your father than you're admitting." Dr. Bazerghi was an extraordinarily perceptive man. He zeroed right in on the most painful and troubled relationship in my life. We began much deeper talks about Clay and me.

He seemed concerned about my sexual experimentation, not so much that I was dating both guys and girls (HIV and AIDS weren't in the lexicon in 1979), but regarding what he perceived as my lack of personal and psychosocial integration. He wasn't at all judgmental about sexuality, but it did seem to me he was hoping I'd eventually choose to play on the hetero team. I already knew I dug guys better, but the lingering Catholic-infused guilt and socially imposed shame about living an alternative lifestyle prevented a full acceptance of my self-identifying as a homosexual. Why take on all the weight and baggage of the sissy boys if you still liked to kiss girls? And I knew I'd never be able to identify with the common, everyday cliché homo anyway. Limp wrists, heavy lisps? Bitchy posturing? No thanks. Gay pride parades? Not really that interested. How about antisocial, maladjusted Bisexually Active Appreciation Day? BAAD! I'm in, let's go.

Dr. Bazerghi confided in me that he'd struggled mightily dealing with human sexuality issues. He told me that as a former Jesuit priest for seventeen years, he'd gone through anguished cycles of inner turmoil and analysis trying to decide for himself where he stood on heterosexuality, homosexuality and bisexuality--the full spectrum. He told me that he'd nearly had a breakdown in his quest for the truth of it all.

"And what did you decide?" I asked, fascinated that he had a wife and family by then.

"I concluded that you can take your dick, and stick it into a vagina, and have sex with a woman; or you can take your dick and share it with another guy, and have two dicks together; or you can punch a hole in the wall and shove your dick into that, and fuck the wall. But none of these behaviors has any more nor any less validity than the others. Sex is communication. Sex is *only* communication. Nothing more, and nothing less than *communication*." Talk about having your mind blown, in the best possible way.

His wise words made me wish I could have sent a recording of them to all the bigoted, homophobic redneck preachers who spread hatred and small-minded judgment against sexual free spirits like me. In time, I would wonder if they all had willfully turned their Bibles into a Handbook Of Hatred, relentlessly trying to convince us that Jesus was a spiteful, judgmental, intolerant redneck. It seems to me like all the fire-and-brimstone preachers have forgotten Jesus was never about hate, only love. Period. Dear God, what have they done to your Son? Dr. Bazerghi's temperate, tolerant words of unconditional acceptance about human sexuality and its use as a communication tool--words from an ex-Jesuit priest, no less, yet utterly free of any religious dogma--would stay with me through the years.

In our third counseling session, Dr. Bazerghi introduced me to hypnosis. I proved an easy subject, and he guided me into and emerged me from several hypnotic trance states, to acclimate me to the process before we took the deep plunge. Once we did, I was hooked. I had no idea the hypnotic state of consciousness would feel so pleasant. Like floating down a river. Just . . . enjoying . . . the currents . . . relaxing . . . and breathing. . . .

I also had no idea that utilizing regression and returning to my close up UFO encounter would engage such a zen-like state of awareness. I really was there, kneeling on my bed as the object lunged down at me. The state of shock I had felt when the intense, all encompassing white light absorbed me seemed very much in the here and now. In some ways, I experienced bilocation: the object was so close, right there outside my window, and scary as hell, when in fact I was comfortably seated, eyes closed, in a skilled psychologist's quiet office in South Austin. The mind is a miraculous machine. It's one of the reasons I firmly believe in a Supreme Being, a Creator.

At the doctor's cue, I emerged from the blinding white-out of the UFO's close up intrusion, where there seemed to have been a sticking point, and my eyelids slowly fluttered open. The first thing I perceived was the doctor's serious face, and the soothing sound of his voice.

Glancing at my watch, I was surprised to see that nearly forty-five minutes had elapsed; it had seemed like maybe ten. So this is what they meant by time dilation during hypnotic trance states. Following some small talk about my first hypnosis session, Dr. Bazerghi gently said, "Wade, your subconscious has chosen to lock away from you a lot of information about this experience. It's behind a locked door. That's why we weren't able to get beyond the point of the object's closest approach. There's a block."

"A block? Hmm . . . yeah, I guess it did feel that way. I couldn't get past the white light. It's like I'd blacked out with my eyes wide open. Do you think we could get through it in another session or two?" I asked.

"Possibly. But I don't think that's a good idea. I would advise against further probing right now. You're already under enough stress trying to come to terms with your sexuality, your sexual issues of being straight or gay. I would advise against it."

"No way! We're almost there!" I pleaded. "I've been waiting over ten years for this. I need to know what happened. I deserve to know all of it!"

"I think it unwise to try to open that locked door right now," he said calmly. "You can go to someone else and they might be willing to work at it and force it open for you. But for now I say, trust your subconscious. When your subconscious is ready for you to know more about this experience, it will allow you to unlock the door and let you into that room. But until then, I'm not going to be the one to put the key into the lock and unlock the door. It's not the right time. Not right now. Trust the wisdom of your subconscious. It knows what's best for you. Deal with your other issues first."

The "deal with your sexuality" became three more sessions in which Dr. Bazerghi tried to help me see that I could be happy in the straight world, and leave the gay world behind. Perhaps my times with guys were merely a trial run, enjoying it just for the sexual heat of it. The old college try. After all, I'd had beautiful steady girlfriends in high school, and had even dated my gorgeous blonde French teacher at UT, a dead ringer for the singer Nancy Wilson of *Heart* in her prime. I wasn't gay, I was simply going through an experimental phase--a glorified college try that snowballed into a very long term scientific project, an in-depth, detailed study of All American Muscle Men of the U.S. Sun Belt, Late 20th - Early 21st Century.

In our last session, Dr. Bazerghi shook his head sadly and said, "I see how much happier and well adjusted you could be with women, and being an active part of the straight world. Gay men pay a terrible price in psychic hang-ups and physical, emotional and psychic damage from their lifestyles; it's unfortunate, but the straight world is very unforgiving and judgmental towards gay men. Even though I hate labels, for now you are a gay man. And about this, I say dammit . . . dammit . . . dammit!" His impassioned rhetoric surprised me. But his genuine concern was evident. He cared about me. I didn't divulge that the mind-blowing hot sex seemed like a fair trade off. (In the end, it was really about me being honest with myself.)

Dr. Bazerghi's denial of access to the locked room in my subconscious mind frustrated me terribly. Perpetually curious about UFOs and my close encounter experience, I craved answers so badly that I wanted to rip the door off of its hinges. It would take me years to appreciate what a good deed the doctor had done. His wisdom and caution saved me from a violent tidal wave of profound psychic trauma. It was also a permanent testimonial to his good, big heart and his impeccable therapist ethics: he had favored my mental, emotional and spiritual wellbeing over the wellbeing of his bank account. I will never forget that about him. The true mark of the man. Noble. Wise. And beneath it all, a good heart beating at the core.

Years passed. My wanderlust drew me to Los Angeles, where I became a personal trainer and massage therapist to the film crew of *The Best Little Whorehouse In Texas*, having met some of them on location in Austin. The trainer/masseur gig worked well: an independent, portable income to buy the time to write. L.A. certainly gave me plenty to write about. It was fun at first–once you got used to movie people, the prevalent phoniness, the traffic jams, the insincerity, the smoggy air and the electronic billboards over the freeways warning you to stay indoors tomorrow due to poor air quality–words you read through squinting eyes, your red, watery eyeballs stinging from the smoggy air *today*. After awhile, L.A. had stung me in a lot of ways. But I did have my fun. An up and coming movie star here, a *Playgirl* centerfold there, a painfully handsome top model who was anything but a top. *The Athletic Club*, where hot West Hollywood boys buffed up before hitting the bars: *Rage*, *Revolver* or *Studio One*. Summer, 1984, cue the ever-present gym music: *Hold Me Now* (Thompson Twins) and *No More Words* (Berlin), one song craving intimacy and touch, the other one pushing the lover away for more space. Apropos accompaniment mirroring my troubled psyche's need to have a healthy dose of both back then.

In retrospect, the primary element about L.A. was that I would cross paths with Carlos, a devilishly handsome, jaded, globetrotting Cuban-American player visiting Tinsel Town for the 1984 Summer Olympics. We met at a post-Olympics party in a villa at the top of a quarter mile long paved stone driveway high above Beverly Hills, one of those marble, limestone and glass show palaces with a cavernous atrium entry hall decorated very hotel lobby, wall to wall with pretty people, actors, models, movie stars and movie star wanna-be's, the smiling faces spilling out into the lush backyard beneath an archway of rainbow hued balloons festively tethered across the sprawling infinity pool carved into the edge of the hillside, the seductive fantasy lights of L.A. twinkling down below like a giant fallen Christmas tree that nobody had ever bothered to unplug.

Anyway, no matter that we connected, went to the condo he was crashing at to play awhile, and he left town the next day; and no matter that I visited him in Miami a few months later, and he ended up moving to L.A. to be with me. And no matter that our El Lay relationship quickly devolved into a game of seeing who could play around the most behind the other's back in our West Hollywood muscle boy showboat playground. What did matter is that his parents, in typical Cuban style, missed their niño guapo so terribly that they offered to buy us a house in Coconut Grove if Carlos would only return home to the family. (Cuban-American families are tight, and they kindly treated me to meals and lodging like a family member while we house hunted. Their warmth and generosity were incomparable.) And so it was that in my strange, zigzag, Texas tumbleweed journey, being in Miami in the late eighties and early nineties would turn out to be a very important thing, integral for uncovering the UFO mysteries from my past. Meeting Carlos played a crucial role in my voyage of self discovery, because Miami is where I found the key to unlock that locked door in my subconscious mind that Dr. Bazerghi had declined to open. And fittingly, with a twist.

So there I was in Coconut Grove. The relationship with Carlos tanked barely six months after our arrival, the perpetual presence of powdered nose candy surrounding him like quicksand, the final blow to our tenuous bond. So I hit the discos and bars, played the field, sewed some wild oats, and after one more shot at a relationship, I decided I was better off single. Being a free agent suited me well. I was alone so often that it seemed sometimes that books were my best friends, people merely occasional companions, and sex a combination of a recreational sport and my pain killer of choice.

My dumb luck intact, I blundered across and settled into an enchanted property in the heart of the Grove, a lush two acre spread for rent--practically unheard of. It was met with enthusiastic approval from my 80 pound German Shepherd, Alf, the beloved dog Carlos and I had bought in L.A. as a pup. The property boasted a stone tower with rectangular windows, evocative of a castle turret from a Tolkien book, with an attached screened-in porch serving as my living room/dining room/second bedroom, overlooking a lily pond and lush, jungle like grounds. Behind the tower door, a bomb shelter awaited: a concrete spiral stairway led down twenty feet to the round tower room, then an L-shaped hallway set off by lead-lined steel doors, and finally a large loft-like space with a living area, bathroom, kitchen and storage closets. The owner, Kenneth Brandt, had built the shelter in the early 1960's after the Cuban missile crisis. He lived in a beautiful neo-classical villa around the corner, and slept soundly knowing that a radiation-proof stronghold would protect his family should commie nukes from Cuba have ever rained down on Miami. It even had a gas powered generator for electricity and purified air. I was ready for Armageddon--and what better digs for a solo bookworm?

Mr. Brandt proved to be the most kind hearted landlord anybody could ask for. When I started working on my Psychology B.A. at FIU--Florida International University (with 60 hours from UT)--he periodically allowed me to work off part of my rent by doing painting and odd jobs around his properties. He was a godsend. Anytime I was short, he'd always deflect the news, saying, "Well, your rent can wait until you have it. Your education is the most important thing. Money is secondary. Education--that's the main thing." They really broke the mold when they made Kenneth Brandt; he was a scholar, a gentleman, a man's man, and a true class act.

Having been long vacant, the subterranean living area was crypt-like, a dark, dusty humid space overrun by cobwebs. A pungent limestone scent lingered in the thick air. A few months of renovation later, and I'd transformed it into a choice bachelor pad. Against the longest wall I placed three bookcases, filling them with my personal library. I was accumulating quite a collection of UFO books, easily a third of the volumes on my shelves.

The late 1980's and early 1990's saw Jacques Vallee, one of the more respected names in the UFO genre, crank out what *The Kirkus*

Reviews referred to as his "Alien Contact trilogy" of books, each title in all caps on the cover, and each with a subtitle: *DIMENSIONS: A Casebook of Alien Contact, CONFRONTATIONS: A Scientist's Search for Alien Contact*, and *REVELATIONS: Alien Contact and Human Deception*. I devoured each of them as they appeared, having always found his analyses thought provoking, insightful and sometimes controversial.

One of the things that really jumped out at me in *DIMENSIONS* was a passage about the Barney and Betty Hill abduction, in Chapter Four under a sub-section: **NEW HAMPSHIRE REVISITED.** It refers to an Air Force incident as notated in an official document. From the text:

> Report No. 100-61, in the files of the 100th Bomb Wing, Strategic Air Command, Pease Air Force Base, New Hampshire, was prepared by Major Paul W. Henderson. The only official document concerning the Hill case, it contains a detail of which both Dr. Simon and John Fuller were unaware: *the object seen by the Hills had been detected by military radar.*
>
> During a casual conversation on 22 Sept 61 between Major Gardiner B. Reynolds, 100th B S DC01 and Captain Robert O. Daughaday, Commander 1917-2 AACS DIT, Pease AFB, N.H., it was revealed that a strange incident occurred at 0214 local on 20 Sept. No importance was attached to the incident at the time.[1]

This simple piece of information hit me like a head-on collision. Ostensibly the most fully documented and respected UFO abduction report ever, the first case expanded into a book in fact, and nobody with access to this information came forward to admit that the UFO the Hills encountered had been tracked on military radar? Why hadn't this gotten major coverage? This crucial fact should have been widely disseminated.

Reading a little further, Vallee really started to lose me with his attitude towards the Hill's encounter, placing key words in quotations:

> I have heard the portion of the tapes covering the "abduction" of Betty and Barney Hill . . . One such detail is the recollection by Betty Hill that, after their car was stopped and a group of "men" had come toward them, the creatures had opened the door of the vehicle and pointed a small device at her.[2]

This undisguised condescension marked for me the beginning of what I would call Vallee's periodic descent into "the Land of May-be-So;" his many years of research seemed to make him vacillate between validating the ETH and then denigrating it in favor of a number of others, such as the concept that an as yet unseen alien intelligence, possibly Earth based, could be training us toward a new type of behavior; or a Jungian variant of this, that the human collective unconscious could be projecting ahead of itself imagery necessary to aid us in our long term survival in the troubled 20th and 21st centuries; or a variant of the Gaia hypothesis, briefly, that all the biological organisms of Earth function like a single organism living in a self-regulating system, presumably meaning that the UFO/abduction scenario is emerging from this system to teach us greater respect for the biosphere and all life within it (sorry guys, it's not quite working up to snuff just yet); or that ET travelers are utilizing radical space-time manipulation, such as use of four-dimensional wormholes for space and perhaps even time travel.[3] The first and the last hypotheses resonate strongest with me, but I'm open to them all.

Speculative variations aside, am I to believe that the military radar which tracked the Hill's UFO was getting a solid return from an illusion or a mythical craft? Or was Vallee insinuating that this was a military PsyOp? That was certainly the attitude he took here. Oddly, he denigrates the reality of Betty Hill's onboard medical examination, but then a few sentences later, expresses that we must "relate" the onboard material with the "physical features" of the military radar tracking the UFO. You lost me there, Jacques.

In his trilogy's final volume, using variations of statistical analysis, Vallee concluded that the total number of alien visitations to Earth would yield "a total estimate of 14 million landings in forty years if we strictly adhere to the ETH."[4] He seems to imply that this is an absurdly high number (by our standards), and therefore the data won't allow that these are alien spacecraft (why would they bother with so many landings?). So we toss the data, erase the chalk board, and say "Next hypothesis?" Suppose we're dealing with a billion-year-old alien technology in some cases? I'd wager they can hop-scotch across the cosmos with inter-galactic jumps! Journeys of billions of light years might be simply a vast board game to them.

The too many landings argument led me in a different direction. My belief: the Universe is possibly spatially infinite, but almost certainly infinite multi-dimensionally. Therefore our planet, and all of known interstellar, galactic, and intergalactic space, could be merely parts of one great big time share condo—except in this scenario, all the tenants co-exist simultaneously, multiple universes sharing the same space. I also believe that our initial atomic test blasts (and later H-bombs), followed by the brutal nuking of Hiroshima and Nagasaki, didn't just kill people (and/or cruelly damage their bodies) and vaporize buildings and tons of matter. The nuclear fireballs also tore the fabric of space-time, and in those brief instances of thermonuclear ignition, bored holes from our Universe into parallel universes (notice how UFO sightings spiked following World War II ?), inadvertently opening up interstellar and inter-dimensional superhighways to other dimensions, alternate universes, and therefore multiple alien cultures. So the atomic flares, detectable from space, didn't just signal alien cultures from other star systems that we'd made it to the nuclear age; they also opened up multi-dimensional gateways for journeys from many places, and many times, multiple universes. It then follows that myriad alien cultures and trans-dimensional travelers would visit. And multiple sightings in numerous locales might be the same objects, in transit, up ticking the UFO report counts. Finally, factoring in the U.S. and Russia's 2,000+ H-Bomb tests since 1945, we may have effectively bored open a panoply of Star Gates scattered about the globe, ramping up Sol-3 for staggeringly hellacious numbers of alien visitations.

After all this, Vallee comes full circle and states in his thought provoking book *REVELATIONS: Alien Contact and Human Deception (1991)*--the final volume of his Alien contact trilogy:

> The arguments raised here are not intended as a complete refutation of the ETH. Until the nature and origin of UFO phenomena can be firmly established, it will naturally be possible to hypothesize that extraterrestrial factors, including undiscovered forms of consciousness, are playing a role in its manifestations. But any future theory should constructively address the facts we have reviewed. At a minimum, the idea of extraterrestrial intervention should be updated to include current theoretical speculation about "wormholes" and other models of the physical universe.[5]

Worthy of note here, too, is that Vallee briefly commented on what he refers to as the outright hoax of Eduard "Billy" Meier.[6] Billy Meier first surfaced on the UFO scene when the oversize photo book, *UFO . . . Contact From The Pleaides, Volume I* (Elders &

Welch) appeared in 1980.[7] While attending UT Austin in 1981, a friend gifted me a copy. I was immediately struck by the extremely high quality and razor-sharp clarity of some of the images, particularly those showing the shiny metallic underside of the saucers in flight against a wide open blue sky.

Beginning around 1975, Meier, a one armed Swiss farmer with a 6th grade education living a simple country life with his wife and children outside Zurich, purportedly had over 100 contacts with the beings flying the discs, human looking aliens who claimed to be from a planet in the Pleiades star cluster about 425 light years away. Although the story is regarded by some as one of the best crafted UFO hoaxes of all time, there are nevertheless interesting elements which remain, among them being the mystery of how, if this is a hoax, some of the extraordinarily clear, close up photos were taken by a solitary one armed man. My biggest beef? If it was all for real, the photos and the story, then why didn't Semjase (the beautiful Pleaidian female cosmonaut) consent to having Meier photograph her "beamship" after it had landed on the ground and settled onto its landing gear in the lush meadows of the Swiss countryside outside Winterthur and Hinterschmidruti? A couple of clear photos like that, with, say, a tractor parked next to it for scale, would have established solid credibility. A Super-8 movie of the whole process, even better (Meier did shoot movie footage, but only with the objects in flight).

An intriguing side note to the Meier case: while living in Houston between early 1982 and Spring 1984, a buddy named Mark Dandridge came by for a tennis date. He scanned my bookshelves before we left, and when he saw the Meier photo book, opened it up and said, "Oh. Interesting. My dad has this book too." It should be noted that his father was a respected real estate broker and pillar of the Houston business community, and as a private pilot sometimes flew his family around on vacations. It seems that in the summer of 1981, the family had been vacationing in Monaco, and Mark's father was out jogging early one morning on the mountain loop. The sun had just risen, and the air was clear, with no clouds. Suddenly a disc shaped craft came into view, flying over Monte Carlo. The metallic disc hovered silently, slowly oscillating until it steadily rose along the mountainside, then disappeared at high speed without a sound.

When I finally got to meet him, Bob Dandridge told me that the UFO looked nearly identical to one of the classic saucer shaped craft that Meier had photographed, even down to the highly polished metallic underside of the vehicle, which in this case was reflecting a near mirror image of the Mediterranean below it. He said that he felt like the pilot of the craft was showing off for him, putting his high performance vehicle through its paces. Gazing along the coastline, the only people up at that early hour were a construction crew hundreds of yards away. He shouted out to them to try to share the UFO sighting, but to no avail. He told me, "Wade, I've been a pilot for years, and I know aerodynamics well. And I can tell you for certain that this thing outperformed by far anything we have." After a minute or two of the aerial show, Mr. Dandridge said the disc suddenly stopped, gently tilted up at a sharp angle, and took off in a flash, disappearing almost instantly without a sound. When he came across the Lee Elders and Thomas Welch UFO photo book of the Meier case in a store, naturally he grabbed a copy.

I took Bob Dandridge's sighting at face value, knowing he was a private pilot with a solid reputation, and had a very respectable business and social background firmly established in Houston. I found it fascinating that he had witnessed this UFO in the same general vicinity as Meier's saucers (Zurich, Winterthur and Hinterschmidruti are practically next door to Monaco), and it looked virtually identical to the discs in some of Meier's photos, like "Semjase's beamship," which had so often been deemed models or fakes.

Discussing Mr. Dandridge's UFO encounter with its intriguing coincidences made me more inclined to take another look at the Meier material, so shortly thereafter I bought a copy of Billy Meier's so-called contact notes, a thick hardback book that recorded in journal format what he'd ostensibly learned from his many meetings with the beautiful blonde Pleiadian cosmonaut, Semjase. The bulk of the notes were purportedly dictated to him by her. From the hardback book, *UFO Contact From the Pleiades: A Preliminary Investigation Report* (Stevens, 1978), I call this passage "Semjase's Warning to Mankind:"

> So there exists sorts which have attained advanced development and knowledge and have made themselves free from their planet and life regions and travel the universe, and occasionally come also to your Earth. Some of them are rather nasty contemporaries, and live in a barbarism even worse than yours. You must take care before these because they fight and destroy everything that comes in their way. They have even destroyed whole planets and beaten their inhabitants into bondage. This is one of our missions, to

warn Earthman of these dangers. Let the Earth humans know this, because more and more, the time approaches, when a conflict with these becomes unavoidable.[8]

Nineteen eighty seven was a banner year for new cutting edge UFO books, and two of them fairly jumped off the book seller's shelves and into my personal library. One of them, *Light Years: An Investigation Into The Extraterrestrial Experiences Of Eduard Meier*, allowed me to delve even further into the background of the Meier material. The author, Gary Kinder, spent a lot of time and money researching the controversial case, devoting many hours and days to Meier, his hangers-on and other researchers. In one of my favorite passages, again spoken by Semjase to Billy Meier, she explains that their beamships use a two-drive system, one which accelerates them up to nearly the speed of light, and the other, a "hyper-drive" in our terminology, which simultaneously paralyzes space and time, collapsing them into "null space and null time." Once this is achieved, they are able to traverse light years in fractions of a second, but the process is engaged in such a manner that the living beings onboard are unaware of it. Semjase continues that it takes them a mere 7 hours for a one-way trip from the Pleiades star cluster and their home planet of Erra to get to Earth, a distance of about 420 light years (next door neighbors in the cosmic scheme of things); upon departure, they take 3.5 hours to accelerate up to near-light speed, getting out into open space beyond their home planet, then engage the hyper-drive and make the jump, and in the wink of an eye they're on the outer edges of our solar system. Then they take another 3.5 hours to get from there to Earth, making the remainder of the journey in "normal drive." She confides in Meier that she's not allowed to divulge specific details of the 2-drive system, but that, "I can tell you that your advanced scientific circles are already working on systems known as light-emitting drives and "tachyon" drives . . . the light-emitting drive serves as the normal propulsion system to move the ships to the limits of space and time. Once there, the tachyon drive is brought into action. This is the hyper-propulsion system, which is able to force space and time into hyper-space. We use other names, but the principles are exactly the same."[9]

This fascinating discourse is further amplified by the fact that Kinder and his research team never found any evidence that Meier, a man with a sixth grade education, had ever collaborated with any-body on these somewhat technical contact notes. And most of the physicists they'd contacted didn't even know what a tachyon was back in 1978. So for Meier to speak of a space propulsion system whose operation used theoretical and not widely known particles was a huge step. (Tachyons are now broadly accepted in theoretical physics as faster than light particles; the existence of the tachyon, though not experimentally established, appears consistent with the theory of relativity.[10, 11])

The Meier contact notes make for fascinating reading. Despite many skeptics adamantly branding this case a hoax, I still keep my mind open about it, partly because of Bob Dandridge's sighting. And striking elements abound in the information imparted in the contact notes; the material Semjase dictated to Billy often reads like it was written by someone with a sophisticated outsider's viewpoint, an off-world being with an extremely broad and detailed knowledge base.

However, because I seek truth above all else, it must be mentioned that Kal K. Korff's heavily researched *Spaceships Of The Pleiades* (1995) unabashedly calls out Meier's UFO case as being an outright hoax, and he backs up his assertion with some powerful evidence, photographic and otherwise. The book is well worth reading. I won't quote from it, but what I still find an unresolved question for me is: how is it that the model that Meier purportedly used as Semjase's beamship in faked, double exposed and/or pasted in photos (the classic "sport model") looks like it's made of a dull pewter metal, and yet some of Meier's clearest photographs of the saucer, when viewed from below looking up at it in a clear blue sky, reveal a highly polished, almost mirror-surfaced metal on the underside, which re-flects the ground below it? The fact that Mr. Dandridge had viewed in plain sight a clear daylight-disc sighting of a saucer in Monaco that looked virtually identical to Meier's polished metal beamships, and which reflected the Mediterranean from its underside's seem-ingly polished chrome finish, will always intrigue me, particularly since I see Bob Dandridge as an utterly unimpeachable witness. His background and reputation are as solid as granite.

The other cutting edge UFO book of 1987 was Whitley Strieber's *COMMUNION*. When I saw the bug-eyed creature depicted on the cover, I laughed out loud in the book store. But after skimming the blurb and thumbing through the pages, it seemed worthy of reading, so I bought a copy.

Strieber's aliens, and the rooms they took him to on their vessels, seemed straight out of a Grimm's fairytale to me. Not to say I discounted the potential veracity of his reported encounters, but rather, that both the fierce little creatures and their onboard environments had about them elements that seemed more like mythical gremlins and their dark, musty lairs replete with stale smells—more organic and dirty than the prevalent abduction reports with sleek, high tech environments and gleaming craft boasting bright lights and seamless, stainless steel tables and walls. The contrast intrigued me.

I was quickly taken in by Mr. Strieber and his tale; he is an abundantly gifted wordsmith, and I found his book an engrossing read. His frankness about his use of M.D.'s, psychiatrists, professional counselors and hypnotherapists I also found compelling, and I plowed through *COMMUNION* in just a couple of days. Immediately I wrote a letter to him, praising his book, sharing my UFO experience and asking if he might grant me an interview for a freelance article.

A few short weeks later, I received a neatly composed letter from Mr. Strieber, full of encouragement about my encounter and cordially agreeing to an interview. I was floored that he'd taken the time to write me, and I phoned him at the number he'd enclosed. He told me that I'd be his 200th and last interview in America; next up was a promotional tour of the U.K. We set a date for a Friday in April of 1987, and on that morning I flew from Miami to La Guardia.

It was a brisk, overcast early afternoon when I arrived at the café in Greenwich Village. I took a corner table and drank hot tea as I watched a steady rain spatter the large windows. It seemed like the perfect setting for a tanned Miami boy to connect with a Manhattan intellectual. When Whitley and his wife Anne arrived, I recognized him right away. Their female companion introduced herself as Mr. Strieber's amanuensis (I looked that up--she transcribed his notes).

After exchanging friendly hello's, the first thing I noticed about Whitley was his enigmatic countenance: a youthful face with almost boyish features, but with eyes that looked like they'd had about a million years of experience. Not to say old or tired by any means; more about an exposure to raw terror. The man had seen some very scary things.

I tape recorded the interview, which lasted over an hour. Whitley quickly impressed me as a very cordial, sincere, exceptionally articulate man; he spoke to me in a calm, clear, steady voice with frequent eye contact. Herein are the high points of our talk:

Wade Vernon: The intense nature of your alien encounters is such that these experiences would tend to stay with you. Do you still find yourself distracted on a given day, your mind replaying images of the abduction?

Whitley Strieber: Not really. I've gotten used to the whole thing. First of all it's happened quite a number of times in 1986. I've become intricately involved with them, whatever and whoever they may be. I wouldn't say I'm used to it–when the encounters occur they come unexpectedly, there's no forewarning–not much forewarning, anyway. If anything, they're more intense now than they were a year or a year and a half ago. I found them to be something I couldn't change, couldn't do anything about the continuations–the whole experience.

Most of the people I've encountered who have any kind of professional competence as scientists just flat out didn't believe me. The others tended to think I was a fraud or took some other emotional viewpoint. The result of this is that I've met, over the past year or so, some really super scientists who are so good that they remain open-minded even in the face of the most extremely strange experiences. And that's wonderful that there are people that good. I found that these people were generally not involved in the UFO community, that their minds had not been essentially closed to the extremely high level of strangeness that is being experienced by people.

The people involved in the UFO community want it to be a sort of nuts and bolts thing. They want it to turn out to be real physical visitors from another planet that has a name, and eventually we can go there. That may be at the core of this experience, but if real physical visitors are at the core of this experience then they are very different from us–so different that it might be that we will never really be able to communicate with them in a manner that is entirely meaningful to both sides. In any case I don't think that the scenario implied by the people who believe that UFOs are actual physical devices and their occupants are as physical as you and me is a real scenario. I suspect that this is something much different and much more complex than that. In part, it involves a quite inexplicable external experience with a very real intelligence. And in part it involves internal spiritual and mystical experiences on the part of the person affected. And when it's narrated by that person, the person tends to narrate it in comprehensible human terms that may or may not be what happened.

WV: Although the public gradually has grown more accustomed to UFO encounter stories, particularly with humanoid contact, much of the media is still extremely skeptical, for example, the unfavorable review *Time* gave your book. Did you have some misgivings about going public with your story?

WS: Sure. I knew I'd be given a hard time by places like *Time* magazine. *Time*, Inc., lives in the past. It belongs to a structure of mind that's dying. By that I mean that it's essentially linear, rationalistic and humanistic. And that's not the way the world will be in fifty years, no matter what happens. Before the next fifty years is over *Time* magazine will cease to exist, and they know that as well as I do. When they see something that is clearly of the new mind, they lash out against it because it's like telling them they're dying, and that's scary.

WV: Aside from the rather nasty comments *Time* gave you, how have the media and public reacted to your book?

WS: It's been fascinating. The public reaction has been overwhelmingly positive. We've gotten nearly 1,000 letters. Two have been negative. Two out of a thousand have been negative. And this is a book that has gotten very wide distribution and my address is in the back of the book. So there's no question about the fact that they can get to me. I even have a listed phone number in New York so they can get to me that way if they want to. The electronic media have been very open minded, for two reasons. One is that they perceive correctly that the public is very much on our side. And they want to serve the public. Two is, I've found a rather disproportionate number of people in the electronic media who've had abduction experiences or UFO encounters of an unusual type. I've encountered now eighteen people, either producers or on-air personalities, who have reported such experiences to me. Among the print media people I've been involved with none have had such encounters, which perhaps reflects a perception on the Visitor's part that the electronic media is more important (laughs). I don't know. No–that's a joke.

I've been on quite a number of national shows. Except for *The Phil Donahue Show*, I was never treated with disrespect–but for one other show, *The Jim Altoff Show* on King Radio in Seattle. They treated me with great disrespect.

WV: The callers or the people behind the show?

WS: No, no, the people at the station. Callers hardly ever do. I occasionally will get a skeptical caller, someone who says they just can't believe it, but they're not ever nasty. Never. Nastiness basically comes from people whose ideologies don't fit this experience, in other words from fundamentalists and from conservative scientists. I think Carl Sagan in his attitude towards this has much more in common with the creationists than he does with real scientists.

WV: Maybe the bottom line of that is just fear. They fear it so they have to lash out at it.

WS: Mmm-hmm. The whole thrust of the interaction I've had with the Visitors in 1986 has to do with learning to fear them less. The future not only of my relationships but of all our relationships with them depends upon our ability to do this. And they've said that it's going to be very hard for us to adapt, and I think they're right.

WV: What would you say was the most terrifying or negative aspect of the experience? The most positive?

WS: I don't even divide it into negative and positive aspects because that's very artificial. In fact some of the things that seemed most negative were the most useful and the most insight-producing. For me the experience has been about the gaining of insights that have helped me to understand both myself and my relationship with the world, the visitors and ultimately also with God. It's been a deeply spiritual experience. Sometimes enlightening things can be very hard. In the old animist religions, enlightenment is achieved through terrifying and painful initiatory processes. When one confronts the Visitors it's terrifying and painful, but if one can get beyond the fear it usually also can be tremendously enlightening. One encounter with the Visitors is worth half a lifetime of meditation. (laughs) I can assure you that's true.

WV: A quote from the book: "I did not have the feeling they were hostile so much as stern. They were also at least somewhat frightened of me. I was certain of that. . . . In some sense, their emergence into human consciousness seemed to me to represent life–or the universe itself–engaged in some deep act of creation." Can you explain what you meant here?

WS: What I meant is this: I visualize them as a force of nature. I strongly suspect that they have been present throughout our entire history, and probably throughout the entire history of the universe. I think that they may represent mind penetrating into physical reality, and we may perceive them as creatures and so on and so forth because of the way our perceptual system works, and what we may actually be encountering is something very much more inexplicable than meets the eye.

WV: When the leader told you, "You are our chosen one," your immediate reaction was, "I don't believe that for a minute. It's ridiculous . . . sing that song to somebody else. I want to go home." Then she asked you, "What if we don't let you go home?" Judging from what transpired onboard, are you under the impression that they chose you for some sort of breeding experiment?

WS: I think when they choose people they choose them for dozens of different reasons. I wonder about the breeding experiment business. I know a lot of UFO investigators believe that's what's going on. But my gut feeling is that that's somehow or another not what's going on, that's only a smokescreen. I have a feeling that wasn't what it was for. It seems to me to be too intimately related to spiritual life and development to involve anything like breeding experiments. It's almost as if they're trying to disguise another form of activity behind that. Just like I think that the UFOs and so forth may be a disguise for a very different type of apparition. It might be that there is a reason why they cannot at any cost allow us to know exactly who or what they are, especially if they are real beings of some kind and somehow emerge from us and are therefore dependent upon our minds for their existence. They may be vulnerable to us in ways that we can't even imagine. So that may be one of the reasons for all the secrecy, the obfuscations.

WV: When you say it almost appears they're using a cover, does that indicate to you they've decided we're not ready to understand the whole thing, so they have to feed it to us in small doses?

WS: They call me *child*. And I think they think of us as children. And it might be that we're in the position of children getting ready to walk via the great mind change I spoke about a moment ago. You can't teach a child to walk. There may be nothing more they can do to enable us to do what they're hoping we will do. And what exactly that is, of course I don't know. Being human, I've never done it before, obviously.

WV: Have you ever read *Childhood's End* by Arthur C. Clarke?

WS: No.

WV: What you've just said reminds me of its storyline. Aliens with a highly advanced technology come to Earth and make their presence known, but they can't let us know all of their plans at first. Finally at the end they reveal that they're guardians or cosmic midwives. Our children, the last generation of mankind on earth, are transcending, mutating into something beyond human, and the aliens oversee mankind's dramatic transition to the next level of evolution.

WS: I ought to read that, because Arthur C. Clarke's had a lot of insights, and I think that one might be one of the ones that's closest to the way I feel about what could be happening.

WV: At one point you discuss how, "if Visitors are really here, one could say that they are orchestrating our

awareness of them very carefully." The 1940's brought us long-distance sightings which gradually phased into meetings and/or abductions in the sixties, more so in the seventies and now eighties. You said that your, "whole experience had been designed in detail by insightful minds engaged in a slow process of acclimatizing humanity to their presence." You further stated that "maybe . . . we have all been chosen–and we are all being tamed." Tamed for what?

WS: The visitors are real, but this is all a speculative thing; I'm not making assertions, but speculations. That scenario certainly does fit the experience on one level.

WV: But tamed for what?

WS: Well . . . you see an elephant dancing in the circus. You wonder how it sees what it does. The elephant's conception of what it's doing may be as different from what it's really doing as our conception of what we'll be doing once we are tamed. Of what we are really doing. It may be very hard for us to know what's really happening, and in fact, as this experience emerges in life more and more, we're going to have to get used to living with a very high level of uncertainty. It may be that we could face a situation where the most important issues we have in life all rest in the unknowable. In other words, if the Visitors' presence becomes a more commonplace and accepted fact, that doesn't mean that we will begin to understand what's happening to us, and it also doesn't mean that we necessarily can understand. There's no way the elephant who's dancing can understand the meaning of the circus he's in. No way. No matter how hard you train him. His mind doesn't work that way. My impression is that no matter how this experience is generated–internally or externally–the external, linear mind, the if-then thinking we do in the circus, is not adequate to address this. We're simply never going to understand it at this level; I don't think it's comprehensible at this level.

WV: So using the analogy of the elephant, we may already be learning to dance for the aliens without even knowing

that we're doing it.

WS: Exactly. An awful lot of abductees report being taught inexplicable things that they can't remember, and then are told they will remember them when it's time.

WV: There is a well-documented contactee case from Switzerland, one Eduard "Billy" Meier, who purportedly has had over 100 meetings with humanoid females who say they come from the Pleiades. In one of their meetings they told him that at any given time the earth is being visited by many different alien cultures–and thousands of visitations per year. Some are here to observe us, since they realize we've reached a point in evolution in which global self-destruction is quite possible. Others refuel here, while still others are here for distinctly unfriendly purposes. The Pleaidians claim they're here because their forefathers are our forefathers. How do you feel about these concepts?

WS: I know a lot of the truth behind that. It's terribly complex and it's going to take years to make it emerge in a correct and useful way. To say that some of them are here for unfriendly purposes is also to cast opprobrium on the lion because he eats gazelles. Not sensible, since both are part of nature. As far as Mr. Meier's case is concerned, I've heard so many contradictory things about its veracity that I don't know what to think, except that, from reading the things he has written, it seems obvious to me that he's had contact with someone or something that knows a lot of the same things that we've been talking about. There are things hidden in that statement that do suggest it was made by someone who knows more than Meier knows himself about its real meaning.

WV: So are you open to the idea that multiple alien cultures are visiting?

WS: I've asked them where they were from and they said to me, "Everywhere." I've also seen seven different types that were not at all human looking, and one type clearly was human, entirely human. So you figure it out; cast sand into the stars, and I think you'll find them that way most easily.

WV: I think I'm going to walk away with more questions than I came with.

WS: Mmm-hmm. You're not alone. Every time I think about this I come up with three or four new questions. Every time I get a handle on it, the scientists I work with and I say, "Ahh, we've finally got this part figured out," then another of the things comes in that throws that into a top hat. The problem is, we don't know how to think about what's happening to us. We think about it in a linear, structural way, and I don't know that that's appropriate. I have a feeling that we just literally don't know how to think about what's happening to us, and that's why we can't figure this. The scientific approach to it is as inappropriate as the superstitious one. They're obviously not Gods from outer space, and they're obviously not nothing. Those are essentially the two approaches that seem to be most current. And they're both just junk–intellectual junk. Nothing more than that.

At one point last July, when I asked them what the Earth was, the Visitors replied, "This is a school." I do not know exactly what this meant, but I have a feeling that life on Earth is not what it seems. I think that they are much more involved here than we realize, and human life could be something along the lines of an experiment wherein mind is attempting to penetrate the physical reality. We are here to learn what this reality means. What we are is a kind of hyper-sophisticated machine which is penetrating this level of reality. Our so-called Visitors are nothing more nor less than the technologists who keep us going.

WV: So we might be an experiment in transcendence?

WS: Very easily could be the case.

As the interview drew to a close, Whitley inquired about my Houston sighting. He asked me if I thought I'd been abducted. I told him I felt like I would have remembered it if I'd left my bedroom that night. He asked how close up the object came. When I told him I blacked out with it barely twenty feet away, he blurted,

"Oh, well then you're in. If it came that close to you, *you're in.*" He seemed certain that I'd been taken. I laughed and shrugged it off. What could I say? His firm conviction fascinated me.

When Whitley had replied to my *learning to dance for the aliens* line with the concept that some people are taught inexplicable things while onboard, to be retrieved later, I didn't elaborate to him about what I'd actually meant by it: that the *learning to dance* might be about Skinnerian conditioning, along the lines of one of Jacques Vallee's themes. More about our Masters grooming their monkeys for performances. Perhaps even for some performances that await in the wings, as yet untried.

I was also struck by the surrealistic nature of discussing his "time onboard" as though the encounter represented a real physical environment with sentient, physical beings (which I don't rule out), but then drifting into and out of his proposed concept that this might represent "mind penetrating our physical reality." Pondering this evoked in me the feeling of being trapped in a lucid dream, struggling to navigate uncooperative astral realms, while in the background knowing that I'm also a physical man in bed with his eyes closed, eyelids fluttering in the REM state.

The whole non-physical, mental projection scenario Whitley proffered conflicts with my conviction that the blazing device of wicked technology I saw and felt and witnessed silently dancing and darting and flashing and hopping and dematerializing and rematerializing and hypnotizing outside my bedroom window in Houston was a manufactured, flying machine—one that emanated intense waves of electromagnetic energy, oscillating fields that made my radio go dead and placed me in suspended animation. Very physical, solid and real. Not from a mythical land. In fact my conviction has always been, and remains to this day, that if the hyper-energized disc that dove down at me, wiped me out and whited me out and shut me down cold like a cryonic deep freeze was not a solid, mechanical vehicle of transport, then neither are Greyhound buses.

* * * * * *

THE NEW AGE ANGEL OF SOUTH BEACH

To put myself through college at FIU, personal training remained my primary income source for awhile, until I had my hypnotherapy certification. Synchronicity was clearly at work when I landed a beautiful female client with a decidedly alternative-reality mindset. Training Angela DesJardin was like getting paid to rehearse with a flamboyant exotic dancer, complete with a voluptuous, sex goddess figure, long honey-blonde hair and a million watt smile. She had the most outrageously uninhibited personality and I never knew what to expect from her. Angela proved to be as open and free spirited as me, and completely on the same page about UFOs, metaphysics and paranormal realities. We'd both read and loved the Seth Material books by Jane Roberts, and myriad other mystical works. From the start, it was easy to feel a strong kinship with such a striking, genuine and dynamic lady. We both had a cockeyed view of that nebulous thing called reality.

When we first met in the living room of her airy penthouse with the view across the bay to Star Island and the downtown Miami skyline, her darkly macho husband Derek by her side, we connected instantly. The rambunctious boy in me loved the mischievous girl in her. Her enthusiastic laughter and playfulness evoked images of a grade school Angela clowning around and passing notes and cutting up behind the teacher's back. Right away I could tell she had a natural irreverence for the staid and conventional, another plus favoring our harmonious interplay. She saw in me someone who could be just as adventurous and open to the new, and the fact that we both had metaphysical interests sealed the bond. Becoming her trainer was an easy sell, so with Derek's blessing she pulled out her checkbook and paid for a month in advance.

Several weeks into our fitness program, one day before training we met for lunch at *Fresh*, the health food café of her swanky gym, the *Skyline Athletic Club*. Punching the elevator button in the parking garage I felt a sense of elation. Angela was one of those rare birds who never brought you down. Even if she wasn't at her best, her exuberance shone through. As the car glided up to the top of the building, I anticipated a fun afternoon with my client and new gal pal.

The steel doors opened to a lobby festooned with mirrored walls, chrome, pricy art and black leather couches--apropos accouterments for our body obsessed, kinky culture of narcissism. The icy receptionist waved me through the etched glass doors, leading to a maze of club chairs full of chubby executives in sweatsuits reading the Wall Street Journal, and leotarded Stepford wives exchanging the latest party scene gossip. Floor to ceiling windows looked out on waterfront reflective glass skyscrapers shimmering in the sunlight bouncing off of the glistening water, a cigarette boat and two cabin cruisers inscribing white wakes onto the iridescent greens and blues. Too bad about the view.

At the far end of the lounge I spotted Angela inside the *Fresh* café at a window table in the back. She flashed a face splitting grin and enthusiastically waved me over. "Hey!" she bubbled. "Great! I've been wanting to do this for a long time. How ya doin', Wade!" She sounded like a giddy six year old girl trapped in a bombshell body. I gave her a quick hug hello, tossed my gym bag on the floor and sat down across from her. Today she sported a full-body beige leotard covered in pale pastel flowers of pink and blue. "You're looking as beautiful as ever, Angela-in-the-garden."

Her face lit up and she giggled when I called her that. "You're lookin' mighty awesome yourself, buddy. Damn, those arms get me," she grinned, sipping on a smoothie. "You must have all the women falling at your feet."

"Not really. I'm pretty shy." The waitress interrupted and Angela insisted I try today's smoothie. Done deal.

Angela smiled and almost started to say something, then stopped. "You know, we, uh. . . ."

"Hmm. What?"

"Well, I . . . I mean. . . ."

"What is it. Spit it out." I extended a cupped palm to her mouth and said, "Go ahead. Pit it out. Pit it out in Daddy's hand."

She cackled and said, "O.K. Well, we've been training for what, about a month now, right? And I've noticed we never talk about your personal life. I mean, I love talking about UFOs, and your interview with Whitley Strieber, and all our shared psychic stuff. But we never get around to your off hours. What's happening with you when we're not training. And if it's none of my business, then that's cool. But you know all about my life, so . . . " she smiled demurely.

"No problem. My life's an open book. What can I tell you?"

"I don't know, I'm just curious about who you're dating, who the lucky gals are."

"Actually, I haven't been with a gal in a good while. I am living . . . what you would call an alternative lifestyle."

"No. You mean gay?"

"Actually, I prefer the term hormone-suckular."

"No!" she said, and began to guffaw gleefully. "God, with a body like that, and–well jeez, why didn't you say so before?"

"Well, I guess first of all, I don't even identify with the word gay to begin with. I mean it sounds like a bunch of guys who would go pirouetting down the street in ballet toe shoes--not my scene at all," I laughed. "I've never looked at being gay as an asset, especially in a society as full of hatred towards my alternative tribe as ours is. And it doesn't define who I am. So I usually keep pretty quiet about it with clients. And I guess I was sort of afraid of what Derek might think, since I've massaged him a few times. Plus, y'all have a newborn baby. I wasn't sure if Derek would feel comfortable about it, especially with AIDS going around and all."

"Ahh, hell!" she bellowed, waving a hand dismissively. "That kind of thinking is for the eggs! I'm not one of them! Listen," she said, lowering her voice confidingly and easing her face close to mine. "Let me tell you what my father told me when I was a little girl. He was dropping me off at school one morning, I think it was fourth grade," she said, brushing back her blonde tresses alluringly. "I was in the front seat next to him, and we were stopped at a crosswalk. The crossing guard held out the flag, and all the little kids crossed the street to the school. My daddy sighed with disgust and shook his head. And he almost spit out the words, 'Look at 'em! Just look at 'em! A bunch-a fuck-in eggs! Angela! Don't you ever become one of those fuckin' eggs, you hear me!!'."

I laughed so hard that I just about turned blue and choked for air. She laughed and said, "It's true. He said it." Once I'd caught my breath I said, "You've followed his advice admirably, Angela." My smoothie arrived and we both ordered. When the waitress left, Angela gazed out the window. She sighed. "Oh, dear, I'm so glad we got that out of the way, not an issue to me at all," she said. "No, doll, Derek and I are very open-minded. You see . . . before we got married, I'd done my own exploring. Hell, what's the big deal about it anyway, right? That's what life is for, exploring, growing, being open to new adventures, pushing yourself, finding your limits."

When I asked her to elaborate, I was treated to a blunt re-counting of her adventures spurred on by her first husband's fixation on seeing two beautiful women together, a non-con-ventional bonding with a wild musician she was madly in love with. Their waterfront property in Jamaica and the tropical atmosphere lent an exotic flavor to her tale of the time Pete brought home a beautiful island girl from the recording studio and made clear what he had in mind, details of which made a suit sitting at the *Fresh* bar put down his fork, turn his head and stare. Angela made a mock funny face, batted her eyelash-es theatrically, poked out her tongue defiantly and winked at him. He smiled and went back to his chicken salad. God, how I loved her. "O.K., now it's your turn," she said.

I shared how I knew about the duality of my attractions as far back as first grade, during my dad's brief Key Biscayne work assignment. I had a crush on my gorgeous, skinny blonde 1st grade teacher, Miss Carlton, a sweet young lady who could have been a runway model with her striking beauty and poise;

and I also knew I had serious chemistry, even as a six year old, for our handsome, muscle-bound 1st grade gym coach. He always wore skin tight gym shorts and tight athletic shirts that looked like the short sleeves were going to burst from his biceps bulging out. But my strongest memory of early awakenings was with neighborhood kids back then. Around the corner from us was this wild family, the Schraders. Their skinny blonde daughter, her shock of sun bleached wild hair atop deeply tanned limbs, self-assuredly escorted me over to her house when the family was gone one day. We staged a fake marriage, and her baby brother was the preacher; then we did the wedding march down the hall to her bedroom, got naked, climbed into her bed, and I had my first taste of how even a young girl could French kiss like nobody's business. We about chewed each other's faces off under those covers. We were both six, so that's all we did, but that was plenty. Her baby brother stood off to the side and kept shouting and throttling his arms, 'I wanna! I wanna!'

"It really didn't occur to me back then that not everybody is attracted both ways like I was," I added. "Not until almost my early teens, I guess. Funny how it's a non-issue when you're so young and innocent, before you're taught about how important it is to hate people who are tuned in differently. . . ." We ventured into wide ranging talks, discovering we were both freelance writers, and how she wrote for a short lived South Beach arts and entertainment magazine called *Cue.* "It was a couple of years ago, back when South Beach was just starting to attract investors. I had my own column, *Audacious Angela Alights.* I was outrageous!" she cackled. "I'd run around to openings and events, have a ball. I could make or break a new boutique or restaurant or club. I had stainless steel balls!"

Our sandwiches arrived and we dug in. I felt relieved to know Angela and I could level about anything now. It also rang true for me, that a woman who was so open to personal explorations would naturally be completely attuned with the UFO and alien realities that were an integral part of my life path and who I was. After our light lunch and a few minutes to

85

wind down our talks, we retired to the weight room.

Wrapping up a full fifty minutes of resistance training with machines and dumbbells, we finished with the weights and trekked to the cardio room, a long narrow space of stationary bikes, treadmills and step climbers arranged in front of another vast picture window looking out on the skyscrapers and multi-hued water Miami is famous for. I put Angela on a treadmill and programmed it. As the machine sped up, she glanced down the line of aerobic torture machines at the snotty skinny broads gabbing away on the Lifecycles and treadmills. "Look at 'em, Wade," she said, "just a bunch of titless wonders! You know, I hear if you do too much aerobic exercise you can lose your tits. Just burn 'em right off! You won't let that happen to me, will you Wade?" she said with her eyes full of mock seriousness.

"No," I laughed. "I promise not to. Though God knows, with those amazing girls of yours that could take years and years." She cackled. The machine sped up and she worked up a sweat.

Later, after our abdominal floor routines, she unwound on the floor mat and we discussed tomorrow's program. We made a date, I jotted it in my book, then I stood and helped her up. "Where you off to next, big boy?" she smiled playfully.

"Actually, I'm going to a psychologist, for hypnosis."

"Hypnosis. Far out. What for?" As we headed for the elevator, I related how the Strieber interview got me thinking more about my UFO sighting, especially his confident, deadpan assertion that I'd been taken; and about how my last hypnosis session eight years ago was stymied by the subconscious block, and the things Dr. Bazerghi had told me. She hung on every word as I walked her to her car. "Remember, I want a full report tomorrow," she said as she climbed into her old, beat up Buick Electra 225, oddly incongruous wheels for one with such a lavish lifestyle. She lived like a princess in a posh penthouse, dressed like a high fashion model, dined out at four star restaurants, and she and Derek were forever jaunting about the globe like the idle rich. So why the clunker? She lowered the window and laughed. "This is so exciting! A real close encounter! Just don't let them take you away, doll. At least not until after I hear all about it tomorrow!" she said, blowing me an air kiss. She backed the wreck out, slammed it into drive, and smiled and waved as she floored it.

THE HOUSTON UFO: ONE MORE HYPNOTIC QUEST

The most beautiful thing we can experience is the
mysterious. It is the source of all true art and science.

–Albert Einstein

Driving to Dr. Jared R. Dixon's office in Coconut Grove, my mind tumbles around, with thoughts and visions of Dr. Bazerghi and Angela fading in and out. Her comment about not letting them take me lingers. These days it seems like anyone who is familiar with UFOs thinks in terms of alien abductions (if they entertain the reality of alien spacecraft at all), perhaps partly fueled by the close encounter UFO books released in 1987 and the resultant media saturation.

Pulling into the parking lot of the small professional building off of South Bayshore Drive, I feel excited and relieved, knowing I'm returning to confront one of my life's biggest mysteries. Dr. Dixon was referred to me by a client, and on the phone he seemed open minded and receptive when I recounted for him my UFO material and previous hypnosis session. He'd come highly recommended for his hypnotic skills, so I made an appointment.

Dr. Dixon's office is tucked away in a ground floor suite. I enter the waiting room and ring the bell by the frosted glass window. In a moment a distinguished looking man with a warm tan, sun streaked blonde hair and dirty blonde beard slides it open. He introduces himself as the doctor, and apologizes for his receptionist's absence. He hands me the customary paperwork, and when I'm through filling it out he takes it and ushers me into his inner sanctum.

I stand quietly and study the diplomas and certificates on one of his walls as he reviews my paperwork. When he's finished reading I sit down in an overstuffed chair and we make small talk for a few minutes. He asks me a lot of questions about what my youth was like, especially around the time of the encounter. I give him a quick rundown.

"And at this time, everything was good between you and your parents? No beatings? No abuse?" he says with a glint of humor in his eyes, obviously not accustomed to a patient telling him that a flying saucer had almost landed on his head.

"No sir-ree, doctor," I laugh. "No arsenic in the food. We were your basic all-American thermonuclear family."

He smiles. After five minutes of questions and answers, it's time for the hypnotic induction process. He leaves his desk and sits in the matching club chair opposite me. He doesn't mind that I've brought my micro cassette recorder, but requests that I don't play the induction for anyone. I agree, and close my eyes and relax.

After a soothing ten minute narrative, Dr. Dixon says, "The elevator progresses downwards quickly, and then, as the doors open you step out. It's like being back in your bedroom, you know exactly where the bed is in your bedroom, don't you?"

WV: Mmm-hmm.

Dr. Dixon: That's right. And there's other furniture in the room, isn't there?

WV: Yeah.

Dr. D: Mmm-hmm. And perhaps, I'm not sure at this point, it could be around midnight, perhaps a little past midnight, you might be listening to your radio or watching television. What do you find yourself doing?

WV: Yeah, I'm listening to the radio in bed.

Dr. D: Mmm-hmm. And what are you listening to on the radio?

WV: Some AM station, just some music.

Dr. D: Listening to some music.

WV: Mmm-hmm.

Dr. D: That's right, just listening. Pretty quiet in the house right now?

WV: Completely, yeah.

Dr. D: Everyone's gone to bed?

WV: Everybody's asleep. Except me. (laughs)

Dr. D: That's right, you're up and you're listening to the radio, aren't you?

WV: Mmm-hmm.

Dr. D: Just enjoying that, lying in bed?

WV: Yeah, I am, I'm lying in bed with it up by my ear.

Dr. D: Mmm-hmm. Very comfortable, aren't you?

WV: Uhh-huh.

Dr. D: Very much at ease?

WV: Very much.

Dr. D: Just absorb that for a moment, just enjoying that feeling, very much at ease, very relaxed. And notice the passage of time as you're enjoying that music, just listening to the radio. A few minutes go by—five minutes, ten minutes, 12:10, 12:15, 12:20, 12:25. I wonder if something begins perhaps to get your attention, are you not listening to the radio anymore, or. . . .

WV: Yeah, I'm listening, and I hear static . . . and strange vibrating noises like the radio's not tuned in right. So I try and tune it with the tuning dial and I turn the volume up a bit because the volume seems to be going down. And, it doesn't seem to help, it seems to be getting worse. So I change the station to see if I can find something else. And everywhere I go there's whatever—music or reports or whatever on the radio—but it's all the same, everything's got the static and the vibrating noises. And the volume keeps going down lower and lower, and so I keep turning the radio up higher. And it seems like maybe thirty seconds longer and finally the volume goes out completely with it turned all the way up.

Dr. D: Mmm-hmm.

WV: And, all the way across the dial everything's just gone.

Dr. D: Rather puzzling, isn't it?

WV: (nervous laughter) It's really strange.

Dr. D: Mmm-hmm. So the radio's no longer working.

WV: Yeah, it was working fine up until then, for the past few days; everything was fine. And I had just bought a new battery a few days before.

Dr. D: So what do you do then with the radio?

WV: I set it on my night table but I leave it on with the volume up. Next to my bed, I set it down on the table.

Dr. D: Where? Was it on the right hand side or the left hand side?

WV: On the left side, 'cause the right side of the bed's against the wall.

Dr. D: Mmm-hmm, that's good. All right, so you set the radio down on the table. . . .

WV: And then, I realize I'm not sleepy enough to fall asleep yet, so I just kneel up in bed and pull the curtains aside to look out the window to see what stars are out . . . and it's a very clear night, no clouds. And, I don't remember what constellations I saw or even if I could recognize any, but I just remember a lot of really bright stars . . . and I'm just looking around very peacefully.

Dr. D: Pretty clear evening?

WV: Very.

Dr. D: Mmm-hmm.

WV: Very clear. And I just . . . suddenly decide, well, I think I'll look straight up (laughing nervously), up over the eaves of the house. It's almost as if I hear the thought *look straight up*, and I sort of laugh to myself, like, how funny I would think to do that. And I do, I kind of turn my head and put my head up against the window and look up. An-n-nd, it seems like just barely a few seconds after I look up there I suddenly see this thing. And, it's this brilliant white ball. White, and I mean just so bright, and very high up. And I notice there are some lights flashing on it, some red lights on its surface.

Dr. D: Did you feel like you had to shield your eyes?

WV: No, it wasn't close enough to be that way but, it was bright like a moon, or brighter, a full moon. And, it almost seemed like it was coming from above, coming down, and then leveling off at a high altitude, like that's why it suddenly appeared to me, because it was coming down and then all of a sudden it's there, and then it, so . . . ahh, my first thought is, what is that, ahh ha ha, what's that? And it puzzles me because it doesn't look like anything I've ever seen before. And then when it proceeds to stay there and keep moving–and I mean it doesn't disappear–then it, I have more time to try and figure it out, but I go through this list in my head really quickly and it doesn't fit anything. So I just–it's unknown, and it, it kinda scares me. And, so it's moving, I guess to the east, yeah, it's moving to the east. Away from my house and towards the next door neighbor's house. And it seems to be about the same altitude and just really gracefully moving along, kind of zigzagging. And then, it seems like it stops beyond the far edge of the neighbor's house, maybe a hundred feet away, at the opposite edge of their house. And just, like in a flash, it dropped altitude, like from point A to point B instantly, with no movement in between.

Dr D: What do you feel?

WV: I feel very scared now.

Dr. D: Mmm-hmm.

WV: (laughing nervously) And I also feel . . . watched.

Dr. D: Mmm-hmm.

WV: I have an overpowering feeling that . . . I've been spotted . . . And . . . then my next recollection is that it moved closer. More towards the middle of the roof next door, and therefore more towards my window. And, it comes even lower, although as it comes down, it's going in like squares, and circles . . . and in fact I'm thinking, this is almost like being hypnotized. Because it's moving in these strange patterns. And . . . I'm a lot more scared. And . . . I feel like there are things about it that I don't wanna know . . . because if I can pretend that I don't see some of the things I see then it won't have to be . . . a ship, and I, I feel like I don't want it to be a ship. Because if it's a ship then, then there's something onboard and it's unknown and it's real scary to me. And it just, umm. . . .

Dr. D: Okay, staying with it, as I count to three, as I count to three, really experience yourself very, very clearly at that window sill. Looking out, really clear, one, two, three. Let it get more and more clear.

WV: It still feels really vague, but, I mean like it's a circular ball of light. And there's, there's lights on it that flash. And as it comes closer, I see that it looks like a disc, it looked like a ball of light before, but closer up it now looks like a disc. And it looks like there are some things extending out of it, like antennae, but that's still unclear. And I feel like . . . like I said, I don't wanna know certain things about it. And I do remember that, as it's coming in really close, it seems almost like it's going into my head. Like it's overwhelming me; I'm fading out, and I'm being absorbed by it. I can feel its energy, and I hear a high pitched hum in my ears. And I'm just grabbing so tightly to the window sill, and I say, 'Oh My God!', and my jaw is dropped open. And I'm thinking that I should either run or–or I don't know what, maybe yell or wake somebody up (voice cracking, very distressed). But I can't move. And I'm really very aware that I can't move. 'Cause my arms are really tight, and my whole body feels like it's just frozen and I just cannot move and I'm looking up at this thing–

Dr. D: Okay. When I count to five, as I count to five, for one moment–just for a moment–you're going to have a very

clear glimpse, a very clear glimpse, as if you're going to see beyond and through your fear, your fright, your anxiety—you're going to see right through it. Just for a moment, a glimpse. One, two, three, four, five. Just a glimpse.

WV: (pause) I really want it. I want a glimpse. But it just seems the same, it's very close up and it's scary.

Dr. D: What are you seeing? You're very frightened, you're by the window sill, you're looking out the window. You think you can't even move. What are you seeing?

WV: It's a brilliant object. And it's the brightest white I have ever seen. And it's the strongest white.

Dr. D: Mmm-hmm. Do you see anything on the surface of this object?

WV: (deep sigh)

Dr. D: It might not be a visual seeing. Sometimes when we speak to somebody, someone says do you know what I mean, and someone says I see what you mean . . . Is there something you know about this, something that you see?

WV: In that regard, I feel like I see that . . . there's some kind of communication going on, and I don't understand what it is.

Dr. D: What kind of communication?

WV: I don't know unless it's uh . . . I don't know, sometimes I've had this feeling that there was some kind of thing planted in my head. Not thing, object, but, I don't know, information or, ahh, a transmission almost. 'Cause this thing was so close. And, like I said it felt like it went into my head, it came so close, it just . . . and even though I just can't quite see it yet, when it's real close, I remember that after it was there, like it was inside my head, then it really slowly moved back, almost like it finished doing what it was supposed to do. And it just really slowly pulled back and up above the roof.

Dr. D: Mmm-hmm.

WV: I just saw flashing lights–I saw something.

Dr. D: Mmm-hmm.

WV: (laughing) I can feel something uncovering a little bit.

Dr. D: Mmm-hmm, stay with it, what are you seeing?

WV: I just saw a glimpse like it, it definitely had flashing lights around the rim . . . and it seems like a disc shape close up, and it looked like a ball when it was farther away–with red, white and silver flashing lights.

Dr. D: Mmm-hmm. And then it pulled away, and you. . . .

WV: It pulled away, and once it was up above the roof of the house again, it moved off to the right, to the south, really slowly, and pulled way up high behind this tree.

Dr. D: And then what did you do next?

WV: Well, I watch it for a few seconds, and it was just silently floating there without moving, behind the tree. And I remember thinking *what if they're monsters?*

Dr. D: Mmm-hmm.

WV: And, I felt like I was supposed to have a meeting or something but I was so scared I didn't want to know anymore. And so I pulled away really quickly, and jumped back down onto the bed, and I pulled the covers up, and I just lay there, wondering what I was gonna do.

Dr. D: Okay, that's fine. And lying there in the bed, wondering what you're going to do, makes perfect sense. As time continues . . . that encounter . . . draws to a close. You're just lying in bed, very relaxed . . . letting it settle . . . just letting it settle. . . .

95

Dr. Dixon then gives me the suggestions to return to normal waking consciousness. My eyes slowly open, and he says, "How do you feel?"

"Whew! Very relaxed," I say.

"Good. Interesting . . . what do you think?" he queries.

"My first impression is that this uncovered more than previously, with the psychologist in Austin . . . particularly that flash that I had, or that, just that quick little look–it wasn't when you said I would have it. Just all of a sudden I remembered, just like when it came in really close it was just like cruising, whoosh, it looked like a disc.

"Flashing lights around the rim," he says.

"Uh-huh."

"First you weren't sure about the flashing lights, but then you saw them. . . ."

"Yeah. Very clearly," I say.

"And the whole concept of communication, perhaps something was imparted to you–"

"Mmm-hmm."

"I mean, certainly if there was the possibility of interaction between . . . yourself and . . . the presence with this disc, there's no guarantee that it would be understood. It's like, you know, if you were going to communicate too, how would you–you'd be very limited. But it's interesting that you had the sense like something was put in your head. You had a little bit of that before from the Austin session, but perhaps this time it was more of a communication. The last thing you said is that you were supposed to have a meeting, " he says.

"Yeah, I remember saying that."

"That could have been very frightening," he says solemnly.

"Yeah. I mean, I felt like, like if I had been outdoors, there's no doubt–I wouldn't have been able to escape it. Whatever it was."

"Yeah, it could really be that your fear, encompassing the experience–which would be natural in an eleven year old–God, it'd be natural in a thirty year old--"

"Mmm-hmm."

"Might have diluted or affected your perception. You know, because sometimes when you're really frightened you just don't see things like you would if you're relaxed–you don't take it all in. So it could be very real, that your fear really did, umm, distort or inhibit your experience somewhat. And so part of it is getting through the fear, and not being so frightened in recollection, and seeing what's going on," he says.

"Mmm-hmm."

"So this feels like it went a little further than what you'd experienced before, in your previous hypnosis session?"

"Yeah, I think so. After the part where I feel like I blacked out, my next vivid memory after the disc darted back up and behind a tree is that I lay in bed awake the whole night scared to death, listening to the radio when it came back on. As far as the content of this encounter, I still don't think I have it all. I do have a gut sense that there's more to this," I say, reflecting.

"If we explored further, you might be able to clarify peripheral elements of it, but the thing is you might be trying to make something clear you're never going to make clear."

"Hmm."

"I get a sense of presence. It reminds me of the movie *Close Encounters Of The Third Kind*, you know the third kind is a case of contact. And you may have been contacted, you know, going with this premise. And if you were, then you may never have an understanding of that–it just may have been something you couldn't have comprehended," he says.

"Mmm-hmm."

"And that's–maybe that's been going on for centuries, I don't know–the lack of comprehension. And maybe that feeling like there's still something there–you might be right. But there may be no amount of hypnotic exploration that's going to make sense out of something we can't make sense out of. That is, we can't give you a framework to understand something that we don't understand. You might get clearer–"

"Details."

"Visually, details, things like that, but it might not be what you're looking for. You might be looking for something a little bit more significant," he says.

"Yes. I do have a strong sense about that."

The session was fascinating and frustrating. Fascinating because I could really feel myself in my old bedroom in Houston, completely in touch with the fear and wonder instilled by that thing outside my window. And frustrating because I knew there was something more, something I was missing out on.

I decide to let it go for awhile. I pay Dr. Dixon, thank him, and leave. Resignedly, I accept that Dr. Bazerghi may have been right. I may have to wait until my subconscious is ready to open up and let me into that locked room. It feels like I've been digging way into the back of a dresser drawer, shifting its contents around, and it will only open part way, because something large is jammed in the back, keeping the drawer from opening. Something large is back there.

THE ADMIRAL

Evening. Dining at a fancy Italian bistro, *Tino's*, with one of my massage clients, Admiral Walter Shawnessy, and his friend Rick, a handsome, clean cut officer who did some tours with him on an aircraft carrier. Walt and I connected through another client who lives in his high-rise, and I've been his masseur for about eight months now.   Scattered around his condo among the antiques and objets d'art are photographs, many inscribed, of some of the most handsome, masculine, muscled up men I've ever seen, some of them famous male models and minor celebrities. Walt had a taste for hot looking models and bodybuilders, and even had a nationally ranked NPC titleholder serving under him on one tour. From what I've seen of his straight and married Navy pals, nobody minded his lifestyle, and it's never alluded to at dinners or functions. I guess "Don't ask, Don't tell" works just fine once you're at the top of the heap.

I'd heard stories about how many gay men there are on aircraft carriers (and probably a fair number of straight men occasionally wanting to blow off steam on the down-low). A few off-color tales came from a buddy who'd served on a carrier; he'd shared stories regarding the large, all male sex orgies that he'd partaken in on a weekly basis while at sea, in a secreted, cordoned off area of the floating city (let's be real, over 5,000 healthy men in their prime in close quarters for months at a time--the male sex drive is what it is, and boys will be boys). But Walt was the first time I'd ever known of a gay admiral. He occasionally made references to having a very high security clearance, which I've gathered means he has access to a lot of Top Secret files, probably even those classified a lot higher than that. I'm happy that he's climbed the ranks despite his sexual proclivities; a respected, decorated admiral who is gay, you just gotta love it.

Born into money, an Ivy league education and good breeding honed a man of great pride and dignity. He seemed to know something about just about everything, and if he didn't he would act as if he did anyway. His almost overbearing bravado was actually one of his endearing qualities. The enthusiastic, ballsy swagger of a leader of men. His gregariousness and polished manner opened doors to friendships with some of the most wealthy and influential people in the world–Palm Beach billionaires, a skin care maven in Manhattan, an Italian fashion designer with a household name, and myriad titled nobles and dignitaries in Europe and the Far East. If you were traveling anywhere in the Northern Hemisphere, Admiral Shawnessy could get on the phone and find you a fun dinner date and maybe even a place to crash. The same charms that helped him establish global connections also enabled him to bed a slew of men, many of them straight but subject to the distractions offered by an attentive, distinguished and cultured gentleman of means. Walt did just fine for himself. He was a real pistol.

Then in his mid-fifties, I could see from many of the framed photographs in his entry hall gallery that Walt was quite a head-turner in his prime. Shamefully, he had let himself go, and his passion for men spilled over into a passion for gourmet food and the high life. I tried to get him to let me train him and burn away some of his bulky girth–which more than anything made me worry about his cardiac health–but it was more of a pipe dream than a possibility. We are all creatures of habit.

Always a gracious host, when he had house guests he often included me in dinner invitations. So far tonight Walt and Rick's conversations have ranged from what kind of opera season the Met would run to who their favorite pro bodybuilder du jour was. I've been silently enjoying the softly lit ambience and the gentle buzz from a Tanqueray and tonic. The overhead pin lights of our fancy café transform it into a night under the stars.

"You've been awfully quiet, sport," Walt says. "What've you been up to lately? Been keeping at the writing?"

"Not as much as I'd like, but several times a week," I say. "And, what else . . . oh yeah. A few days ago I went to a psychologist for time regression hypnosis. He regressed me back to a really intense, close up UFO sighting I had back in Houston when I was eleven. It was a pretty far out session."

I immediately regretted there was no hidden video camera to capture the expressions on Walt's and Rick's faces. Before my declaration, both men looked relaxed, happy, content. We were having a great time, the booze was doing its trick and conversation flowed. The atmosphere was jovial, full of smiles and laughter, echoed by the happy murmur from surrounding tables of faceless, fawned over diners in the shadows. Now Walt's eyes opened noticeably wider, almost too wide, and he turned his head to survey the bubbly crowd as he sipped his Rum Runner. His deeply tanned face looked somber, funereal, and all the blood seemed to rush out of it. Was that fear in his eyes? Bizarre. Robotically, Rick turned his head to gaze in the other direction, like an opposing bookend. Staring off into the distance, there was a distinct look of fear, or something beyond it, in his eyes too. The two officers looked and acted as though they'd just watched a scary scene from *The Exorcist*, like when little Regan MacNeil started bellowing with an angry old man's voice. Their sudden, dramatic transformation surprised me, but only for a moment. Despite my alcohol induced tranquility, I quickly connected the dots: high ranking Navy officers, Naval Intelligence, high security clearances, UFO files, Need-To-Know basis, deadly serious material, National Security, must maintain cover up. Time to dummy up. Load zombie program. Program loaded. Run program. Running. So this is how it works. Just put yourself somewhere else. Dummy up. Remain neutral. Let it pass.

Walt sipped on his cocktail, eyes still averted, and mumbled hoarsely, "Mmm-hmm." Rick was completely in a daze. Still no eye contact. Long seconds of awkward silence ensue. I'm wondering if my dinner pals think I haven't noticed that my mentioning "UFOs" has pulled the plug on our casual conversations. Clearly, it doesn't matter anyway; they've been conditioned to respond like this, and that makes experiencing the whole scene even creepier, watching them run their pre-loaded programs like two drones. Presently, turning his attention back to the table, Rick mechanically squeezes a lime wedge into his rum and Coke. Finally, the program ends.

"It sure is nice to be in a city where shorts and t-shirts are comfortable this time of year; D.C. is freezing," Rick murmurs dreamily. The spell is broken. Walt takes the ball and runs with it, launching into a south Florida welcome wagon spiel. My story is shelved. I'm tempted to confront them both with what has just ensued, but the lure of fine dining and keeping the peace with a friend override any chances of that. I run my own program, called friendly conversationalist. The carousel of casual banter resumes its lazy spin.

Ahh-ha, I say to myself. I file this scene away for another place and time. We'll be talking about this again, Admiral Shawnessy.

A few years later, the right time arrived when the Admiral phoned me to ask a favor: could I help him unpack a shipping crate, and rearrange some heavy furniture in his high-rise? We had become good friends, and helped each other out whenever the need arose, so I was happy to oblige. I paid him a visit the next afternoon.

After a good solid hour of work, we took a break and sipped some iced tea. I marveled again at the spectacular view he had to the south from the 27th floor; you could see all of Key Biscayne and beyond. The colors were mesmerizing. On impulse, I decided to ask a question that had begged for an answer for a long time. I recounted the dinner we'd had at *Tino's* with his Navy buddy Rick, and how odd I thought it was when everything changed after my mentioning UFOs. I told him I was really clear about that night's events, and I had a question.

His countenance went from happy go lucky to cloudy, so I pressed on. "Look, Walt, you've known me for a couple of years now. It seems to me you'd realize by now that I would never do anything to jeopardize your standing with the Navy—or hurt you in any way otherwise. I've made it clear that I've been researching UFOs for a long time, and that I had a close up sighting in Houston, and so now I want you to do something for me. I understand the implications of your high security clearance, and I know you can answer this. I'm certain that the government and the military are hiding a huge amount of information about UFOs from the public. At the very least, I'd like for you to confirm that for me."

Walt's face flushed red, and like before, his eyes looked spooked and he averted his gaze. He took a big gulp of his iced tea, and stared intently out the south-facing window at the sun drenched water and sky. He seemed a million miles away. When he finally spoke, I was surprised by how curt and angry his words sounded, as though he had to struggle against years of conditioning to get them out. It dawned on me that maybe their superiors really did threaten these guys with death for breaking their security clearances, their oaths of silence. He spat out the words, in an angry and defiant tone, "There is a cover up. It's a matter of National Security. That's *all* I can tell you. And don't *ever* ask me about this again!" The *ever* was emphasized by a vigorous sweep of his hand. He never made eye contact, throughout. Whatever the covered up material was, it seemed to have scared the crap out of him. The steely resolve in his eyes was chilling. Case closed. The massive vault door clicked shut with a heavy clang of bolts. Steel against steel. Impermeable without the proper credentials and pass code.

I didn't see the Admiral again for months.

I can't say I was surprised by Admiral Shawnessy's response, but the finality of it was startling, as if a grenade had suddenly detonated in the room. It was disappointing that all I got was a confirmation, but that told me a lot. The microcosm represented by the information blackout between the Admiral and me was barely a fragment of the macrocosm, the global UFO Cover-up that the government, the Pentagon and the military-industrial complex maintain. Mustn't let the eggs know about the saucers. Keep them dazed and confused by the hypnotron, all dumbed down and doped up and in the dark. The big picture is only for the insiders.

* * * * * * *

WELCOME TO *PSYCHIC CENTRAL*

Sunday, August 11th, 1991. Evening falls at 3636 Hibiscus Street. I'm seated on the couch of my breezy screened-in porch, the ceiling fans blowing air like a twin engine plane. Alf has just been fed, and happily patrols the twilight grounds of my Coconut Grove domain, his paws making sifting, hissing sounds as they swiftly pad across the gravel trails. The timer activates, and my dense jungle wonderland of royal palms, oaks, ficus trees, areca palms, hibiscus and poinciana trees comes to life with red, green, blue and yellow Malibu lights suspended from the tree limbs high overhead. Beneath the stone bridge's wrought iron railings, the bullfrogs happily croak at the shore of the lily pond. Olivia, one of my favorite hot Latina lipstick lesbians, once said that nobody could appreciate my property as much as I do, and she was right on target about that. I feel blessed that Alf and I have been able to enjoy the incomparable serenity of this tropical idyll for years. It's been like a dream.

I've just finished my B.A. in Psychology at FIU, and with no immediate plans for grad school, my degree fully qualifies me to pump gas or flip burgers at Micky D's. I take a break from the *Miami Herald* classifieds, suck on a Coors Lite longneck, and kick back to reflect. My two acre sanctuary is great for that, all the living things in their lush environment inspire contemplation. My landlord Mr. Brandt says I contemplate my navel too much. He may have a point.

I am a man who has remained true to his obsessions. My vast library harbors virtually every book ever written about UFOs, alien abductions and paranormal phenomena. Boxes of adult VHS videos and bodybuilding magazines take up the slack. Both ventures are apropos distractions for a wounded loner with a strong sex drive who often finds himself good company. Living in the midst of my dreamy jungle compound with a castle turret overseeing the grounds only seems to have enhanced my guarded, monastic tendencies.

My teenage rampage in New Orleans, what sometimes I'd called my crackup at 13, has remained a mystery to me--as does the padlocked room in my subconscious that hides the secrets of the white out, when a saucer came too close for my eleven year old self to handle, and fried my circuits. Sometimes I feel stuck in a permanent state of recovery. The Catholic brainwashing obviously didn't help matters any--all that angst was a killer. So much wasted energy, years of guilt about my attractions, idolizing the competition muscle gods of the muscle magazines. The men I wanted to be like. The men I wanted to have. And once I'd built up my muscles with regular gym visits, the ones I did have. The hidden power of the flying saucers, and the implied power of the muscled up über-men. Still obsessed by hidden, mysterious power. Safe to say, I am enamored with high performance machines, flesh and blood or off-world nuts and bolts. Beautiful engines: Potential energy, waiting to perform. My path. My way. Big muscles make me happy, as much as all of the alien machinery. Marching to a different drummer's beat, indeed.

Now the contest level muscle men of the Joe Weider magazine ads have become an extended procession of real, live South Beach muscle heads. (He only cares about them from the neck down, is what I've heard some say about me. The boys do love to talk.) And when a soldier from the beach battalion isn't available, my expansive video library takes up the slack. In my sheltered youth I suppose I would have thought a man like me was sort of twisted, a real piece of work. Maybe that's the truth. But now I see my routine as a healthy release for my raging libido. And it keeps me off the streets--most of the time, anyhow. (Hell, at least I'm not a monster like the Catholic priests who have repetitively ruined young boys, trashed their souls and left them scarred for life, wolves feeding on the baby lambs from their flocks. These black robed dudes wanna save *me*?)

And in the midst of my habitual patterns, huge changes loomed on the horizon, as though riding in on the winds of an unseen tropical storm.

Transformational experiences, like love, often come from unexpected places. This has always been true for me--and was to prove itself as an axiom when I scanned the Help Wanted section of *The Miami Herald* that evening. It must have been fate or destiny that my training and bodywork businesses had slacked off, and I'd have to job hunt in Miami in the summer of 1991. Otherwise, I never would have met Kimberly, the beautiful psychic who held the key that would unlock and open up the door in my subconscious. Kimberly Lee's gift would change my life forever, unleashing bottled up truths with the explosive force of a nuclear warhead detonating.

Sandwiched among the typical fare seeking bartenders, waiters, bellhops and cab drivers, one ad jumped out at me. It read:

Psychic Telephone Hotline

Seeking gifted Tarot readers, Clairvoyants, and
Astrologers for national (900) phone line. Must
be psychic, or will train.   Serious callers only.

The "or will train" had me laughing out loud--it had to be a joke.
Even so, I needed a job, and my curiosity was piqued, so I called
them. The man who answered said the ad was legit, they really were
a 1-900 phone service providing psychic readings to callers for a fee.

"And how would I get hired by you?" I asked.

"Have you ever had any experience doing readings?"

"Yeah. It's been a long time, but when I was in my early teens I
used to read the Tarot and I was really good at it. I guess the main
reason I let it go was because I started getting so accurate with pre-
dictions that it sort of freaked me out. And being raised Catholic,
they pretty much have you convinced that the Tarot and all the other mystical arts are works of the Devil. So I guess I got scared off."

"Mmm-hmm, yeah, I'm familiar with that one. Do you still have
your Tarot deck?"

"Yeah, packed away in storage somewhere. I'm sure I could dig
it up."

"Well, then all you have to do is brush up on your skills for a week
or two. When you feel like you're ready, call us back and we'll have
you do a phone reading for one of our people here. Based on that,
if you show promise we'll have you come into the office and do two
more readings for two more of our staff. And if that goes well, we'll
hire you. How does that sound?" he asks jovially.

"Fair enough. Should I ask for you?"

"Sure. My name is Tim."

Intrigued, I hang up. I didn't have a problem accepting the idea of
absentee phone readings; I used to do Tarot readings on people not
physically present, and regardless of whether they were next door or
thousands of miles away, the information came through clearly. I
remember my first exposure to the cards. My cousin Nathan McAllister from my father's side of the family—the son of my dad's sister—
was studying architecture at Tulane while we were in New Orleans.
He showed up for dinner one night, a sandaled, easy going, long
haired hippy, and brought along his acoustic guitar and Tarot cards.
He gave me a fascinating reading, and in fact accurately predicted
the deeply troubled summer awaiting me at the upcoming family re-

union in Virginia. Eerily prescient. Captivated by the colorful card's highly stylized look and the information behind them, I bought a Rider Waite deck of my own shortly thereafter, and some books on the art of laying out the cards and interpreting them.

As for distance readings, I'd always had a gut level belief that the transmission of thoughts and emotions is not limited by space or time, which are merely constructs we've created to help us get a grip on our perceptions, anyway. So why should it be a problem when you have a distant stranger on the other end of a phone? Once the connection is made and the third eye opens up, the information flows freely. It all made perfect sense to me.

The following week I put aside the phone line ad and did a round of job interviews. Unable to connect with anybody, I dug out my old cards and re-familiarized myself with the Major and Minor Arcana. After a week or so of doing practice spreads, and regaining a feel for the cards, I call Tim and give him an impromptu reading. Something really clicks when I mention the presence of a legal matter; he perks up and tells me to zero in on that. I tell him that the Page of Wands falling next to JUSTICE, the eleventh card of the Higher Arcana, signified that whatever legal battles he might be involved in would have a favorable outcome, especially since they fell in the position of Future Environments, and Hopes and Fears.

He seems pleased with what he's heard, and we sign off. Barely ten minutes later he calls back and asks if I can come by that afternoon for my next two readings. Prospective employers weren't exactly beating down my door with offers, so I figured why not. What the hell.

Psychic Central is housed in a suite of rooms in a non-descript white concrete office complex, part of a sprawling utilitarian office park off of Bird Road near a freeway off ramp. The sunlight almost blinds me on this nearly cloudless day in late August. The invigorating, clear blue sky looks like wet paint. I park and find the entrance, chastely signified by the address stenciled on the mirrored door. I check out my reflection one last time before entering, tuck in my shirt, and catch a glimpse of my tarot deck and handbook clutched at my right hip, like a cowboy grasping his holstered gun. So this is how I'm going to put my Psychology B.A. to work. Counseling unseen strangers from distant cities with Tarot cards. As I punch the buzzer I wonder what I'm getting myself into. True to form, I'm diving into strange waters and starting to swim with the currents.

The door buzzes open, and I enter an ante-room with a beat to death couch and a sign on the wall that says PLEASE WAIT FOR ASSISTANCE. I plop down on the godforsaken sofa and thumb through my handbook. The stench of lingering cigarette smoke poisons the stale air. Definitely not my favorite smell. After about five minutes the inner door opens, and a stocky, attractive man with penetrating eyes steps out with a smile. "Hey. I'm Ryan. You must be the new guy Tim screened. Welcome to *Psychic Central*."

I introduce myself, we shake hands, and I'm shown into the main office. After a brief introductory discussion, Ryan answers a buzz on the intercom, picks up the phone, and lets me know that they're ready for me upstairs. I follow him through a back hallway. Half way down it, a beautiful redhead with a showgirl figure breezes up to us on high heels. Smartly dressed in a white ruffled shirt, tight designer jeans and a short camel blazer, she exudes confidence. Ryan introduces me to Donna Lee. We shake hands, and I follow her down the hall. She walks in short strides and subtle glides of the arms, accentuating the swivel of her perky butt. I feel like I'm following a top tier real estate agent through a show home as she points out the snack room, restrooms, a vast multi-booth phone room with an imposing master control desk, and stairs leading up to more suites.

We retreat to an upstairs office with nicer furniture than the other rooms. We sit down on opposite sofas, and gazing intently, she queries, "So how long have you been doing psychic work, hon?" I can tell I have my work cut out for me.

Placing my cards on the table between us, I tell her I'm just getting back into it, having given good readings in my teens. "Mmmhmm," she smiles, and her heavily made up emerald green eyes are telling me let's make this as quick as possible, hon. As I shuffle the cards, Donna Lee relates how she is the staff astrologer, and gives mini-astrological readings to the callers. I tell her that I think that's really cool.

I cut the cards a final time and spread them. I don't think Donna Lee is impressed by how many times I refer to my handbook. As I tell her what the cards show, her eyes stare at me, but psychically I can see that she's busy grocery shopping, running impending errands, doing chores at home. When I nail her on having artistic outlets beyond her profession, she zooms back into the room and it breaks the ice. Human nature being what it is, everybody perks up when their egos get stroked. By the end of the reading she says she isn't in charge of hiring, but will give me a favorable rating. I thank her and gather my cards.

Donna Lee departs, soon replaced by my final reading hostess, Anna. Her short strawberry blonde hair frames a face almost untouched by make-up. She has the open, innocent face of a freshly scrubbed farm girl. I'm always impressed by women who can look appealing without piling on the war paint. She smiles easily and giggles like a little girl. We make small talk about the folly of working on a 1-900 psychic phone line. Now I'm comfortable, and the reading flows. She gives me pointers and a lot of positive feedback.

Back down at the front desk, Ryan assures me that he'll be in touch soon to let me know if I made the cut. He holds the office door open and smiles again. A handsome young Latino boy of about twenty, with black menacing eyes like liquid tar, wanders out of an adjacent dark room from which emanate male and female voices. Tar Eyes stands close to Ryan, possessively so, and they watch me walk out. Anna smiles and waves.

A week later, I'm summoned to the office to fill out paperwork and W-2's. Voilá. I'm a phone psychic. Ryan congratulates me, issues me a headset, clipboard and stopwatch, and sends me back to the upstairs office to rehearse my phone patter with Beryl. Before I go, he says, "One thing, Wade. If you work here full time, that means you'll be giving eighty or more readings a week. Don't be surprised if, after a month or so, your Third Eye suddenly opens a lot wider. You'll be amazed at how much your psychic abilities will grow from that point on. You'll be able to see things you couldn't see before." I thank him for sharing that.

Beryl is so skinny I can't see how she manages to balance her head on her shoulders with the wild Afro of streaked blonde hair billowing down her back. She has the clear complected face of a little girl, and she'd be genuinely cute if it weren't for her scary eyes. At first they look normal enough, but once her words start flying and she hits overdrive her eyelids periodically open so wide she looks like she's scared to death or in shock, like those classic black and white stills of good ol' Charlie Manson in full psycho killer mode. I try to stay focused and not let my mind stray to troubled musings of the sinister things Beryl might be capable of doing in her off hours.

We sit down and I put on the headset. Her eyes regard me tranquilly as she says, "Now ahhm not sure how much Ryan has told yew. Did thay mention yew need a phone name? Fer example, mah phone name is Airy-ad-knee." Her heavy southern drawl helps me to relax despite her axe murderess eyes.

"Uh, yeah. I'm going to go by Seth."

"Seth. Oh, ahh like that name."

"Yeah. I chose it because of the Seth material. You know, the entity Jane Roberts channeled?"

"Ye-e-e-ss!" she exclaims with shocked eyes. "Ahh love those books!"

"Yeah. Me too."

The ham in me takes over, and the mock readings go smoothly. I'm given pointers about our required spiel, use of the stopwatch, making sure the caller is over eighteen, employed, and holding a major credit card. My main question about duration of readings is answered with, keep it at roughly twenty minutes. At $2.99 a minute, we're looking at a mighty high-end shrink visit for an hour, $180.00 a session in 1991 dollars, give or take. So I'm told to use my best judgment: if the client seems well off (or I can psychically tell they're loaded), we can let the clock run, but generally we're advised to wind it down after fifteen or twenty minutes. *Psychic Central* wants to avoid phone charge-backs and billing disputes. Even soothsayers have to pay the rent.

Barely half an hour later, Beryl says I'm ready to start, so I'm told to go downstairs and plug right in at a vacant booth. Navigating the spacious, gloomy phone room, I choose a cubicle off by myself. Five or six ladies are scattered around me, and it seems that I'm the only man working today. I get comfortable, set up my cards, and as I'm about to plug in my headset I notice on the far wall there's a pin-up poster of a beefy muscleman wearing a big smile and a skimpy towel. Feels like home. I overhear soft moaning from a nearby operator, and she quietly purrs, "Ooh, baby, go slower, you're so big . . . it's *too big* for me . . ." The control desk lady laughs as I gather my gear and walk out. "Psychic room's up front," she says, pointing. Shame on my Third Eye blindness, anyhow.

I enter the dim, candlelit haven. Phone sex wins the popularity contest, twenty eight booths to nine. Tar Eyes occupies the far corner, his desk aglow with flickering black candles, and crystals of varying shapes, sizes and hues clustered around them. He wears all black, matching his hair, sideburns and eyes. He methodically pulls and lays out cards from the Crowley Tarot deck and mumbles into his receiver in a deep baritone. Figures he would use a deck designed by a man who was allegedly one of the most infamous Satanists of the twentieth century. Aleister Crowley's cards always gave me the creeps. I'm happy with the more traditional Rider Waite deck like my cousin Nathan used.

Instinctively I select a vacant cubicle away from Tar Eyes. I pull up a chair next to Sherry, a slightly zaftig but otherwise very attractive blonde with a pageboy haircut. She swivels briefly and smiles at me while chatting away, sounding as smooth as a seasoned stockbroker with her nonstop patter–fitting since she is selling people their futures. I smile and nod back and set up my station.

Each of our cubicles has walls of industrial grey carpeting, with enough desk space to spread out our cards, and ample room for personal items. I sit down, switch on the lamp, and unplug the trim line phone on the desk and plug in my headset; then I adjust the earbuds and mouthpiece to my ears and mouth. I pull out my cards and handbook from my backpack, jot notes on the Psychic Report sheet, place my stopwatch at the ready, then throw the switch marked Available.

Momentarily, Ryan buzzes me to say he'll be sending me a call in a few minutes, and that he'll be silently monitoring my first few readings. I tell him that's fine. He says he's confident I'll do just great.

About two minutes later, I hear a double tone and see the red Incoming Call light flashing on my desk. I take a deep breath, click on my stopwatch, grab my cards and begin shuffling them, and punch the Accept Call button. A green light goes on. It's show time, folks.

"Good afternoon," I begin, "you've reached *Psychic Central*. This call is being billed to you at the rate of $2.99 a minute. My name is Seth and I'll be giving you a reading today. May I have your first name, and city and state you're calling from?" The caller complies, and I jot it all down. "Now Julie, I want you to understand that the information I'll be giving to you represents only my impressions and the things I might see and feel from talking to you. I'll be getting impressions from my Tarot cards, and sometimes I can pick up things from your voice. Please remember that you always have free will, and you are free to accept or reject any information I'll be giving you, and your free will enables you to create any reality in your life that you see fit. . . ."

Thus begins my new role as a card reader for *Psychic Central*. As the day wears on, new faces working staggered shifts enter our dark cubbyhole, and soon we have a full house of preachers in the dark. After a break, Scooter prances in arm in arm with Sherry, and they soon establish themselves as a comedy team like Laurel and Hardy, though in gay slang they're more like the nellie fag and the fag hag. Scooter is a puckish, diminutive, effeminate fellow, and a couple of hours of his loud, affected and shrill acting out make him my least favorite card reader. The poor bloke speaks with profoundly affected sibilance; like a buddy of mine used to say, how many s's are in *truck?* Sherry could be a real knockout if she trimmed down a bit. Too bad she chose such an offensive creature to bond with.

Late afternoon. The lines quiet down, and my fellow psychics exchange greetings with me. Tar Eyes introduces himself as Ramon, and we shake hands. When he smiles his eyes aren't as threatening as before. His large silver pentagram pendant hanging from his neck choker glistens in the subdued light.

Pandora—amazingly, her real name—quietly introduces herself in a soft spoken, shy manner. Her angelic face and plump, petulant lips make her a dead ringer for Angelina Jolie, years before Angie's superstardom. I gradually come to realize that Pandora favors feminine ladies like herself, which grows more clear in time.

Scooter and Sherry launch their rolling swivel chairs in my direction. They slow to a stop a few feet away from me and stare like twin mimes. I'm standing near Ramon and can feel Scooter's little rodent eyes chewing their way up and down my body. "Hi, I'm Scooter," he says with a half dozen s's, extending a limp wrist. Why do boys like him have to revel in their sissified energy? "You simply mu-u-u-st give me a reading sometime."

"Yes, and me too," Sherry says, introducing herself. Anna and Donna Lee roll their eyes and exchange knowing glances.

"Aww, Chuh-rist, you guys," Tim grumbles from his desk. "This is his first day here. Don't corner the poor guy!" He chuckles, shaking his ample belly. I see Tim as my ally, he's very easy going and fun and grounded, and was instrumental in my being hired.

"What's wrong, Tim, just jealous of all the a-ten-shun?" Scooter needles.

"Not likely," he laughs throatily. "Not for the likes of you, Tinkerbelle!" The room erupts into laughter. Scooter trots over to Tim like a little girl in ballet tights, bends with hands at the waist as though grasping a tutu, curtsies, and gives him a big loud raspberry.

In the coming days, though I gradually gain confidence in my

113

readings, there are times when I'm not connecting at all with the caller. That's when we're allowed to tell them they can try someone else for a better connection, and send them back to the control desk operator. Other times it's like I'm sitting next to the person, leafing through their scrapbooks, like with Robert, from Manhattan.

In the background, the hubbub of machinery and many voices echoed from a very large space. He spoke in very assertive, confident tones, and jumped right in. "I have a stock in front of me," he stated. "I'm considering buying a lot of shares. Tell me everything you see." The bulk of the cards I throw show a huge amount of money at stake, possibly a family dynasty, perhaps even a highly innovative invention in need of investors. Judging from Robert's reactions, I'm hitting some bull's eyes. The suit of Pentacles comprise seven of the thirteen cards, indicating money and worldly goods, and The Sun, The Magician and The Chariot fall auspiciously (the more Higher Arcana cards, the greater the impact.)

"The Ace Of Pentacles is the final card," I say, "representing the beginning of great wealth or material gain. Man, I'm almost wishing I could have a piece of this stock myself. If you can get back to me sometime, I'd really like to hear how things turn out."

"So what you're telling me is that the stock looks good," he ponders.

"From what I can see here, if I were you and had the information you already have, and it agrees with what I've been telling you, then I'd have to say, 'Go whole hog,'." Robert thanks me and abruptly hangs up.

Weeks pass. My initial impression was correct, Ryan and Tar Eyes (Ramon) are lovers. By now Ryan openly flirts with me, at times even coming up from behind me to massage my shoulders and traps while I'm on the phone. It's made things uncomfortable for me at the *Psycho Collection* (a nickname bestowed via group consensus), and good ol' Tar Eyes habitually shoots me looks of utter malevolence. The cards really fell into place after Anna told me that Ryan, Ramon, Sherry, Beryl, Pandora and Scooter are in a coven together, and Beryl of the Manson Eyes is their High Priestess. Some of my friends have told me to get the hell out, but I need the money. And flexing my psychic muscles is fascinating. So I stay.

Back when Ramon and I were on speaking terms he commented that it was nice I wasn't judgmental about them being Wiccans. I'd met practicing Wiccans at UT, discovering that it didn't necessarily indicate witchcraft. But time brings clarity, and the spells Sherry casually speaks of casting with her covenites don't always sound benign. Beryl's repetitive assertions that casting an evil spell on someone brings back the same dark juju to the caster times the power of three makes me that much more apt to believe their jolly bunch invokes them anyway. The lady doth protest too much. And my gut sense is that Ramon's unsettling eyes have seen plenty of threefold dark stuff boomeranging back to him, too.

By late September, every day I sense an oppressive atmosphere at the *Collection*, a heavy blackness hanging over us in our dark room, more so all the time. One evening Tar Eyes took a break during a lull, strode to the dry erase board and wiped away the details of chronic phone addicts we should cut off–psychic reading junkies who didn't pay their bills–and used the space to write a truly frightful paean to the Devil. I couldn't bear to look at it for long: stanzas about how he couldn't wait to die to reunite with his Master in Hell, and happily serve his sweet Satan. And his foreboding outfits, damn. Straight from *Fredericks Of Hell*. Today, the usual black ensemble, though he wore a white pattern to break the monotony–scores of white skulls painted on his shirt, accented by the pentagram. Cheerful.

Getting ready to start my shift, I take a moment to listen to what eight phone psychics chatting, babbling, burbling and voicing their insights sound like. To play along, assign separate actors for the parts, to be read aloud simultaneously:

Scooter/Morgan: "Honey, trus-s-st me, men are all alike; dump him, dump him, dump him . . . I know it hurt-s-s-s, honey, bel-ie-e-eve me I do."

Sherry/Charmaine: ". . . The DEVIL card doesn't mean you're going to be possessed! No! Of course not! It means that there might be some bedevilment types of disruptions going on in your life . . . No! Don't worry, you silly thing. . . . "

Pandora/Alta: ". . . but I'm not getting that she's the one for you . . . no, I'm definitely not getting that, more like a passing fling, enjoy her and then just move on when you can tell that it's the right time to go. "

Donna Lee/Angel: ". . . and are you aware that this call's costing you two ninety-nine a minute, hon? Okay, now I'm an astrologer and although I can't do an in-depth chart in the short time we have, I can give you a detailed overview. . . . "

Beryl/Ariadne: "Ah would. Yes! If ah was yew I'd march straight back in there and tell him 'Ahm not gonna take that kinda crap no more.' And the cards say you have the upper hand here. . . . "

Kimberly/Zandra: "Mmm-no. It's not going to happen right away, but you can tell your boyfriend Brian–" (animated screaming and shouting heard through the headset)–"what, now calm down Mary Ann, I can usually get names of people and places. . . . "

Ramon/Rennie: "Umm, and now I see the Two Of Pentacles. And the Two Of Pentacles is not necessarily a bad card, it can mean things are bouncing back and forth, like a scale. And now I see the Eight Of Wands, and The Eight of Wands can mean a trip. . . . "

Tim/Adam: ". . . it'd probably be okay to sell the Firebird right now, but be sure to hold out for a better deal than you got with your '65 Mustang. Whaddya mean how'd I know what your last car was? I'm a psychic! I didn't get this job with my looks. . . . "

Kimberly brings light to our dark den. She first caught my eye as I was leaving the *Psycho Collection* for the day and she'd just arrived, sitting on the office sofa chatting with Ryan. I wondered what such a striking lady was doing in our midst. She looked as out of place as a Blackglama mink at a yard sale.

Kimberly's tousled head of Billy Idol white-blonde hair fell to her mid back. Even seated I could tell she had a model's body, lean and with a great figure. In that first brief glance I saw that her eyes had witnessed their share of sorrow, but they also seemed to radiate an iron will. She held her head high and proudly, and dressed in smart looking pants suits. I could tell that I'd either really hit it off with her, or else she'd be a distant ice princess and we'd steer clear of one another. And I decided that if we were to become friends, she'd have to come to me.

A few weeks passed, and Ryan started scheduling me evenings. On my first night shift, it actually felt better knowing that I wasn't missing out on a sunny day. Besides, mystics say that night time is when one's psychic powers peak. It's always seemed to be that way for me.

I enter our shadowy realm and spot Kimberly setting up her desk. I grab a cubicle at the other end of the room. Everybody's too busy to say hello. It's a little past six, and the lines are ringing steadily.

Later, around eight, there's a lull, and a lot of us disconnect for a breather. Just as I hit the break switch, I turn to find Kimberly at my side. Thankfully, Scooter and Sherry are busy gossiping over cancer sticks in the break room. A little quiet time.

"Hi," she says sweetly. "I'm Kimberly." She sports a black pantsuit and white blouse that look custom tailored. Her huge smile lights up my dark corner.

"Hey, Kimberly. I'm Seth--er, Wade." This doll actually has me flustered. She giggles, and the playful sound is evocative of Elizabeth Montgomery's Samantha Stephens of *Bewitched* fame. She even looks a bit like her. I wonder if she can twitch her nose and whip up a pile of cash for me. "I noticed you a few weeks ago talking to Ryan but I wasn't sure if you were working here," I say. She grins shyly at me, and folds her arms together. Lord she is such a pretty lady. "What. What is it?" I ask.

"I'm sorry. It's just that I've been getting impressions from you from across the room, even while I was giving readings. Do you mind if I tell you what I'm seeing?" she asks.

"No. Not at all. Have at it."

She stands close up and stares at me, then looks sideways, then back at me. Damn. I can see a big blue ocean in those clear bright eyes. They are as beautiful as Ramon's are unnerving. A few seconds pass. Then she says matter-of-factly, "Architecture was very important to you for a long time." I try not to help her out, but I'm sure my eyes widen a bit. Of course it was my childhood obsession and my first major at UT. "Yeah," she smiles bewitchingly and nods without doubt, "in fact I can see you drawing lots of things as a boy, as a teenager . . . designing houses, floor plans, even inventions . . . aircraft . . . spaceships?" she laughs. I tell her she's amazing. Her arms still folded, she looks away, as though at some invisible screen. Then she closes her eyes and puts her thumb and forefinger to the bridge of her nose. A few moments of silence. She sighs. When she opens her eyes she moves closer and stares deeply into mine, like she's scanning my soul. "Did you ever have a black sports car?" she queries. I shake my head no. "Are you sure? 'Cause I'm seeing you driving one . . . let's see, ummm, it's very sleek . . . a, oh, some sort of Toyota, yes–it's a Supra. Are you sure you didn't used to drive a black Supra?"

"Believe me, Kimberly, I'd remember if I had a car like that–wait a minute," I break off, laughing nervously. "What did you do, check up on me?" That Samantha giggle again. "Wow! I never owned one, but I used to drive one a lot. A friend of mine bought a black Supra and he had me teach him how to speed shift. I drove it so much and got so attached to it that it felt like it was mine." She smiles and nods. She's used to being right. "You're really good. But how funny that you'd jump from my design interests to a car I used to drive," I say.

"Yeah," she shrugs, "I never know what images will pop up when. I just try to allow the stronger ones to grow clearer."

In the ensuing weeks I learn that Kimberly has two children and an abusive husband who beats her periodically. He suspects she's having affairs, and resents her psychic gifts, and that she's trying to develop them. He cheats on her, and when he returns home from his flings she shakes her head at him and describes who he's been with. I'm dumbfounded that a lady with a figure like hers could be a mother to two children. Already I'm feeling protective of her, and grow to resent her husband. I wonder if she's savvy enough to know that I'm a gay man who finds her attractive. But if she knows my situation then why does this feel like a mutual flirtation?

My feelings are obviously mirrored by Pandora. More than once I've caught her ogling Kimberly like a shiny new bauble she covets. With her whole Angelina Jolie arsenal of looks, the long, flowing hair, sexy body and angelic face, she was another of our psychic girls who probably could have modeled or been an actress if she'd wanted to. (A hot straight boy psychic could have really cleaned up here.)

Pandora looks and dresses like a total package, except for the frumpy old witch shoes she wears. Clunky, like high heeled clogs, only homelier, creepier somehow. Very Wicked Witch of the West-ish. Maybe she shops with Ramon at *Fredericks Of Hell*. I first noticed them when she sat at her desk with her long legs crossed; I envisioned her on a ratty old broom flying high over Miami at night, her wild mane of hair swept back in the wind, and I had to stifle a laugh. A fitting fantasy, since she is also a member of the *Psycho Collection's* coven. And right now my psychic abilities tell me she's an enchantress who would sell her soul to call Kimberly her amour.

I can picture her at the witching hour, whooshing in a huge arc over the waterfront pastel lit Miami skyline on her broom, scrolling in cursive script across the heavens, *Surrender, Kimberly*. . . .

* * * * * * *

Dusk. I'm reclining in a chaise on the west terrace of Angela DesJardin's penthouse, nursing a glass of buzz-inducing Merlot. I'm hypnotized by the yellows, roses and burnt oranges of the sunset as they bleed over the water and bathe downtown Miami's fantasy skyline. The fancy, Emerald City skyscrapers and shiny towers ignite in a blaze of neon lights to greet the falling darkness and compete with the sunset's splendor. Atop the tallest one, an iridescent green laser beam pierces the sky, zigzagging back and forth like Luke Skywalker's light saber. All is reflected on the tranquil water surrounding Star Island. The sight never fails to move me. The surreal shimmer of carnival colors conjures up scenes of my outings on South Beach. My blurry, escapist neon nights of finding a new big dumb muscle head to happily trance out with, bring home, and then part ways with at dawn. In my gut sense, I believed we might really come around to the physical plane multiple times to explore a wide spectrum of life experiences for our soul's portfolio--but we're only young once each time around. This was the season for sewing wild oats, of sharing the fire of youth, and I knew I wouldn't always be able to be as wild and free as my early thirties allowed me to be. So I happily indulged the merry-go-round of the South Florida triangle: the gym, the beach, the bars. Lather, rinse, then repeat.

"Wa-a-a-de!" her sing song voice calls out to me. "It's time to come in and get started. Everybody's here." Angela is hosting a metaphysical get together, a psychic house party high in the sky. I've brought my cards to give free readings. Derek is away, overseeing the construction of their dream house on their recently purchased waterfront acreage on Martha's Vineyard.

Angela came clean with me about their situation, right about the time I'd pieced it all together--although I'd never asked questions about their finances. Early on, she'd told me that one of her ex-boyfriends used to deal huge amounts of cocaine, and shuttled it around the country on Lear jets. They ran in the fast lane and partied with high rollers from Manhattan to Palm Beach to Malibu.

When her beau was busted and jailed, his attorney sprung him on a technicality after serving a few months. Turns out the swashbuckling ex-boyfriend was her hot rogue hubby, Derek.

Derek and Angela will have a very comfortable life with the millions he socked away. Angela once told me she sets aside time every day to walk around and say out loud, "Thank you, dear God, thank you, God, thank you, thank you, thank you. . . ." Wary of IRS goons who battle it out with their attorneys every April for access to their hidden assets, they eschew flashy purchases—like the Rolls Royce Corniche convertible Angela said she'd rather drive than her beat up old Buick. But trips to Europe and the fine homes and the dinners at four star restaurants more than compensated. Life is sweet for the DesJardins. I'm happy for them.

Angela looks spectacular tonight in a sheer white gown tied at the waist with a thin gold belt. She wears her hair pinned up by gold ornaments embedded with crystals and gems, and gold sandals on her feet. She radiates like a Grecian goddess, aglow in the soft light of tall white taper candles. She deftly maneuvers the crowd into a large circle on the plush carpet of the living room. Our enclave boasts a neurosurgeon, a couple of psychics, curious jet setters, a few South Beach models, an advertising executive and his realtor girlfriend, a deli owner, some gal pals from the gym. A burning smudge stick to cleanse the air is passed around after our hostess quietly says a prayer. A relaxed atmosphere pervades. We're all happily floating on Angela's cloud. A blissful night with memorable scenes you cherish later.

One by one we share why we've come tonight. When it's my turn I tell them I was opened up to other realities when I had a UFO sighting at eleven. "It tends to open up your mind to them when you have a flying saucer almost land on your head," I say. Others share similarly alternative, mystical, paranormal experiences.

We do a fifteen minute group chanting of OM, holding the note for a full ten seconds. It's amazing how it makes the room feel. Still. Centered. Clean. Pure. Afterwards, Angela has us queue up and approach her, one at a time. While humming loudly, she takes a pointed chunk of quartz crystal the size of a grapefruit and scans over our bodies. She says she is restructuring our DNA. No one questions this, so I don't either. At various points she'll stop moving the crystal, zero in on a spot, and her voice rises to a crescendo of "OOOOOOOOOh, KIAH!! KIAH!! KIAH!!" I guess that's when the quartz is really doing its stuff. Her five year old son Alexander wanders in and asks why he can't play with us. Maybe the sound effects remind him of Cowboys and Indians, too.

I give some readings on the terrace and Angela sets up her massage table in the library, giving short, clothed sessions of what she calls Healing Touch. Between the brie and other delicacies in the dining room, and the champagne and spiritual happenings here and there, everybody mingles and samples it all, a New Age smorgasbord. Later when the crowd has thinned out and just a few of us linger, I get on the table, lie on my back, and Angela works me over quietly, in gentle rocking motions. Her fingers barely touch me and their warmth soothes me. Suddenly, she says, "Oh, honey, your adrenals are fried. What happened to you?"

I laugh bemusedly. "Hmm. I don't know. Just a fast life in the nineties, I guess."

When she gets to the back of my head, she slows down, gently palpating my scalp and the back of my neck, then stops. "Oooh, Wade," she murmurs as she cradles my head. "I'm not sure I'm supposed to probe around here." Her tone is soft and maternal, evocative of the Earth Mother persona I'd always seen as one of the many facets of her warm hearted, benevolent presence. She has a quizzical look on her face. Her warm hands gently rub and soothe the back of my head and neck. Her fingers slow to a stop, and she seems dazed.

I look up into her eyes. "Huh? What's wrong. What is it?" I ask.

"I don't know what to say," she says softly, her palms gently cradling my head. "It's like there's something stored away in here and I'm not supposed to mess with it."

* * * * * * *

KIMBERLY'S GIFT

All great things must first wear terrifying and monstrous masks
in order to inscribe themselves on the hearts of humanity.
                                        –**Nietzsche**
                                        Beyond Good And Evil

"I'm really sorry you couldn't make it to Angela's get together last week," I say.

Kimberly and I are sharing a candlelight dinner of grilled chicken and pesto pasta on my porch, the garlic in my homemade pesto hot enough to sear the tongue. She missed Angela's metaphysical party because she'd filed for divorce from her husband, and with Pandora's help shuttled her kids up to her brother's home outside Ocala until the smoke cleared. They'd be safer there in case of repercussions. She has temporarily moved in with Pandora and her family in a spacious waterfront home in North Miami Beach. I offer her alternative shelter in case Pandora's flirtations get too aggressive, or she just wants to get away.

We finish our dinner. Kimberly helps me clear the dishes and take them down to the kitchen to be washed. Afterwards we return upstairs to the porch and finish our wine on the couch. Late November is still as humid as a Thai jungle, so the ceiling fans cosset us. I tell her how amazed I am by the psychic skills of some of our readers at the *Psycho Collection*, and we both regret the increasingly ominous vibes in the phone room. I mention that during a recent break, Tim out of the blue told me that I'd been right, the lawsuit settled in his favor. He laughed at my startled reaction. "It's why we hired you, buddy," he said. But it was understood by all that Kimberly was the star psychic, and no surprise to any of us that the call center began to get deluged with requests for readings by Zandra, her psychic phone line alter-ego. She could open up a line of her own without a hitch.

Her gift blows me away, and I ask how she does it, does she practice. Yes, she said, like recently when she'd gone out to dinner with Pandora and they played a psychic skills game. They alternated on who would sit with their back to the entry. When it was Kimberly's turn, Pandora would say, for example, "Ok. New subject entering. Begin reading now." Kimberly closed her eyes and described in painstaking detail the man's physical characteristics, height, weight, hair color, color and style of clothes, distinguishing features. Accuracy? Nearly 100%. She has eyes in the back of her head.

Another time, on a girls night out at a disco, they sat at a table enjoying some drinks, grooving to the music. A big, muscle-bound player strutted up to Kimberly, put down his beer and put the moves on her–no introductions, no names, just a fast one-liner. She puffed on her cigarette, took it all in, smiled up at him, nodded and said, "You know, Billy, you think you're really hot stuff cruising around in your fire engine red Porsche Carrera, workin' all the bars and playin' all the girls in town. But your flashy stuff doesn't do it for me at all." Billy, slack-jawed and stunned, gathered his wits and his drink and silently skulked away like a spanked puppy dog. Pandora loved that one.

I've told her over and over she should do some volunteer work for the police. She could do a world of good with the missing persons cases. Most of the time she tells me she wants to, but fears the exposure. Once people discover the scope of her skills, they call her day and night and drain her dry. Even Pandora sometimes keeps her up at all hours asking for free readings, bombarding her with interminable questions.

Tall white candles flicker in the ceiling fan's downdrafts. Alf contentedly sits at attention in his classic pose, tongue protruding, face smiling, his radar dish ears vigilantly scanning the dark grounds outside the porch. Sometimes it feels like we're kid brothers. Man's best friend.

"He sure is a beauty," Kimberly says.

"Yeah. I've always loved German Shepherds. He's one of the reasons I wonder if I had a past life in Germany. I seem to have such bizarrely karmic relationships with Germans and Jews, like they're marching through my life, Germans and Jews, Germans and Jews, on and on. Lovers, friends, and especially employers. And I'm half-Kraut myself."

"That's so funny you'd say that," she remarks.

"Why?"

"Remember on those night shifts, when I'd sometimes lean back in my chair and look over at you and give you those strange looks?"

"Yeah, like your eyes were gonna pop out of your head. I used to be afraid you were seeing some things that . . . well, that I didn't want you to see. All my nasty boy stuff."

She chuckles in her Samantha Stephens laugh. "Relax. I don't allow myself to intrude like that. I don't probe into personal things unless asked. But what I did see is flashes of you in an SS officer's uniform. Standing in the snow with a big German Shepherd at your side. The picture was so clear for me it's uncanny. Every time I looked at you funny, that's what I was seeing. Every time."

Her revelation has a strong sense of truth to it. "Wow! Really? Damn . . . I can see that. I'm getting goose bumps. . . ."

"Mmm-hmm. I didn't want to tell you because I was afraid it might upset you."

"No. Not at all. It wouldn't surprise me at all. I've had so many troubled relationships with Germans and Jews in my life it's like I'm paying off karmic debts this time around. Payback time. Especially with Jewish employers--big time. Did you see anything else?"

I have to encourage her to open up, convince her I can handle the scope of her visions. Her protectiveness is touching. She gazes into her wine glass a few moments, then looks at me with her ocean eyes.

"Mmm-yeah. I saw you at a concentration camp. You were an SS guard, standing at your post, a German Shepherd always at your side. You did what they wanted you to do out of duty, but deep down inside you were against it all, like a conscientious objector. You helped Jews escape when you could. You were against all the torture and killings. You weren't always a good guy, but you helped the Jews escape whenever possible, and tried to protect them other-wise. I know it sounds pretty heavy, but it's what I see."

The whole scenario triggers flashbacks to Dad and my family's no-madic six month assignment on Key Biscayne, Florida, summer of 1963, and scenes with our neighbors the Schraders, and their huge Shepherd, Wolfie. I had such a powerful connection with him, a control over him, like I'd bonded with large German Shepherds be-fore--but he was the first Shepherd of my youth. He was a beautiful wolf-ish animal, maybe he was part wolf, and I was really taken by him. Wolfie seemed to sense a connection to me too, and the com-munication resonated between us. The local pound often came by and picked him up, as he was notorious for chasing kids and biting them. His jaws could have shredded my little six year old body, but he somehow recognized my authority and he never challenged me.

He strangely respected me somehow. I'd even have to walk Natalie home from school to keep Wolfie at bay, as he knew she feared him, and the devil would be waiting for her on a corner across the street from the school grounds when the final bell rang at 2:45pm. Maybe my mastery over German Shepherds was something I'd brought along with me from the darkest days of another place and time. It was part of my energy essence. Wolfie saw that, and he connected with it. I even sometimes stood in our front yard and studied him perched on the Schrader's driveway as he gazed over his domain, and I smirked at him, like *I dare you.* He would stare back at me calmly and keep his place. He didn't mess with me, ever.

I share all of this with Kimberly. She smiles and nods in assent. "Wow . . . it makes so much sense," I say. Her revelations silence me for awhile. The crickets and bullfrogs play us a little night music. My Grove domain is hypnotically tranquilizing at night.

"You know, Kimberly, I've been needing to tell you something for a long time. And, well, I guess tonight's as good as any. . . . "

"Ye-e-es," she smiles mischievously.

"I don't doubt your psychic gifts at all. But sometimes I wonder . . . how much you can see. I mean, I . . . I haven't been with a lady, in--"

"In awhile. Yeah, I know," she smiles warmly.

"Yeah. And I haven't exactly been like Mother Teresa, if you get my drift."

She giggles her Samantha laugh again. "Yes, I did figure that out. For one thing, your dining table here has qu-i-ite--a--story to tell, mister man!" she smirks and pokes me in the ribs.

I toss back my head and let out a great big horse laugh. The dining table is one of my favorite things about the porch. I'd snagged it from a yard sale. It had been an oversized butcher block lab table, and I'd used a whole can of paint stripper to reduce it to the raw oak, then bleached it with Clorox, and finished it with clear lacquer. It was great for dining, spreading out lots of books for studying, and . . . hot grappling with one of my bad boys. Incredibly, Kimberly had picked up on my nasty interlude with Sean, my occasional Ft. Lauderdale muscle buddy. He'd dropped by one night in late August, and I'd just gotten home from the gym. In a matter of minutes we were on the porch and all over each other. The dining table/lab table was handy, and the grounds of my Grove domicile provided for us an ample jungle ensuring visual and auditory cover for a late night outdoor nature-boy tumble in stealth mode.

Once we'd had our fun and collapsed into each other on the table top, I'd gazed down into his eyes and said 'Do you realize that if my landlord had walked by on the street out there and caught us in the act, I'd be out of here in forty-eight hours?' 'Try twenty four,' he'd said, ever the witty rogue.

Earlier, after I'd poured the wine, when Kimberly picked up her glass, she'd acted like she'd gotten a static charge from the oak table top, and she looked at me funny and laughed. So she had gotten it all in a flash. Unbelievable. She wasn't just an amazing clairvoyant and psychic empath; she had skills at psychometry, too; but I still had to call her on it. "You mean you picked up vibrations from this table, too?" I asked.

"Sure. Psychometry. Lots of psychics can do that. Everybody and every thing has its own story, recorded and stored in its energetic field. You know that as well as I do. And what a story-we-have-here, mister!" she teased, tapping my table and cackling.

"Wow . . . your gift is phenomenal." She beamed and blushed. A sudden thought arises. "You know, I hate to ask you for a freebie like all the vampires at the *Psycho Collection*, but there's something you might be able to help me with, to clarify for me, if you wouldn't mind. I know it's late, but would you mind doing a quick reading?"

"No, of course not, my dahrlink." I loved how she sometimes lapsed into a faux Romanian/Bulgarian accent, lampooning gypsy fortune tellers in movies. "What is it?"

"Well, when I was eleven I had a really intense UFO sighting when we were living in Houston. Very close up, very frightening. I've been to two psychologists for time regression hypnosis, and I've always walked away feeling like there was more to it. Like I'm missing something--besides a few screws, right? You suppose you could scan back to the sighting and see what you come up with?"

"Sure." We leave Alf upstairs to guard the grounds, and retire to my downstairs sitting area. I dim the lights. Kimberly has me lie on my back on the sisal rug in front of the sofa. She sits in the lotus position and cradles my head in her hands. "What I want you to do now is close your eyes and relax. Go back to the very beginning of your sighting, maybe a minute or two before. Then silently replay it in your mind, from the very beginning to the end. Just relax and let it flow. . . . Good . . . that's good."

In seconds, I'm there, an eleven year old boy at a windowsill

127

watching something in the night sky. Immediately Kimberly oohs and aaahs, but says no words. I allow the memory to play in my head like a silent film. Soon I come upon the moment when the object makes its closest approach, coming down at me, down . . . down . . . down. Both in my bedroom as a boy and during my hypnotic regression sessions as a man, I uttered, "Oh My God!" At that very moment of the visual replay running in my head, as though she was reading subtitles under my visuals, Kimberly exclaims, "Oh – My – God!"

I bolt upright with the shock of it as she releases my head. She pulls away as if from an electric shock, and leans against the sofa with her eyes closed. "How the hell did you know I said that?! And right then! I can't believe this!" I exclaim.

"Oh my God, Wade," she whispers quietly, her face in her hands. "Oh my. . . . " she fades out.

"What? What, Kimberly. Kimberly, what is it? You're scaring me. What is it? What did you see?"

I nudge her a little bit. She pulls her hands down and places them on her lap. Her eyes look distant, glazed. Swallowing hard, she says, "I saw everything--the way the thing zigzagged and darted, and the bright lights. So bright! How it came down at you. I'm just not sure how much of this I can tell you." After a minute or so of cajoling, I get her to spill the whole story. "Well, for one . . . the object was outside your window for between two and four hours. And . . . I don't know how else to say this. . . . There was a transmission of some kind. It--or they--transmitted some kind of information to you. Into you." Her last words hit me like I'd been clubbed in the head. Transmission. Information. The words that surfaced in both regression sessions. I tell her this and she nods, unsurprised. But around for a few hours? "What kind of information?" I ask.

"I'm not sure you'll ever be able to find that out, no matter how much hypnosis you do." Dr. Dixon's very words echo in my head. The stark bold-print accuracy of her words grows eerie. "But my sense is . . . that they've used you the way that they've used a lot of people on this planet. As a living library book. A place to store information."

"Huh?"

"Yes. When I scanned the object, what I got was a deep, deep sense of urgency, desperation, sadness . . . loss–like time is of the essence. And if I'm right, then their race or planet is dying off and they want a record preserved--of their history, their culture, their planet, what they're like. So they place it inside of the heads of people like you."

"But if this is really what happened, why? And if they died out, how would anybody access the information?"

She shakes her head. "I don't know. But I do get some strong impressions that you're going to be working with them somehow, someday."

"Wow. That's funny. That's definitely one of the things I've always wanted to do. From the first time I ever heard or read about UFOs . . . I've always felt a connection." We sit on the floor quietly for awhile, side by side against the sofa. "But Kim, I've still got this feeling that there's something more. Maybe something further in the past. Could it be something that happened when I was younger and I don't remember it?"

She takes my hands in hers and closes her eyes. She's got to be one of the most beautiful psychics in the world, the delicate, sweet face, the shock of long white-blonde hair. A long moment of silence, and then she says quietly, "Six . . . and three."

"Huh? "

"You had encounters when you were six and three years old."

"Yeah? Really, that's what you're seeing? Damn." I swallow hard. "Wow. . . . Okay, I'll buy it, how about six. What do you see at six?"

She closes her eyes again. "I see a huge cabin in the woods. Two. No. Three stories, very big, massive, solid, made of logs and stone. A big two story entry and a large wooden staircase. Lots of trees around. Lots of leaves on the ground. Out in the woods, in the outback . . . hmm . . . could it be North Carolina?"

Knowing precisely the place she's found, I say, "How about Virginia?"

"Yes! I was going to say Virginia. That's where it is."

"And?"

"I see you running through the woods at night, as though you're being pursued. And you are. You're being chased. Some kind of light in the trees. And you make it to the cabin, and run inside and then . . . I see you on a bed, cowering, with all the covers scrunched up around you. And there's a strange light shining through the window near your bed. And . . . you're mumbling to yourself, 'Not again . . . not again,'."

"Not again? Shit. Damn, Kimberly. That was the cabin we used to have family reunions in when I was a kid, way out in the back-woods of Virginia. Not again? "

"Mmm-hmm. You were referring to what happened at three."

"At three. I can't imagine what that could have been. But okay, I'll bite. What do you see at three?"

She closes her eyes and holds my hands again. A minute of si-lence. "I see you out in the middle of a huge field," she says dreami-ly. It's daytime. There are wildflowers and huge, wide open areas of grass and weeds growing wild. And beyond the open fields, trees . . . a huge forest of tall trees, miles and miles of them."

"Mmm-hmm." Age three. Tyler, Texas. Living on the edge of a vast pine forest.

"And you're pretty far away from your house . . . but still, I do see the frames of new houses in the distance. Houses under construc-tion. And . . . you're out in that field, and suddenly. . . ." her voice trails off.

Abruptly, she pulls her hands away from mine and wraps her arms around her bent knees. She starts to shiver, first subtly, and then in a more pronounced fashion. "Hhhhshhhshhhuhuhuhuh," she mutters, shaking and shivering, then rubbing her hands up and down and over her arms.

"Kimberly. Are you okay? What is it?"

"I'm c-co-old. It's c-cold inside. I'm onboard." She continues to shiver.

"What! Onboard?!"

"Unhhh. Ohhh. My God. UNHHH!" she exclaims. Her eyes open wide, incredulous. She shakes her head and her hair goes flying. She runs her fingers all the way through her long blonde locks.

"You're telling me I was abducted when I was three?" I almost whisper it to her. "Come on. Is that what you're telling me? Is that what you're seeing, Kim?"

She catches her breath and rubs her arms vigorously with her hands. I can't believe what I'm hearing. "Yes," she says solemnly, looking me right between the eyes with a steady, soulful gaze. "I'm afraid you were taken, Wade, as crazy as it may sound. They got you when you were playing, way out in that field. Depending on how you look at it, you were in the right place at the wrong time, or the wrong place at the right time."

This is too much for me to accept. I'm thinking I've finally discovered that she does make mistakes like everybody else. "I'm sorry, Kim. No. No way. No way." I shake my head. "You've been right on target before--with pretty much everything. But this time you're way off. I'm sorry. I don't mean to doubt your gift, but everybody makes mistakes. If I was really abducted when I was three, I know I'd remember something. That's the way my mind works. I'd have some kind of memory trace. Something."

"Mmm-hmm," she nods gently at me. "You do have memories of it. The experience terrified you. It was so traumatic for you that you had horrible, scary nightmares about it . . . recurring nightmares . . . for years and years." Her eyes are moist, her face radiating pure compassion as she steadily gazes deeply into my eyes.

"No I didn't. I'm sure I didn't. I--" A sudden realization flashes to life, an epiphany that surges like an unexpected jolt of electricity from a live wire. It was as though a blinding theatrical light suddenly winked on where before there had been a dark stage. Like a klieg light illuminated a place so dark and distant I didn't even know a stage was there, or that secret scenes played out on it in several acts, to a limited audience.

A powerful chill runs through my body, from my head to my toes, as though my blood has turned to icewater, and my body is challenged to adapt to the sudden shock of it. It's the chill of recognition, of what I now call Body Memory. My body is rocked with waves of cold fear, and I roll over on the floor and fall forward, with my face against the cold terrazzo tile. I begin to beat my fists against the hard stone surface. "Oh fuck! Oh my God! OH NO!" I choke on the words. I was finally confronting the mystery presence lurking in the darkness of my childhood, crouched down low in the shadowy edges of memory, and it was staring back at me. It did not give me a warm feeling. Nothing about it resembled warmth or kindness. It brought with it the death of innocence. My childhood's end had come much too soon. With brutal abruptness.

With tears in my eyes, the words come flooding out. "Oh my God, Kim, I did! I had recurring nightmares that started in Tyler, back when I was a little kid--right around three! I was on a table; I was always surrounded by these living things . . . blurry figures, they were never clear, but I don't think they were human, and I didn't want to see them clearly. It seemed like they were observing me. And it seemed like they weren't like us. I was pinned down somehow–I couldn't get up, I couldn't move, I was paralyzed. And I remember

I always woke up screaming bloody murder, and Mom would come running in and comfort me. And I had that same nightmare for years and years . . . and the sleepwalking! I sleepwalked for years, too, starting in Tyler. And sometimes Mom would find me wandering around the house, or she'd almost trip over me lying on my back on the big area rug in the den, flat on my back, arms spread wide, staring blindly up at the ceiling. Jesus. It's all starting to make sense. Oh, my God. . . ."

Kimberly brushes my face with her hand consolingly, her eyes full of tears. "And the doctor. It all makes sense now. I was so afraid of going to the doctor in Tyler that my mom and sister had to make up stories that they were taking me to get an ice cream sundae or a banana split. And by the time I knew where we were going and we'd get to his office, I'd be in a panic. And by the time he'd be putting me up on an examination table, I'd just . . . be . . . flipping out and screaming at the top of my lungs, 'No!! What's he gonna do to me? What's he gonna do to me!!?' And the doctor would look at my mom like he wondered what the fuck my problem was . . . like what the hell's going on with this kid at home? Damn. It all makes sense now. Oh my God."

Kimberly hugs me and I hold her tightly for awhile in silence. With my head resting on her shoulder, I say, "So what did the bastards do to me, anyway, Kim? What did they do? Can you see that too?"

"That's why I looked so shocked, once I realized I was onboard . . . what I saw was, they had you--"

"No! Wait." I put my hand gently over her mouth. "Don't tell me. Not yet. This is such bizarre timing . . . but I went to a dinner party a few weeks ago, and met a psychotherapist who does hypnosis, and works with people on past lives. We were planning on starting some past life regression sessions next week, bartering, with me providing massages for some of her clients. But she's so interested in UFOs that I'll bet she'd be glad to take me back to Tyler instead. Let me see what I find with her first. See if we can go back and find this stuff in Tyler. How about you write down what you've seen, everything, with as much detail as you can, and then seal it in an envelope and put it away for awhile; we can compare notes later. Is that ok?"

"Sure. I think that's a good idea."

"Damn . . . Kimberly . . . and you're sure about all of this? As sure as anything you've told me before? As sure as you were about seeing me driving the black Supra?"

   She smiles and nods slowly, emphatically. "Yes! I promise you it happened. I can see you in the field, the thing coming down, even what it looked like, and how scared you were. You were so scared. You were so afraid they weren't going to let you go," she says dreamily, squinting at me. "And that's one thing I don't get. Why they let you go. 'Cause I'm not getting that letting you go was part of the original plan at all. And there's one other thing, Wade. I also get that it's not over yet. . . . They're coming back for you."

* * * * * * *

THE MAESTRO OF DEEP TRANCES

> The mind loves the unknown. It loves images whose meaning is unknown, since the meaning of the mind itself is unknown. The mind doesn't understand its own *raison d'être*, and without understanding *that* (or why it knows what it knows), the problems it poses have no *raison d'être* either.
>
> **--Rene Magritte**

Ellen Lyon has an office at the end of the hall on the third floor of a large crescent shaped building near a busy intersection in Coral Gables. The plaque on her hall door says *Ellen Lyon, M.S.W.*, and under that, *Psychological Services.* Sitting in her waiting room, I reflect on the events that have brought me here, and on the remarkable synchronicity that has drawn us together with such perfect timing.

Barry, the host of the dinner party where we'd met, said Ellen was a gifted hypnotherapist who sometimes worked with Dr. Brian L. Weiss, the celebrated Miami psychiatrist and author of the metaphysical bestseller, *Many Lives, Many Masters* (1988). Having read it three times, that in itself was enough to capture my interest.

When Barry introduced us, I noted that Ellen had deeply penetrating eyes--so intense that they seemed capable of inducing a hypnotic trance without words. Over drinks before dinner, we discussed hypnosis, psychotherapy, the rigors of her training, past life regression, and UFOs. She was fascinated by my sighting, and told me that Dr. Weiss was becoming more interested in the issue of alien abductions. That I was studying practical hypnotherapy at an institute in Ft. Lauderdale piqued her interest all the more.

Our immediate connection led to a dinner at my house the following week. She loved my funky bomb shelter, and the lush gardens of my jungle retreat. I showed her through my library of UFO books, and she was intrigued by the doggedness of my sleuthing over the past twenty-five years. I gave her a copy of my B.A. Psych thesis, *The Presence Of Specific Personality Factors And Their Effect On A Person's UFO Belief Systems* (see APPENDIX--includes a self-designed 15 Question UFO questionnaire with a 7-point Likert scale; a Content Validity Test-Retest Reliability of the questionnaire revealed a Pearson's r =.93 --wv). By the time dinner ended we were like old friends, and she had agreed to the barter. Alf really took to Ellen too, always a good sign to me.

I thumb through some magazines to pass the time and take my mind off of what Kimberly and I seem to have uncovered. In a few minutes her door opens and Ellen steps out with a man in a well tailored suit. She pats him on the back as he leaves, then walks up to me and we exchange greetings. She shows me into her private domain. I follow her, carrying my notebook and micro-cassette recorder. She's dressed very casually in faded jeans and a green corduroy shirt, but it doesn't compromise her professional aura.

I sit in an armchair across from her; reclining on a sofa with her sandaled feet tucked under her, she looks more like a happy kid at a slumber party than a therapist in a session. "How's everything?" she says, and smiles girlishly. Somehow her shoulder length stringy brown hair gives her the air of a tomboy, and I smile to myself.

I fill her in on the latest developments with Kimberly's reading; she nods and arches her eyebrows. Her interest is comforting. After sharing a brief overview of my Tyler material with her, I ask, "So do you give credence or credibility to information coming from a psychic, if that psychic seems unusually gifted?"

"Mmm, yes, I would have to say that I do," she says softly. When I tell her of my desire to switch tracks, and forego the past life work to focus on Tyler instead, she says, "Absolutely, we can do that. But I want to be sure you understand . . . that what we find back there might be kind of unpleasant . . . and possibly very disturbing and painful."

"Yeah. I realize that."

"And you're prepared to take that chance--I just want to be sure you understand what you might or might not be facing," she says with a very serious look on her face.

"Yeah. I know . . . But I have to know. I have to. Because the thing is, I've always had beautiful memories of Tyler. It was such . . . an idyllic place to spend childhood. I mean, it was such an innocent life. We almost never locked our doors. One of my most vivid memories is of my parents taking Natalie and me to the woods near our house, and it was such a beautiful sunny day, and we had these tin pails and we picked blackberries and raspberries from the vine. And I remember how happy I was that day, happy from all the simple things--the way the sun felt, the feeling of having all that space, how the wind blew through the fields, the way the berries tasted. We ended up eating a lot of them, smearing the juices all over our faces. And when our pails were full we came home and Mom baked blackberry and raspberry pies. Those kinds of memories are typical for me.

"But the thing is," I continue, "Kimberly's been so accurate about all the other stuff before Tyler. Almost beyond belief. And she was dead right about the nightmares. And I started sleepwalking in Tyler, too; and I continued to sleepwalk for years, almost until I was twelve. Isn't sleepwalking supposedly a sign of some inner turmoil going on?"

"It can sometimes indicate unresolved conflicts," she says.

"And I started getting sick a lot around then, and it was when my allergies first appeared. And those scenes in the doctor's office really hit home--the way my mom and sister had to trick me into going, and the way I would absolutely flip out when the doctor put me on the examination table. And I guess the way Kim described the setting, the field I was in, sort of spooked me. Like a trigger. So I called my folks a few days later. I casually mentioned to my dad that I'd had a lot of vivid memories of Tyler recently, and didn't we have a lot of open land around us? He said, 'Oh hell, yes! We lived on the edge of town. The Smith's house three doors down was the last house on the block, and beyond them were open fields of grass and wildflowers, and then a pine forest that went on for *miles and miles.*' When he said that, and emphasized miles and miles the way that he did, I got chills. It was exactly how Kimberly described it, and how I remembered it too. And as hard as it is for me to accept . . . I have this strange feeling, almost in my bones . . . that there's more to Tyler here than meets the eye."

"Mmm-hmm. Okay," she says with her head tilted up to one side like she's listening for something on a frequency beyond my words. "I'm curious about Tyler. What your family life was like."

"Sure. Well, I had an older sister, Natalie, and my little brother, Aaron, was born just before we moved away. Dad was a research scientist with an oil company, and Mom was a housewife and part time artist. Sometimes I played with Natalie, and other times I played with a little girl named Bree, and other times I played Cowboys and Indians with the boys across the street, or with some of the other neighborhood kids."

"Mmm-hmm. And how did your parents get along? Which of the

two were you closer to? I'm curious about a few positive experiences and a few negative experiences you might have had."

"My parents seemed to get along fine back then. I think I was always closer to Mom. Both our parents used the back of a wooden hairbrush to spank us on our behinds if we disobeyed or were bad. I got along fine with my parents, I guess. My dad was pretty emotionally distant from me, though. And he went away on business trips pretty regularly. But when he came home he always brought Natalie and me presents . . . if what you mean is you're wondering about child abuse or sexual abuse, I can't imagine that. Is that what you're looking for?"

"It's something we can't rule out at this time," she says gently.

"Well if there was any, I wouldn't believe it'd be my mom or dad--I mean, not in a million years. And I don't have any relatives who lived nearby, no sleazy uncles who might have done something. I can't even think of anyone around us then who might have done something like that."

We discuss Tyler for awhile longer, and then Ellen decides we have enough time for a short hypnotic session. We leave the office and enter a door adjacent to the waiting room. She uses a separate room for hypnosis, she says, because it prevents the client from inadvertently slipping back into trance during the post-hypnotic discussion.

I make myself comfortable in the recliner, setting up my tape recorder on the lamp table. Ellen says the fact that I've been trained as a hypnotherapist and have used hypnosis with some degree of proficiency—and have therefore been hypnotized myself numerous times while learning (aside from my previous UFO hypnotic sessions)—should make me an excellent subject. I prove her hypothesis correct.

I switch on the tape while she dims the lights; then she has me close my eyes, relaxing me with her induction patter. "Remain relaxed while you explore the part of your unconscious mind which holds memory and knowledge that is important for your healing and your growth. Feel a lightness. Listen to my voice. Each time you swallow, go deeper and deeper. Just let it go. Drift deeper and deeper. Moving . . . in space . . . as if you're carried by a hot air balloon . . . drifting . . . drifting through beautiful clouds . . . deeper . . . (she taps my forehead), deeper . . . good. I'm going to count backwards from four to one. When you reach one, you'll begin to see images of your life flash in front of you . . . at your speed . . . at the speed your unconscious knows is important and significant. Pictures and experiences in your life, going back through time. But before doing that, let the part of your unconscious that protects and takes care of you to be aware that at any time that part opens the door to a memory it would not be helpful to know consciously at this time, then after this session, the door can be closed again, and healing can take place. The door can be re-opened at any time in the future–as your hand moved to the right, that part of your mind that protects you became aware. Start to allow to begin to remember what you might have forgotten, and what might be helpful for you to know. Is that okay? You can talk to me."

WV: Sure.

Ellen Lyon: Do you feel that you're deep enough now to begin going back?

WV: Sure.

EL: When I reach one, begin to allow pictures in your mind to come into awareness. Four, deeper with each picture, three . . . deeper . . . two . . . deeper . . . one. Begin now. Back. Picture after picture. Going deeper and deeper. Only pleasant memories. Now, back to when you were eighteen.

WV: I was already in Tyler.

EL: In Tyler?

WV: Yeah. I was seeing all kinds of things very, very clearly.

EL: Okay. Go back to Tyler.

WV: I can see the den. It's just so funny . . . I can see the TV . . . so what, huh, it's just a TV, but it's really clear, it's a funny looking TV, it's an old black and white.

EL: Black and white?

WV: And I see the wicker covered bottle my mom made, she was sort of a frustrated artist. She painted a wicker thing that was a bottle holder, a great big wicker thing this really ugly color, sort of like a bright pinkish purple almost. We all kind of enjoyed it because she did it.

EL: How old are you?

WV: Somewhere around four or five, I just know I'm pretty young.

EL: Hold onto that picture for a moment. I'm going to count backwards from ten to one. With each number, go deeper, and put yourself into that picture. Okay?

WV: Okay.

EL: Ten, nine, deeper, deeper. (counts down, gives deepening suggestions). . . . One—be there.

WV: I can see the couch. (laughs)

EL: Where are you?

WV: Back in Tyler, in the den. I guess that was a really comfortable place for me.

EL: Go. Be there. Be there . . . be there . . . in the picture.

WV: Okay.

EL: Where are you?

WV: Sitting on the couch.

EL: What are you feeling?

WV: Happy to be there.

EL: What are you happy about?

WV: Just simple things, I guess. Enjoying being young.

EL: How old are you?

WV: I guess I'm four.

EL: It's okay to not be happy all the time . . . when you're four. Did you know that?

WV: No. I don't feel like I have to.

EL: Where's your mother?

WV: (pause) In the living room with my dad.

EL: And you're alone in the den?

WV: I guess . . . my sister is there.

EL: And you're happy?

WV: Yeah.

EL: Move back in time, in that house, to your birthday party, when you're four years old. Go there, now.

WV: Okay. I got a lot of presents.

EL: What are you getting? Open one of your presents.

WV: Um, sort of like a shop vise, but it's made of plastic, with a little crank handle. It's fun.

EL: How are you feeling?

WV: Having fun.

EL: Go back to when you're three . . . what's your first

memory?

WV: Being outside a lot. There's a lot of land.

EL: Go on to that lot of land outside, when you're three. I'm going to raise your arm up (lifts left arm up, elbow bent, giving deepening instructions). Remember the parts that are significant to remember about Tyler . . . your nose knows . . . go in to the picture . . . when your thumb touches your nose, you'll be there . . . your nose knows. (long pause) Where are you?

WV: I don't know. It just seems to be outside. I see a bunch of different scenes, but not all at one time. I can see all kinds of different flashes of things.

EL: What do you see?

WV: Umm. My neighbor's house. The Smiths. There's a pothole in the street in front of our house. A strange little kid across the street sometimes--he's really young and still in diapers--and sometimes he pretends to go to the bathroom in it, and I think he does (laughter). Umm. . . . I feel, I remember there's so much room.

EL: Be there. Move from seeing to being there. Feel you're there. Is there a breeze? Is it hot or cold?

WV: It's really sunny. But it's not hot, it's like--the air is just right.

EL: The air is just right. And the sun is shining on you. Move. Go there.

WV: I can see a trail . . . lots of wild grass and weeds, tall, just everything's running wild. And it was fun to have it nearby. Room to play and hide. Just have fun.

EL: There. Be there. On the trail. Feeling your little legs moving along the trail. Are you walking or running?

WV: Doing a little bit of both. Skipping and running along. I can see my shoes.

EL: What kind of shoes?

WV: A kind I haven't seen in a long time. Keds. Little kid's shoes. Almost like deck shoes . . . like Keds or something.

EL: Move forward, into that little kid's body. Go be there . . . skipping . . . running.

(long pause)

WV: Seems like I'm a little bit farther in the field.

EL: Farther. What happens next?

WV: I'm not sure.

EL: Into the woods. . . . What's happening?

WV: I don't know. It just seems kind of blank . . . just sort of like I walked into the woods and then, I don't know where I am.

EL: If you could know, if you could know, what might you know?

WV: What might I know?

EL: Yeah. If you could know.

WV: I don't know. I mean . . . I know what Kimberly told me.

EL: Go back. Go back. . . .
(pause)

WV: I'm not sure I see the trees yet. I see . . . the field is so . . . wild. And I see the trail is muddy . . . not muddy, but the dirt is smashed down from people walking on it. Almost like . . . the weeds almost look like stalks of wheat . . . some of them are flat on the trail.

EL: Are you there alone?

WV: I think so, yeah. Sometimes I went alone, sometimes I went with friends.

EL: Who might they be?

WV: I sort of see this girl living next door, I can't remember her name. She's a little bit older--there's a girl and a boy next door to them and they're both a little bit older. The Smith's kids. Sometimes she would take us, me and my sister and a few other kids, just for a walk.

EL: Look back, back away from the field. Look down. Look at yourself . . . as you move back, feel yourself go deeper, deeper . . . back up, back up . . . I'm going to count from one to ten. As I do, I want you to count from ten to one out loud. With each number, going deeper, until you reach a place where you can see yourself clearly. Begin now. (deepening process begins)
Go back . . . looking at yourself, at three. Skipping and running. What do you see?

WV: I see my haircut is a burr. (laughs)

EL: You have a burr?

WV: My dad always cut it almost to the scalp. I see one of the shirts I used to wear. It's really loose fitting, and shorts, I wear shorts a lot. And it feels really . . . uh . . . easy. Don't have to do anything. There's no school, haha. It's just fun being a boy . . . the days seem so long . . . I'm a little bit farther in the field. There's always seemed like there's some sort of mystery about the field.

EL: Be there.

WV: Something really unusual happened . . . it's almost like I might have hidden something from myself.

EL: Something? What kind of thing?

WV: I don't know. I would think that something might have happened.

EL: Are you willing to go back?

WV: Yeah. I want to.

EL: Back into that three year old. Are you willing to do it now? I wonder, if you would do that when I count backwards from three to one . . . to discover (voice trails off) . . . three, two one–go there now.

WV: (silence)

EL: What's happening?

WV: I don't know. I really wanna go back.

EL: Where are you?

WV: In the field.

EL: I'm wondering if your unconscious could share with us now, what it is that you're going back for . . . would your unconscious do that now?

WV: (long pause)

EL: What's happening?

WV: Just trying to remember . . . (long pause)

EL: Where are you now?

WV: I get some distance in the field, and I just, I don't know where I am.

EL: Good. That's good. Your unconscious is protecting you. And right now, the important thing to find out is what is the positive intention of your mind's remember-

ing. **What is the positive intention of your unconscious right now?**

WV: I don't know, I guess it's protecting me.

EL: **From what? Ask. Ask right now. Are you being protected?**

WV: Am I being protected? (long pause)

EL: **What's the answer?**

WV: (long pause) Something happened.

EL: **Something happened.**

WV: Yeah.

EL: **If you were to discover consciously what happened, what is your unconscious afraid might happen to you?**

WV: I don't know unless, somewhere in there, it thinks it'd . . . it'd be hard to deal with.

EL: **What is it that would be hard to deal with?**

WV: If I could see it I'd tell you. Looks like there's just a gray wall.

EL: **A gray wall?**

WV: Yeah. Like at the edge of the field there's just a gray wall. But I mean I know there's no wall in that field.

EL: **Your unconscious is protecting you.**

WV: Yeah. But I really want to know.

EL: **When our unconscious wants us to know . . . and before we break through the gray wall, it seems that the point is to begin to earn the trust, with great appreciation,**

of that part of your unconscious which has protected you.

WV: (plaintively) I just feel like I have a right to know.

EL: You do know in your unconscious ... I'm going to ask ... begin to ask the unconscious what you need to be doing, consciously, in order to begin to access memories of importance which have been forgotten. Ask.

WV: Ask now?

EL: Yeah.

WV: What is the importance of what I can't remember, and what do I have to do to remember it? What do I have to do to remember it? I know that if I lived this far ... I mean, I lived through this, so it couldn't hurt me. So a remembering shouldn't hurt either.

EL: It shouldn't.

WV: No.

EL: There's a part of your unconscious that doesn't want you to get something.

WV: Are you asking me?

EL: Ask yourself.

WV: Is there a part of my unconscious that doesn't want me to remember? ... I can see myself playing alone.

EL: Good.

WV: I think I was on the ground. I did a lot of make believe stuff--I didn't even need toys, I could make things out of rocks and stuff.

EL: Make things out of rocks?

WV: I used to make believe a whole lot. I guess I was on the ground a lot, on my hands and knees just playing in the dirt, just pretending things.

EL: What are you pretending?

WV: I think I built a fort, or something. I love building things. I'm just sort of in my own little world. I didn't care about anything else that was going on. I was just having fun.

EL: I'd like you to leave yourself there, having fun, in the dirt on the ground with the fort. Move back. Move back. Back. Away, up, allowing Wade to drift. You can just barely see you. Put a circle around you.

WV: Put a circle around me?

EL: Yeah. Put a circle. Frame it. Frame around you in a circle.

WV: Okay.

EL: And pull that circle—everything in there safe and warm—back into your unconscious, storing the memory, to be opened up again another day. Done?

WV: Yeah.

EL: I want you to take a moment and search your unconscious; ask if you've been programmed in any way by anything outside yourself not to remember. Ask aloud.

WV: Have I been programmed not to remember? (long pause)

EL: What's happening?

WV: I don't feel like I have an answer, but I feel like . . . kind of like I'm pushing at the edge of my protection, or whatever.

EL: Okay, good. I'd like your unconscious to know that any of this work we do now or in the future will be done at

148

a speed and with gentleness, to protect Wade's wellbeing. And that nothing will be forced that might or might not cause any harm. Is that agreeable?

WV: Sure.

EL: Okay. It's important and it's significant to remember even if remembering to forget consciously, that Wade is in charge of himself, and that anything that was implanted can be looked at gently, safely, and that whatever keys there are can be utilized with care and that anything that will be done will maintain and be aware of possible triggers to something else.

WV: Possible triggers?

EL: Yeah.

WV: Like what?

EL: That we need to ask; we will ask at a later time.

WV: You mean like a memory trigger?

EL: Yeah.

WV: Yeah, I feel like there must be some.

EL: It seems as though what your unconscious is saying is we need to move slowly and cautiously. Is that correct?

WV: Yeah, I think so.

EL: Okay.

WV: It's really intense.

EL: And we'll take adequate protections for you along the way. Is that alright?

WV: I didn't understand what you said.

EL: As we work, we'll incorporate new protections, built in, to assure and maintain your safety, on your journey.

WV: Yeah.

EL: You feel that need to, right?

WV: I don't know, I guess part of me does.

EL: Okay. Now I'd like you to move back to this room, this space.

I emerge from the trance and switch off the recorder. We return to Ellen's office. We take our seats, and I say, "So, what did you think?"

"What did you think?" she replies.

"Mmm, well . . . I found it fascinating that there really did seem to be a gray wall out in that field. And it makes me see that what I've learned about the subconscious is true--that the subconscious mind is the protective mind."

"Mmm-hmm."

"I guess I'm a little impatient with the process."

"Yes, it sounds like Wade would like to get in there and pry things open with a crowbar," she smiles wryly.

"Yeah," I laugh, "you're probably right about that. Another thing that . . . intrigued me, was when the words *something happened* popped up, it didn't feel like I was saying them. It felt like the words came from some very deep place, like the subconscious was speaking . . . without the conscious mind getting in the way."

"Mmm-hmm."

"And I found it interesting that upon re-visiting the field I would come up with the idea that there was something mysterious about it. And yet, when I think about it now, I feel that that's a valid memory. It feels right. Because something happened."

* * * * * * *

"Angie?" On the phone with my favorite client and gal pal.

"Hey-y-y-y, handsome. What's new and exciting in your life?" Angela's vitality is always a potent elixir, especially for a man of moods like me.

"Well, actually, I, ahh . . . I think I'm on my way to the *Twilight Zone*."

"Ooh, sounds like a great getaway to me. But how'd you manage to get tickets?" she cackles good naturedly. "Bah-dum-bump!"

"Exactly. Bah-dum-bump!" Well, I think I got issued tickets years ago." I fill her in on the latest developments with Kimberly's reading and the hypnosis session with Ellen.

"Oh! How exciting! Keep me posted, love. I want to hear all the details--I mean, provided you're comfortable sharing all of it."

"Of course, darlin', that's why I called you."

"One suggestion, if I may be so bold?" she asks with a gentle lilt to her voice.

"Sure."

"I'd be very careful who I told this to if I were you. Because, I mean, a lot of people aren't ready for this kind of information . . . and they might start saying things about you. You know what I mean?"

"Absolutely."

"Well, sorry to cut it short, but Derek and I are due at a gallery opening. Keep me posted, darling. I love you."

"Love you too."

My week at the *Psycho Collection* drags on interminably. The high point is on Wednesday, when three different women phone in and request a reading from our little nellie boy Scooter/Morgan because, "she's so understanding when it comes to girls with man problems." Choice. Kimberly keeps an eye on me but gives me space. She has plenty to deal with on her own, what with Jerry's attempts to win her back, and Pandora's attempts to get her on her back.

* * * * * * *

BREAKING THROUGH THE GRAY WALL

Finally, it's Thursday, December 12th, 1991, and I'm face to face with Ellen again. She's in full hippie garb again, and I'm finding her studied casualness reassuring, like it's her way of saying the clothes don't make the shrink. By now it's abundantly clear to me that I'm in very good hands, anyhow. "So, how's it been going, Wade?" she begins.

"Doing okay. I mean I guess I've been a little bit preoccupied with thinking about the stuff we're working on here. Maybe a little anxious, a little nervous. But I'm okay."

She nods. "That's understandable. Anything relevant pop up this week, any memories, dreams?"

"Yeah. Two things. One, I had dinner with a friend at *ESP*, that new restaurant across the street from *Warsaw*." *Warsaw*, a/k/a *The Warsaw Ballroom*: South Beach's legendary hottest gay disco of the day, a destination for both locals of all persuasions and international jet setters and circuit boys, one of my favorite haunts when on the prowl for fresh beefcake. "And for dessert, I had fresh berries. And when I bit into some blackberries, I suddenly saw myself as a child–I mean I was there, I was the kid in that sunlit field, with my tin pail of berries. That tart burst of flavor took me right back to Tyler. Pure synesthesia. It was really beautiful. It was like I instantly got in touch with . . . an innocence that will never be recaptured."

"Good. That's a good memory," she says, so softly I feel like I'm reading her lips. "And the other thing?"

"For some reason I remembered gardenias. We had huge hedges of gardenias around our house in Tyler; Dad was so proud of growing them, and Mom used to float gardenia blooms in bowls of water on the dining table. Their incredible scent filled the whole house. And when I remembered the smell, it brought Tyler back for me again. The fragrant scent was . . . comforting . . . and safe."

We make small talk for a few more minutes, and then Ellen and I retreat to the hypnosis room. Again, I set up my recorder and get comfortable in the chair. I close my eyes and she begins the induction patter.

EL: Allow your curious mind, and the curious part of your unconscious, to begin moving up the hill that's just before you, walking slowly, and comfortably ahead in a steady pace up the hill. Feeling the breeze . . . feeling the breeze . . . the sun on your back . . . warming, aware of your muscles . . . gently being massaged . . . deeper and deeper relaxed, walking up the hill, feeling the breeze, feeling the sun on your back. As you do, become aware of yourself becoming smaller and smaller. When you reach the top of the hill, what you're about to do, you become aware of the breeze, the gentle breeze, carrying the fragrance of gardenias. And you become aware, allowing the gardenia's fragrance to aid you in reaching, reaching back in time . . . any place you want to be, and any time . . . as soon as you become aware, let me know.

WV: I see a flash . . . of this great big huge wooden spool that my dad got from a trip to one of the oil wells he supervised; it used to have cable on it . . . and he brought it to the house and put it in the backyard for us to play on . . . and I can see us rolling it through the yard, having fun with it. It was as tall as we were.

EL: As tall as you were.

WV: Yeah.

EL: And you are enjoying yourself?

WV: Yeah.

EL: I need you to go be there, in the yard, playing with the spool that's as tall as you are, enjoying yourself, allowing your conscious mind to stay behind, to go be in that experience.

WV: I see some kids around. There's Bree and Frankie. Bree's a little girl I know, and Frankie is her little brother.

EL: How old are you?

WV: Three or four.

EL: Three or four.

WV: And my sister's there.

EL: Your sister's there? And what are you doing?

WV: Rolling the spool through the yard.

EL: Rolling the spool through the yard?

WV: Yeah.

EL: Having a good time?

WV: Yeah, it's fun.

EL: Be there. Be there–be there. Go from there, and move back further in time, to a time when you're in the field. Crawling around there, in the field, feeling the earth, crawling around in the field . . . go there now. (pause) What's happening?

WV: I'm still on the ground . . . playing make believe stuff.

EL: What are you playing? What are you making believe?

WV: I think it's just this fort that I've made.

EL: You're making a fort?

WV: Yeah. It's with rocks and stuff.

EL: You're making a fort. As you're making a fort, be aware of what you're feeling. What do you feel?

WV: (pause) I feel a sense of . . . space. Being very far away from things.

EL: There's a lot of space, and you're very far away?

WV: Mmm-hmm.

EL: And what does that do for you?

WV: It sort of gives me . . . the ability to have more . . . imagination . . . 'cause I'm far away from the house, and the other kids.

EL: And Wade is very curious, and discovers all kinds of possibilities?

WV: Yeah, I'm very curious . . . about everything.

EL: What are you most curious about, right now?

WV: To know about something that happened in the field.

EL: Go on that wave of curiosity . . . on that wave of curiosity, and ride it, ride it wherever it leads . . . to the land of discovery. Ride the wave . . . whenever you get to something, let me know. (very long pause) What's happening?

WV: I just have this feeling, there's something really big there. I keep asking myself, did I make it up or is it really there?

EL: Does it matter?

WV: Well, I guess I just wanna know the truth.

EL: The truth is in your curiosity. What's real is real; whatever comes to us is real. Varied things, varied shapes, and size, and texture . . . there's no significance between what is real and unreal. Real and not real–real is a made-up construct. Whatever is in our consciousness is there for a purpose. And now, ride the wave of curiosity. . . .

WV: (long pause)

EL: Riding the wave . . . riding the wave. Just noticing, not judging, just noticing, be in curiosity, in curiosity. (pause) What are you experiencing?

WV: I can feel myself . . . moving into . . . my body when I'm little.

EL: Go there.

WV: I have the sense of something . . . made of gray metal . . . huge . . . and close to the ground. And I wanna be there so I can really see it.

EL: You want to be there? So you can really see it?

WV: Yeah, I do.

EL: Go there.

WV: I'm not sure I know how. (long pause)

EL: Noticing . . . something that's huge and gray. You're wanting to move closer so you can see . . . you're not sure how. What do you see next?

WV: I'm scared, but I'm really curious.

EL: Scared, but really curious. And then what do you do? Scared, but really curious.

WV: (pause & deep sigh)

EL: Stay with the scared, and curious. You're safe. Allow yourself to move deeper, and deeper. Scared . . . curious.

WV: I just keep seeing, uh . . . sort of like a . . . a door, a black opening, on the side of it. And the whole thing, seems like all I can see . . . is the shape . . . of a lampshade almost–but flattened, tapered . . . it goes out like that (gesticulating the shape in the air,

157

resembling the flying saucer from *The Invaders* TV show, 1967--wv) with a dark opening.

EL: A dark opening.

WV: Yeah. It's dark in the underside.

EL: As you're riding the wave of curiosity, what happens next?

WV: I wanna go towards it.

EL: You wanna go towards it. And what are you doing?

WV: Standing up from where I was.

EL: Be there.

WV: And I was looking at it like I can't believe it's there. There's a light around it too.

EL: Be there.

WV: I just don't understand how it can be there.

EL: You're not understanding how it can be there?

WV: 'Cause I don't remember it coming down; it's like all of a sudden it was there.

EL: Be there. Be there.

WV: I don't understand. (nervous laugh) I just have this feeling that it's real. (laughs) I can't believe it's real, but I think it is.

EL: And what happens next?

WV: I'm slowly walking towards it.

EL: You're walking . . . you're walking slowly. What are

you feeling as you're walking slowly?

WV: I feel mixed feelings. I guess . . . I know it's something really strange, and scary. But I also wanna know more. I wanna know more.

EL: You wanna know more. As you're walking slowly, what do you feel?

WV: Fear . . . and amazement.

EL: You're feeling fear and amazement, walking slowly. Feeling fear and amazement, walking slowly, and I'm wondering, what happens next?

WV: There's something–it's like a hallway . . . leading towards the center. I step up on the edge, and I just wonder if I'm doing the right thing.

EL: You step up on the edge and you're wondering, you're wondering . . . deeper, be there, be there wondering, be there, wondering if you're doing the right thing. Deeper.

WV: I'm scared.

EL: Scared.

WV: 'Cause I can't be. . . . (voice fades out)

EL: Put yourself there, scared.

WV: I kind of feel like I wanna cry.

EL: You feel like you wanna cry. And do you cry?

WV: (sigh) I think I do . . . it's just so much . . . so much at one time . . . so much.

EL: So much. So much. You feel like you want to cry.

WV: Yeah.

EL: Deeper. Deeper. Be there. Be there.

WV: I'm scared about what I'll see.

EL: Scared about what you'll see.

WV: Yeah.

EL: Yeah.

WV: 'Cause I don't know what's inside. It's kind of dark. Something tells me it'll be okay. Part of me wants to run, and part of me wants to go in, but I'm just going in anyway.

EL: You're going in.

WV: Yeah.

EL: Be there.

WV: (deep sigh, pause)

EL: What's happening?

WV: I don't know what's going on.

EL: Describe what you feel, what you're experiencing.

WV: I feel a bit more calm.

EL: You're more calm?

WV: I feel like I need to go deeper to remember.

EL: Okay. What I'd like to do, is for today, have you be at that door, looking down the entrance, being aware of that curiosity, and knowing you can come back and explore further. I want you now to jump ahead to when you

were five, and be in your yard, at four or five, smell the gardenias, smelling the gardenias, the fragrance of gardenia, safe in your yard, knowing that you can look back, and knowing that whatever occurred, whatever you did or didn't do, you got through safely. You got through safely. You survived, and learned, and grew, and can return to the fragrance of gardenias, to your yard, to your family and friends, to good feelings. Can you do that?

WV: Yeah.

EL: Take a moment, and do that now. (pause) Now look back on that younger self, a younger Wade that's in the field, making a fort. Tell him that he will survive, he will be safe. And he will enjoy the smell of gardenias planted in his backyard. Can you do that?

WV: Yeah.

EL: Back. Back to the time on the hill . . . climb back . . . as you get to the top of the hill . . . pause for a moment. . . .

WV: I'm having the weirdest feeling. I can feel it right now.

EL: The weirdest feeling.

WV: And all those nightmares I had–OH GOD!! Ohhh! I feel this pressure on my arms (choking), it's the same pressure I had in those nightmares.

EL: Yeah.

WV: All this pressure against my body.

EL: Pressure.

WV: (dramatically startled) Ah ha! And I'm just realizing this is all tied together. (almost shouting) THIS IS ALL TIED TOGETHER!!

EL: What is it that's all tied together?

WV: Just like I thought. (crying, sobbing)

EL: What?

WV: The experience in the field. It's why I had these weird night-mares, and I didn't know why I felt the pressure, in the nightmares, but I can feel it on my arms and my body right now.

EL: And how would you describe that feeling?

WV: Like I'm underwater hundreds of feet, pressure equal every-where, but just smushing me. It's really smushing me and I always wake up screaming.

EL: And you have that feeling now?

WV: (laughing incredulously) I can feel it! I don't know why, I just feel it!

EL: And when do you feel that feeling in your present day?

WV: When?

EL: Yeah.

WV: I don't.

EL: Only in your nightmares.

WV: Back then.

EL: Yeah.

WV: That's why it's so weird right now, I can feel it.

EL: Yeah. Feel the pressure?

WV: (laughing) I mean, EXACTLY the way I felt it in my nightmares, EXACTLY the way I felt it. (laughing nervously)

EL: So you felt the connection.

WV: (nervous laugh) I feel the connection. I guess . . . 'cause it's the most overriding feeling I had–I couldn't move and I was being smushed, and there were things around me, and I'd wake up screaming.

EL: Okay, good, that's a good awareness. I want you to move back from that awareness, back up to the top of the hill. Back up, and look at yourself with the awareness: pressure on your arms, being smushed, things all around you. Look back on all the memories you've just had, and draw a circle around it–any color that's a healing, safe color, any color of your choosing. Then, with each breath, allow that scene, that experience, as memories to fill up with that color. When it's done, let me know–when you've filled up the memories with color, safe, protective, healing colors.

WV: Okay, I'm done.

EL: How big . . . is that color . . . of memories?

WV: How big?

EL: Yeah. What size?

WV: It's a huge bubble.

EL: A huge bubble.

WV: Yeah. It's all around me.

EL: It's all around you. Where can you put that huge bubble, where it can continue to heal, to be safe, to not interfere with Wade?

WV: Around me in the field.

EL: Around you in the field. Okay. Do that now.

WV: Okay. I have. (deep sigh)

EL: Let go. Now that huge bubble, and all these memories become smaller and smaller and smaller and move back to a place where it's safe in your unconscious mind. As you move away, let your protector self know that you will remember to forget whatever is significant for you now to consciously remember any sooner or any later than you need to forget to remember. Your nose knows what it is that's significant for your eyes opening and closing to know, as your nose knows remembering. . . .

Upon emerging from Ellen's sometimes unique and somewhat jabberwocky hypnotic patter (she'd told me over dinner at my home that she'd learned some of it in her training, and some was uniquely her own), I follow her to the office and collapse into the cushy chair. She sits opposite me on the couch and gives me some time to collect myself.

My mind darts back to the shocking re-experience of the pressure on my arms; I'd completely forgotten about it, the kinesthetic element of my nightmares. The feeling of those sensations while stretched out in the recliner in her dark room was made even more eerie in light of how they'd surfaced, after so many years, somehow inextricably tied to a dim awareness of walking towards and boarding something that came down from the sky in an open field. On some level, it seemed my mind had pinpointed the source of those sensations to be onboard that thing. Though I'd consciously remembered the nightmares, the kinesthetic memory I had just now uncovered and felt again today--for the first time since childhood in Tyler--was evidently in its original form sufficiently terrifying to warrant burying at the deepest levels of my subconscious.

With my eyes closed I get snatches of scenes of being laid out on a table; the pressure on my arms has a tingling quality, not unlike electrical current, accompanied by a sense of a fluid in motion. The tingling in my arms flowed in a line, with a direction. And I could faintly hear the muted, nearby sound of liquid steadily flowing, like water quietly bubbling through a laboratory retort.

"I find that last memory you had significant–of the pressure and

so forth. . . ." Ellen begins slowly when she sees that my silent re-
flections are waning. "There's a good chance that sometime this
week . . . you'll experience some flooding. Do you understand that–
about flooding?" Her eyes can't disguise that she's deeply concerned. My nose knows.

   "Hmm. Flooding. Yeah. Meaning that things, like potentially
heavy stuff, might surface because of my hypnotic breakthrough."

   "Mmm-hmm. . . . Yes . . . and if you do flood, just try to go with
the flow, and let things surface as they will. But if you have any
trouble before next Thursday, and you feel like it might overwhelm
you, don't hesitate to call me, okay?"

FLOODING 101: A CRASH COURSE

The particulars of my latest session, and the uneasy things that emerged in it, linger with me long after leaving Ellen's office. As the weekend looms, I seem to be plunging into a torpor, like an animal readying for hibernation. I feel lucky to be scheduled off Saturday and Sunday; I'm so out of it I doubt I could focus enough to do decent readings for anybody. Knowing that I won't see my family at Christmas puts a melancholic spin on an already tough time.

I reflect on what Donna Lee told me Friday afternoon. We were on break and I asked her for a quick astrological reading of current aspects. She said sure, took my dates and thumbed through her ephemeris, the astrologer's almanac. She said, "Right now, the way Saturn and Neptune are moving in your chart, it looks like you're going through a cycle where you're having to face up to some things about yourself that you might rather not have known about, things that might be kind of tough to deal with. You might be realizing or learning some things about yourself that you weren't aware of, as though things that maybe haven't been clear before are now becoming clear. Things that were hidden are coming to light." Her words are so dead on accurate that I'm taken aback by them; I put on my best poker face and thank Donna Lee for her guidance. Sometimes astrological aspects can emerge from the cloud of misty generalizations and really kick your ass. And on a Friday the 13th, no less.

At a Christmas party on Saturday night, I only have one drink and can't seem to get into the holiday spirit. It feels like whatever gears I have inside of me that keep me running are grinding together and falling out of synch. Shortly into the festivities I feel something's amiss and beg out early. When I get home, I decide the best thing I can do is to turn in early, so after taking Alf for a long walk through the Grove I crawl into bed.

I toss and turn for hours and skirt along sleep's shore but never quite make it to the land of Nod. Alf shares my restlessness and wanders from sofa to rug to the end of my bed and back. I glance at the clock and the luminescent numbers say one forty-nine. My grinding gears are doing so almost audibly now--who can sleep with this racket? Will blessed sleep ever be a possibility tonight? Even the jet black darkness and perfect silence provided by my subterranean refuge are not enough to lull me into dreamland.

Resorting to tactical maneuvers, I clamp down my eyelids force-fully, but it only ignites a phantasmagoria of phosphenes that swim across my corneas; the light show merely imparts an elevated awareness of my insomnia. Defeated, I relax my eyelids and sur-render to whatever foe lurking in the dark conspires to keep me awake. Oh well. If my mind won't shut down, at least I can enjoy the inky blackness and pretend that I'm asleep. And I can readjust the sheets and comforter as I explore new positions.

Though my bed chamber is as silent and dark as a sealed wine cel-lar, I have a sense of agitation that won't let me be. As time drags by the anguish only increases. Gradually I notice that the lights seem to be going up in my room. Before I open my eyes I wonder if the dimmers on the lamps are malfunctioning. A quick lift of the eye-lids reveals the room remains dark and still. Closing my eyes again, the gradual bleeding in of light progresses, so I bewilderedly invite its presence. Great! Let 'er rip! If I can't crash I may as well be en-tertained. But a monster movie isn't what I had in mind.

With a final jolt of lumens, the footlights go up in my head and I'm treated to an interactive, three dimensional scene worthy of a classic horror picture: I am lying on a table, looking up at a group of non-human creatures whose enormous eyes stare down at me; a bright light shines on me, so their bodies lurk in shadows and their large heads are in silhouette. Still, the eyes bug out of the shadows, and I am reminded of Yeats's line, *a gaze blank and pitiless as the sun.*

I do the first thing that comes naturally: I scream, a bloodcur-dling, bone chilling roar that seems to come from my viscera but gushes from places deeper than that. It's a howl of pain and anger that's been trapped inside of me for over thirty years. It erupts with a violence that threatens to shatter the plaster from the walls. My neighbors are lucky I live so deep underground.

Alf flies off the sofa and his metal dog tags jangle as he rushes to my side; he stands bedside at attention as his master thrashes on the bed. He whimpers and nervously shifts his weight from one leg to the other. I continue to shout and scream, bellowing guttural cries like an animal being tormented. Tears pour from my eyes in streams, and when I can muster the words, I venomously shout, spitting out the words, "NO! NO-O-OOO! YOU HAVE NO RIGHT! YOU HAVE NO RIGHT TO DO THIS! I'M NOT AN ANIMAL! YOU

HAVE NO RIGHT!"

My cozy room and its furnishings have vanished. I am on a table. I am not using my imagination to create the monsters. I am reliving a childhood experience that happened. My subconscious mind told me the truth--something happened. In short, I am flooding.

The scene plays out at its own pace, a streaming video clip that runs without pause. Being trapped on the table is terrifying. The truth is, I'm getting what I'd asked for: the time regression hypnosis sessions unlocked the door of my secret room, and unwittingly I have punched the PLAY button. Now there's no turning back, no way to press STOP or REWIND. I founder into my pillows, and cry convulsively for a good ten to fifteen minutes.

Gradually, the scary goons surrounding me fade out, but not before I catch glimpses of machinery and silver and blue lights illuminating the strange space I'm in, bathing it in a metallic bluish haze. The protective mind has been compromised, the dam has burst, and I have flooded. The lock on the door to this room had held remarkably faithfully, bolted firmly shut for over thirty years. I feel a renewed respect for my subconscious mind. All the energy it spent on this task . . . it's no wonder I cracked up when I was 13. No wonder at all.

Finally, I lay still, utterly spent as though from a violent thrashing. My face is drenched in salty tears, and my nose is blocked up. Flooding doesn't seem that different from drowning. I whimper and wind down as the crying fades into steady breathing. Alf slowly approaches, eyeing me with concern. I gently stroke his head and he licks my face. How fitting to be consoled by an animal when you have just relived being treated like one laid out on a lab table.

Presently the scene fades to black and I settle down. I switch on the bedside lamp and reach for a handful of Kleenex. I clean myself up, and Alf licks my face again. I collapse back onto the bed, and he jumps up to lay beside me. A German Shepherd always at your side. . . . Our backs press against each other. I stare up at the ceiling and contemplate my next hypnosis session.

On impulse, I go to the shelves housing my metaphysical books. I grab my bag of Runes, the Celtic stones of prophecy, and the accompanying handbook, and plop down on the sofa. I close my eyes, take a deep breath and focus my thoughts and feelings, centering them around the flooding that has just transpired. I take a minute or two to meditate on it. Reaching my left hand into the bag to feel the stones with my fingers, I say aloud, "Did this really happen to me? Was I really on that table as a child, surrounded by these creatures? Did this really happen?"

Momentarily, a stone separates from the bunch in the bag and pops into my palm. I remove it, hold it up in the lamp light and scrutinize it. It resembles a car's left turn signal. I leaf through the handbook and match it on the chart: #14, *Kano*. Turning to the page, I see it's also called *Opening, Fire, Torch*. "This is the Rune of opening, renewed clarity, dispelling the darkness that has been shrouding some part of your life," it begins. "You are free now to receive, and to know the joy of non-attached giving.

"*Kano* is the Rune for the morning of activities, for seriousness, clear intent and concentration, all of which are essential at the beginning of work. The protection offered by *Kano* is this: The more light you have, the better you can see what is trivial and outmoded in your conditioning. . . ." It continues in that vein, but the last sentence three paragraphs later is the clincher: "Simply put, if you have been operating in the dark, there is now enough light to see that the patient on the operating table is yourself."[12] I keel over into the sofa cushions, cover my eyes, and laugh out loud, almost convulsively.

At four-thirty a.m., I thumb through my Day Runner, find Ellen's office number and punch it on my phone. Her outgoing voicemail voice sounds reassuring. I calmly tell her machine what I've been through, and I ask that she please be willing to push a little harder, if need be, in my next session. I tell her I really need to do this, to be sure to flush out all of this stuff. Then I tell her I think she's wonderful, that I'll see her Thursday, and I thank her for helping me.

Later, exhausted, I finally plunge into a deep sleep. When I awaken in the early afternoon I phone Angela and Kimberly. They both sound concerned and want to hang out with me, but I decline, opting to wander through my private jungle property with Alf, to regroup and recharge. The vitality of all the life forces in my jungle gardens offer me the best therapy around.

* * * * * * *

THE LOCKED DOOR FINALLY OPENS

They cannot scare me with their empty spaces
Between stars--on stars where no human race is.
I have it in me so much nearer home
To scare myself with my own desert places.

--**Robert Frost**
Desert Places

Thursday, December 19th, 1991. A date that will always remain in my memory as a day of pivotal changes. I arrive in Ellen's office a little before eleven, as usual with my tape recorder. When she opens her office door I feel a jolt of nerves, almost like a kid at his first confession. She says "hey" and pats me on the back, I say "hey" back, and I take my seat. She curls up on the couch like a cat. With eyes full of concern, she says, "I got your message from the weekend on Monday morning, and I didn't call you back because you didn't ask me to. Is that okay? "

"Sure, yeah, I wasn't expecting you to."

"So tell me what's been going on. It sounds like you had a really intense time flooding," she says.

"Yeah, you could say that," I laugh. "It's such a funny thing that I live in a bomb shelter. If I'd been screaming and carrying on like that in a regular house, the cops would have been banging on my door."

"Any thoughts or comments? " she asks.

"Mmm, yeah. For starters, all the stuff that flooded out reminded me of the scene in *Rosemary's Baby* when Mia Farrow suddenly looks up from her bed at the coven and sees the Devil and screams, 'This isn't a dream! This is really happening!' That's the feeling I was left with, after this horrible scene played out. This wasn't a fantasy sequence–it really happened."

"What did they look like?"

"I could never see clearly because the light seemed so bright. And maybe my memory was a bit blurry--I was only three. But I saw really big eyes and I got a sense of large heads. I was terrified because I couldn't move. I was immobilized on the table. It also reminded me of the childhood fear I had when I watched *The Wizard Of Oz*. When the witch's flying harpies kidnapped Dorothy and locked her in the tower, I had to leave the room. I couldn't handle it. I wouldn't come back to the den until Natalie promised me that the scene was over."

"Mmm-hmm."

I tell her about my Rune reading, and she arches her eyebrows and smiles. "The timing of that stone . . . jeez, uncanny. And the wording, 'if you have been operating in the dark'--that's basically what I've been doing my whole life. I've been . . . a wandering amnesiac. A seeker . . . but not even sure what I was looking for. And the final line, 'the patient on the operating table is yourself.' Too much! I laughed so hard I almost choked," I say. She nods silently with her Cheshire cat smile.

Ellen's neutrality throughout all of this has me wondering. "So tell me, do you still think this could all be the result of sexual abuse as a child?" I ask.

"It's one of the possibilities," she says thoughtfully. "And one of the others is that this really happened, and you were abducted," she says softly.

"And which do you subscribe to now?"

"I can't really say at this point," she says.

I refer to an index card that I've jotted notes and questions on. "But if this were sexual abuse, wouldn't I have at least a glimmer of a memory trace--something about who was involved, where it was, that kind of thing?"

"Not necessarily."

"But why would it be disguised as aliens onboard a ship? At that age I hadn't even seen any space monster movies."

Ellen shrugs. "Sometimes the mind can cloak traumas like this. It's called a screen memory."

"Yeah, I've heard of them. . . . Oh yeah, another thing. When I first came to you intending to do past life work, for our first, preliminary session, you asked me to come prepared with things that seemed like themes or repetitive patterns in my life. We discussed my promiscuity. The fact that I've been with hundreds of men--probably over a thousand. And you said that might indicate I was, I can't remember the word, but we studied it in school, what was it--something that allows me to have sex and set my heart or emotions aside--"

"Dissociated, or a dissociative personality disorder," she says.

"Right. So if this really was a genuine abduction encounter, are you thinking that could account for my dissociation?" I ask. Most of the members of my tribe think multiple sexual escapades just mean you're having more fun than breeders are--because we can. We certainly weren't all abductees.

"Certainly that could be part of it. Anything sufficiently traumatic can cause us to split apart in order to handle it, to separate ourselves from the pain, the horror," she replies.

Realizing the time, Ellen takes me into her hypnosis room. "We'll end this around noon," she says. "We'll close it, but not totally close it off from your consciousness, so that if you want to allow more material to come up, we'll be able to do that."

"Okay. I'm very confident that even if it's really tough, you know, I'll be ok."

"Oh sure. You're here now, right? " she chuckles.

"Yeah."

"This experience was in the past. And you are okay," she says with assurance.

"Mmm-hmm."

"You already have the proof. But the kid back there doesn't know that."

"Exactly."

"That's something you might use. If you find that kid really getting into terror, you can pull back to the Wade of today, and tell the kid that he's gonna make it through. And you know it, 'cause you're here," she laughs.

"The other night I was almost doing empty chair therapy with an alien in the chair. But I didn't get to forgiveness."

She waves her hand dismissively. "Don't worry about forgiveness--forgiveness is usually bullshit, okay?" We laugh. "Forgiveness means, 'Come in here, you son of a bitch, I'm gonna release you!'," she snickers, making a fist and punching the air.

I switch on the tape, get comfortable, and Ellen dims the lights for me. She begins the induction, delivers her characteristically smooth hypnotic patter, and soon I'm deep in trance, feeling so relaxed it's like I'm floating on a cloud.

EL: Where are you now?

WV: In the backyard, in Tyler.

EL: In the backyard, in Tyler.

WV: All the ground is covered in snow. It was snowing the day before. It was amazing because it never snowed before.

EL: Your first snow.

WV: Yeah, it's the first time I've seen snow.

EL: How are you feeling, today, with the first snow?

WV: I'm watching a little house I built melt. It's like a, a Fred Flintstone kind of thing--for *The Flintstones*. And I'm just watching it melt . . . I have the feeling of knowing that everything kind of fades away . . . just watching it melt. . . .

EL: And everything fades away.

WV: There's something really beautiful about it, just watching the light on the snow. It's just . . . kind of happy and sad at the same time, 'cause it's melting.

EL: Okay, stay with it, happy and sad, as the snow melts . . . go deeper. . . and deeper . . . as you continue, search for the time that you first felt, your arms being smushed, like pressure, pressure underwater. The first time you had this feeling. Go there. GO THERE NOW.

WV: I don't feel it yet, but I know where I am.

EL: Where are you?

WV: (voice breaking) I'm inside of this thing . . . inside of this strange thing.

EL: What are you feeling?

WV: (starting to cry) I'm scared.

EL: Scared.

WV: I can't believe it's happening. (crying and sobbing)

EL: What's happening?

WV: (sobbing like a child) I don't know what they're gonna do to me. (sobbing and crying forcefully) I don't know what they're gonna do! (crying harder and harder) OH NO! I should have never come in!

EL: Should have never come in.

WV: (almost choking on tears) I should've walked away . . . 'cause they've got me . . . and I can't get away.

EL: You can't get away.

WV: No-o-o-oh! I wanna get away! (sobbing hard) I WANT MY MOM AND MY DAD! And they can't hear me!

EL: What's happening.

WV: I'm on a table!

EL: On a table.

WV: (sobbing, gasping) I don't know what they're doing to me! I just know it's wrong! IT'S WRONG!

EL: What do you see?

WV: I don't wanna look!

EL: Your eyes are closed?

WV: Sometimes. (continuing to sob)

EL: And you're on a table?

WV: Yes.

EL: Can you hear anything?

WV: (sobbing) Some strange noises . . . I don't know how to describe it. I think they're making the noises . . . but I'm just trying to pretend this isn't real! (almost hysterical) This can't be real!

EL: Just trying to pretend it isn't real. What are they doing to you?

WV: They're keeping me down on the table

EL: How are they keeping you down on the table?

WV: (sobbing) I don't know—I don't even think they're holding me--I just, I can't move . . . and I wanna move, I wanna get away!

EL: What happens next?

WV: (crying harder, sobbing, not answering)

EL: We can leave little Wade there on the table and back up, and look at little Wade, crying, wanting to get away. Back up. (still crying hard) Back up, look at little Wade on the table. What are you doing? You're far enough away to where you can evaluate this, knowing that little Wade made it through this experience. Knowing that he did, back up far enough to where we can get useful information to share about what's really going on.

WV: I just know that they've stuck things in me.

EL: They're sticking things in you.

WV: Yeah.

EL: Where?

WV: (crying) I don't know.

**EL: Back up further. Back up, back up far enough, know-
ing that you made it through. Knowing that you made it
through, and there might be useful information available.
Back up, far enough to be able to see . . . what's being stuck
in little Wade on the table. Can you see?**

WV: I just know that I'm surrounded and that when I see them
they look scary like . . . some kind of Halloween thing! (sobbing)
And I can see that their colors are sort of like orange and brown . . .
Some kind of strange orange and brown color. . . .

EL: Strange orange and brown color.

WV: And they don't have any feelings (crying), I know they don't
have any feelings! (sobbing)

EL: No feelings.

WV: No . . . and I just wanna kill them all! (sobbing) I just wanna
smush 'em back like they're smushing me.

EL: They're smushing you.

WV: Yeah.

EL: How are they smushing you?

WV: (crying) I don't know, I don't know.

Ellen gives distancing suggestions, that I put a piece of Plexiglas
down to separate me from the experience; and that I put myself on a
TV screen and watch the encounter, knowing that I'm safe, viewing
it from a distance. (long silence)

EL: What's happening?

WV: I can't see where they touch me, but I just know I'm being
poked and prodded, and I'm naked--just like an animal or something.

EL: You're naked on the table?

WV: Yeah.

EL: Being poked and prodded, but you don't know how.

WV: Yeah.

EL: Where do they poke and prod you?

WV: (long silence) I can't tell.

EL: Can't tell.

WV: I know they are, but I'm not sure where.

EL: Not sure where? Okay, let's leave that, we'll explore that further another time. I'd like you to move forward, to what happens next, after you've been poked and prodded. What happens next?

WV: I have a faint memory of walking around . . . almost like a tourist.

EL: And how are you feeling, walking around . . . almost like a tourist?

WV: Numb . . . numb. And I'm still angry . . . but I'm under control.

EL: Whose control are you under?

WV: I'm under their control.

EL: What are they saying to you?

WV: Look around . . . look around.

EL: Look around.

WV: And I'm angry at 'em.

EL: Angry.

WV: I hate 'em. And I just wanna get outta there. But I am curious . . . I am curious.

EL: You are curious.

WV: Yeah . . . 'cause if it's lasted this long, then . . . I feel like it's really happening, even though I . . . want it to be a nightmare. I want the whole thing to be a nightmare.

EL: You want it to be a nightmare?

WV: Yeah, I do.

EL: And what would you get from that?

WV: Then I wouldn't have to accept it.

EL: And what would it mean if you accept it?

WV: That there are monsters . . . and they treat us like animals. And if it's a nightmare, then I can wake up and I'll be at home. (sobbing) 'Cause I wanna be at home.

EL: Wanna be at home.

WV: (crying harder) I'm not sure if they're gonna let me go . . . I'm not sure if they're gonna let me go. (crying harder) And . . . I don't know why this is happening to me. I wanna know why this is happening to me.

EL: What do they say is the reason?

WV: Because I was there. (sobbing)

EL: Because you were there.

WV: I was convenient . . . I was far away. (crying)

EL: What is it that they're trying to find out? What are they trying to do?

WV: (sobbing) I don't know.

EL: When I touch your thumb, I want you to drop that Plexiglas screen . . . (gives distancing suggestions again). Just remember, anytime you need to . . . you can go to a place where you have ALL your adult resources.

WV: Okay.

EL: Just do that. And from there, if you can, right now, see if you can discover what the intention was . . . of laying little Wade on the table . . . poking and prodding, and showing you around.

WV: (long pause) Observation.

EL: Observation.

WV: I feel like they're as curious about me as I am about them.

EL: And if you had the opportunity, would you put them on the table, poking and prodding?

WV: Right now I would. I'd smush all their heads in.

EL: And you're curious too.

WV: Yeah.

EL: So are you saying, they weren't intending to hurt you?

WV: I don't know . . . but I feel like . . . it was a trade off. That if I put up with . . . what they did to me . . . then I would get to see things . . . that no one else gets to see.

Lauren begins her emergence patter, and soon my eyes are fluttering and I'm returning to the present. When I open my eyes, I

179

look at her and say, "You can believe what you want, but now I know that this really happened. This was real."

We return to her office to sit for just a few minutes. Her next client is waiting. "This week, I'd like you to sketch from memory anything that seems pertinent to you--things you may have seen during these sessions that we didn't get to talk about," she says. "Since you're right handed, try to use your left hand; that may help you to tap into the unconscious a little better."

As I get up to leave, I give Ellen a hug. I head out, through her waiting room and out the hall door. I must have been really loud today. The hypnosis room is right next to the outer door, and in past weeks I've noticed that the adjacent office leaves its door wide open. As I head down the hall and put on my sunglasses I can't resist turning to the left to glance in there; I meet the gaze of the secretary at her desk, craning her neck in my direction. I smile at her, wink, and pull my shades down over my eyes. She quickly looks away.

* * * * * * *

Twilight. Saturday, December 28th, 1991. I'm sitting at the porch table sketching scenes from last week's hypnosis session. As soon as Ellen told me to move forward from being on the table, in my mind I saw a vivid sight unfold as smoothly and steadily as film footage. (This phenomenon is fairly typical of deep-trance hypnotic regression sessions, of seeing but not reporting to the therapist every single thing witnessed.) The stream of visual information was constant, and I was so entranced by what I was seeing I simply remained silent and let it flow like a video clip. I'm convinced that what I saw was a memory replaying. It manifested too quickly and smoothly for my rational, conscious mind to think to itself, "Hmm, what should I be seeing now?" and then quickly manufacture this scenario. Furthermore, I was by then in such a deep trance state that I was fairly swimming through a subconscious river of information, flowing virtually independently of the currents and eddies of full waking consciousness. My conscious mind wasn't in charge, and wasn't making this up. There was a steady visual info stream.

I stood in a very large, cavernous space. Though I'm sure it was enclosed, I wasn't aware of a ceiling overhead. The creatures stood behind me as I walked towards a wide hallway that gently curved off into the distance, suggestive of a huge round structure. The entire outer wall was comprised of glass enclosures (or what looked like glass), each separated by partitions, like a vast array of terrariums. Inside the nearest cubicle to my right, I saw a creature I doubt I'll ever be able to adequately describe, but I'll give it a try.

At first glance, I detected undulations, fluid random motions like a sea anemone's tentacles. In the next instant, I glanced from the tentacles to its face, and it became apparent that the creature wasn't in water or fluid, but rested on something like a rock outcropping. It had a furry face vaguely resembling a wolf, but with a flat area where a nose would have been. Its body appeared squat and equally furry. Tentacles of the same dark brown hue as the hair radiated out in a circle from the face--or what could roughly be called the face--in a fashion similar to a sunflower's petals. Each tentacle was thicker at the base and smoothly flared into a more slender appendage at its end. All of the tentacles were in motion, slowly, rhythmically curling and unfurling randomly. Meanwhile, what passed for little round eyes and a small round mouth opened and closed, opened and closed, almost keeping time with the slowly thrashing tentacles. There was a sense that it was gasping for air but I don't think it was in pain. This was simply how it looked at rest. It was a living thing, at once hideous and fascinating. I wondered what more I could have seen had the session lasted longer.
(See copies of both sketches in APPENDIX, marked with "LH": Left Hand--wv)

During my second session on December 12th, I also had a clear visual of the ship before it landed. It came zooming in and hovered near the ground, and as it did it teetered subtly, like a spinning top slightly out of balance before it righted itself. It moved so quickly that it appeared almost to materialize near me. There was no noise. The saucer part had a conical tower in the middle, and I perceived a glow as if from lights around its edge. Now I realize that it was too small to account for the enormity and spaciousness I experienced in the interior scenes. Perhaps it was a shuttle, an elevator to a much larger craft in space, or elsewhere.

Kimberly has just pulled up, and Alf is galloping for the gate. I put the sketches away for now.

"How's it going with the Spooky People?" I ask. I pour us some mineral water from the fridge. I'm grateful that I had the sense to cut loose from the *Psycho Collection.*

"About the same," she says. "Everyone really misses you. I do too. I wish you hadn't quit."

"Yeah, I wish I didn't have to, but Ryan kept after me, and Ramon was fuming with jealousy. The vibes got too heavy, and with everything else I'm dealing with now, it got to be too much. I needed to lay low for awhile. How's Pandora. Still hot on your trail?"

"Just driving me crazy. She comes into me at all hours of the day and night expecting me to give her one reading after another. Her parents are so kind to me, but still I can't wait until I move out of their house."

Kimberly and I go upstairs and recline on the porch waterbed, and I gaze up through the screened in windows at the trees bathed in the glow from the Malibu lights while we listen to the tapes of my three hypnosis sessions. By the time they're over she's teary eyed. "Oh, it was so hard for me to listen to your pain. It's strange, I didn't get that they were evil or malevolent, but they sure didn't sound nice here! I kept getting that they were in awe of you, as much as you were of them. Almost as though they thought you were an adult human, as opposed to a little kid."

"Well, that was really bright of 'em!" I laugh, and run my fingers through her hair. "Another sense I got was that it was sort of a zoo ship, and they collected species."

"Yes! Exactly! " she exclaimed. "That's exactly what I was getting, too."

"And I was just something to add to their collection."

"Yes. And that's one reason I haven't been able to figure out why they let you go. . . . And I've seen them showing you around--that part you didn't mind, you were fascinated by that. They showed you around a lot, showed you things no one's ever seen."

"Maybe that explains my obsessions with designing fusion reactors and starships, not your basic fourth or fifth or sixth grade hobbies."

Kimberly then informs me that they put tracking implants in the back of my head, to find me when they return. She tells me they'll be back sometime in the next three years. My life is so completely *Twilight Zoned*-out by now, nothing she tells me is going to faze me anymore. I tell her how frustrated I was at stopping the last session when we did. "I know there's more to it. I still wonder what the pressure on the arms was all about."

"I can answer that for you," she says, digging into her purse. She produces a sealed manila envelope with *Wade* scrolled on it in her handwriting. She waves it in front of me, and judging by her smile I'd say there's something pretty powerful in there.

"Your impressions from when you were "onboard," during the reading you gave me?" I ask. She nods. "Okay, let's have at it." I take it from her and tear open the flap. I pull out a single white sheet of typing paper, on which she'd typed:

> These are all the things that I saw:
> You were laid out on a table, somehow they were able to put you under, into a very deep trance type of sleep. They began to insert a type of tubing into your arms, the tubing ran from some type of cabinets. There were several of them, and each one of them was very much interested in what was taking place. In each arm was placed this form of tubing, one side for entrance, one side for exit. Ok, what took place next is almost unbelievable, and hopefully by the time you read this you'll have already seen this by yourself. Wow, how do I find the words? You were transfused! The liquid you contain in your body is still blood, but changed in some way. I really am not sure how to explain? It's like a super charged atom or something, I think they purified it, tested it, analyzed it, just totally absorbed all information possible, then proceeded to awaken you. During the whole process your blood was being circulated through some kind of machine. So, that is what I meant when I said you are your parent's child, but in some ways you're not. This is what I saw. I know this is enough to blow you away so I hope you find out all this information by yourself. I LOVE YOU!!!!!

> Sealed this day, Dec. 19, 1991

I read Kimberly's note two or three times. It gives me chills, recognizing that the bizarre, almost electrically pulsating pressure on both of my arms that I'd re-experienced in my second regression session with Ellen (and had instantly recognized as always being a kinesthetic element in my Tyler recurring nightmares, while being on an examination table surrounded by scary figures under bright lights), seemingly fit like a hand in glove with Kim's psychic impressions of being onboard. The recurring nightmares all along had been replaying the full reality of the encounter, including the kinesthetic elements. The paralysis, the pressure on the arms and body, the electrical current, the feeling of fluid in motion, and the sounds of it, too, all had been replaying in my Tyler nightmares for years.

"How bizarre! My God, Kimberly, I think you're right!"

"I can promise you that's what happened. I saw it very clearly. Wade, everything you are now is because of what they did to you then. They made you who you are."

* * * * * * *

When I call Angela DesJardin and tell her about my third session, we agree to get together soon to go over all the notes and tapes. "God almighty!" she blurted out at the outset. "How could Kimberly say they were benign after what they did to you? I don't call terrorizing a three year old and forcing him to lie on a table for experiments an act of friendship!" Then, after a moment to reflect, she says, laughing, "But you were such a curious little kid, you shouldn't have been so far from home, wandering out in a field like that! So tell me, doll, how are you handling it? Are you okay with it?" she asks. Her voice carries with it the weight of years of shared experiences, and love.

"Yeah, I guess so. It's just that some of it has been so brutal. That's gonna take me awhile to work on. And the shock of suddenly having all of these things come to light."

"Uh-huh," she says, pensively. "But do you also see . . . that it was a *gift*?"

I love the way she lilts her voice when she wants a question to really sink in. I give it some thought. "Mmm, not so much now. But I think eventually I will. I'm just a little close to it, the brutality of it right now. But I see what you mean. I guess in some ways it was an honor. I was inducted into the Light Years Club."

"Yes you were."

"And my membership card is still in good standing," I laugh.

I phone my brother Aaron in Austin to share all my recent revelations. Before I can get into details, he interjects, "I guess you realize that I'm the only family member you can tell this to, right?"

I laugh. "Yeah," I say. "That's all Mom would need to hear to push her over the edge. And Dad and Natalie would probably send for the men in white coats." I mail him copies of the tapes and transcripts, and a note with the conclusion that my gut sense is that I am a genuine abductee.

Aaron wrote me back a week later. In his letter, he said,

When I was in third or fourth grade my classmates and I were taught

184

about a few of the great thinkers: Socrates, Galileo, Copernicus, et. al. There was a general pattern these great men's lives seemed to take. They would make outlandish statements (which would eventually be proved true) that would unfailingly bring them ridicule and scorn and even death at the hands of their so-called peers. As a schoolboy I was always angered at the ignorance that had created disaster. I always wanted to go back in time with a modern weapon, say a well-oiled AK-47, and mow down the lot of progress-impeding boneheads. I always thought of how foolish those same skeptics of old would feel if they could look back on what they had done from my vantage point. And I always swore to myself I would keep an open mind and never casually dismiss any new idea no matter how bizarre it may seem.

For this reason I choose to believe that what you believe happened to you over thirty years ago is possible. As you have probably noticed, I worded the previous sentence very carefully. I am not, by any means, fully accepting the notion that extraterrestrial creatures ever actively participated in your life. But neither am I going to ask you to drink hemlock.

I guess at some point in my past I defined the limits of the wonders of reality as all thing appearing in *Discover* magazine and relegated all the things in *OMNI* magazine to the nebulous, unreal universe. But the fact is, as much as I may like to believe I have a firm grip on reality, there really is no such thing. Reality is tenuous. All things really *are* possible. "There are more things in heaven and earth, Horatio, than are dreamt of in your philosophy."

I'm very impressed with your psychic friend. If Kimberly can narrow down her estimate of when your next encounter will take place you can bet I'll be clinging to your back when it happens. If I remember my transporter dynamics correctly they'll have to beam up my molecules if I'm attached to you.

That's all for now. Please keep me posted.

Much love,
Aaron

* * * * * * *

Mid-January, 1992. Ellen took a month off for the holidays, during which time she's read the copies of the transcripts I mailed to her. I've been using the time to decompress from the shock of my newly discovered self, the separate Wade I've been carrying around inside of me for over thirty years.

Sitting before her again, I feel rested from the break. It did me a lot of good to cut loose from *Psychic Central*, despite the financial hardships. Although I regretted having to sever ties to a job that enabled me to expand my psychic abilities, the cost was too high: I really didn't relish being the unwitting third wheel to Ryan and Ramon's triangle, and I sure as hell didn't need Satan worshipers pissed off at me. Ol' Tar Eyes would have to direct his evil juju at somebody else. The *Psycho Collection* was history.

After our hello's and how are you's, Ellen, as ever in her relaxed and casual garb, says, "So what's new with you? Any more stuff surface?"

I refer to my ubiquitous note cards. "Not so much that, but I did have some insights. My training client and New Age pal Angela referred to my adrenals as being fried, while giving me energy body-work at her party. It seems worth mentioning, because it occurred to me that if I was really trapped on a table like this, the "fight or flight" reaction would have been short circuited, since I couldn't have done either one. And that's all tied into the adrenals, right?"

"Mmm-hmm. Interesting."

I show Ellen the sketches I've made, and a copy of Kimberly's visions onboard. "And what's really strange about it is that the idea of a blood transfusion makes sense to me, too."

"Because it explains the feelings in the arms," she proffers, reading Kimberly's note.

"Yeah, that. But get this: back in 1987, when I took my first HIV test, it was under the aegis of the University Of Miami, through a research project of theirs focusing on HIV and the immune system--I think they called it the Huff & Puff Study. In addition to giving the participants a free HIV test, they also ran a full scale blood analysis. A few weeks later I went back and the nurse gave me the results, HIV-negative just like now, and said the doctor wanted to see me for a minute if I didn't mind. I said sure, and waited in his exam room. He came in and said, 'I just want you to know I've never seen blood like yours.' When I asked him what he meant, he said, 'One of the tests we did is called an Absolute Number count. Basically, we take one c.c. of your blood and put it under a microscope and count the number of white blood cells vs. red. For a man of your age (then thirty) in good health, an acceptable norm would be around 250 to 300. Your count was between 1,250 and 1,500. And this doesn't indicate stress or trauma--you simply have a very potent immune system. Whatever you're doing, keep doing it,'." Ellen raises her eyebrows.

"So whaddya think, doc? "

"That's very interesting," she says reflectively. "Any other thoughts, observations?"

I share with her that my childhood pollen allergies, which first emerged in Tyler, gradually grew to such severity that by the time we lived in New Orleans, at the insistence of our family General Practitioner doctor, in the fall of 1969 I was taken to an Allergist and Immunologist specialist at the Tulane University School Of Medicine, a Dr. John E. Salvaggio. "The doctor and his staff did extensive testing," I tell her. "They lined up two rows of droplets on each of my forearms, each droplet with a different allergen source, and punctured the skin so the blood would absorb each one. Another two rows extended up my triceps on both arms, from the elbows to the shoulder joints of my upper arms. I looked like a human pin cushion! I think they called it an allergy skin blot test.

"Anyway, the doctor was surprised that I'd reacted to everything they tested me for, except for salt water. But he thought that I was only mildly allergic, since the reactions seemed mild. In fact, the reaction was so severe that my body didn't know how to handle all of the allergens they'd bombarded me with; when I left the office about a half hour later, I started to see double as my mom and I headed for the elevator. I suddenly developed severe respiratory problems, and I couldn't breathe--it felt like a strong man had a strangulation grip on my throat. I was choking for air. We rushed back to the doctor, and by the time he put me on the exam table my body was going into anaphylactic shock. The whole room started to fade out, and it felt like I was being steadily strangled to death. The doctor administered an adrenaline shot, and that stopped the process. I started to breathe more normally. Once my condition normalized, he told me that if I'd gotten into the elevator, I might have been dead by the time we reached the ground floor lobby.

"It also seemed worth mentioning, that when he was doing a basic physical examination prior to the skin blot test, when I had my shirt off he'd commented to my mom that for a boy of almost 13, I had an unusually sunken chest, and how this sometimes indicated adrenal gland problems. He basically said my adrenals had been severely stressed out. Interesting that Angela's assessment was the same. In her bodywork session, Angela also commented about something stored in my head she wasn't supposed to mess with." I toss at her ideas regarding a tracking implant . . . transmissions from a close up UFO . . . and Kimberly's "stored information, like a human library book." All those concepts seemed intertwined somehow.

"But Dr. Salvaggio's most interesting comment came a few weeks later, after all the lab results were in, and they were developing the allergy serum for my treatments," I say. "The doctor told me, with Mom standing there, and this is an honest to God quote, 'You're severely allergic because your blood is missing a huge number of antibodies that most of us have naturally. It's almost as if all the antibodies have been filtered out of your blood,'." I laugh to myself as I share this with Ellen. "I have to say that this meeting with Dr. Salvaggio is very significant to me now. In light of what Kimberly said she saw, and the hypnotic material that spontaneously surfaced with you, with the pressure and pulsations on my arms, and how intensely I felt those sensations in real time . . . those are all strong markers to me, signifying in an almost spooky way, that this really was an alien abduction. The way that all these different things dovetail together is pretty uncanny, wouldn't you say?"

"Yes, I'd have to agree with you on that," she says thoughtfully. "It might also explain how you were able to remain HIV negative despite the hundreds of men you've been with. The whole concept of your immune system being, like Kimberly says here, 'supercharged atoms'."

"Yeah, that's true. But my negative status could also be explained by the fact that I've had mostly safe sex, and up to now have been more of a top sexually, and haven't been into hard drugs, just occasionally recreationally . . . but I can't deny, the big picture sure is a cohesive one."

I shuffle through my note cards. "A few more things," I say. "Reading over the transcripts, it struck me, how funny, that at one point I said, 'I just wanna smush 'em back like they're smushing me.' And for the past ten years or so I've been schlepping a massage table across the country and smushing people into it. It's like I've been acting out. Working out my stuff, my conflicts—that I didn't even remember having—by taking other people's pain from their bodies."

"Mmm-hmm. A good observation," she says. "That's a good thing to ponder. . . . "

"And, like I told you, Kimberly kept seeing images of me in a past life as an SS officer at a concentration camp. And when I put it all together, it came to me: what better way to pay off a karmic debt of allowing medical experiments to be done on human beings in Nazi Germany, than to come back and have a similar thing done to you at the age of three. Although I guess I got off easy, since they let me go. . . ."

"Mmm-hmm. Yes."

"So what do you think about my experience now. Do you still think I might have been sexually abused?"

"Well, like I told you, reading the transcripts you sent me would be useful, since I sometimes drift into a partial trance myself when I'm using hypnosis--many hypnotherapists do, as I'm sure you're aware. And . . . based on what's transpired here, and on what I've read in the transcripts, I'd have to say that in my opinion there's a very high probability that this was a genuine abduction. And I would venture to say that there's likely a lot more information from it stored away in there."

As we draw our session to a close, Ellen tells me that she's decided it's best if I hold off on further hypnotic exploration for awhile. She says that she perceives my life as being like a train running on a track that has suddenly split into two rails, and that more hypnotic work might only serve to reinforce my sense of dissociation, the mysterious life of the young Wade vs. the present day man striving to integrate himself. She tells me to take some time out to process the material that has already emerged. Seek wholeness.

I follow her advice, and focus more on my hypnosis practice. I run an ad in a local weekly newspaper, plugging my business as the Grove Hypnosis Center. I do ok with it, and get a fair number of calls. Some of them are more interesting than others, especially the clients requesting time regression hypnosis. The fact that *Many Lives, Many Masters* had been a bestseller, and it was set in Miami, meant that numerous clients requested past life regressions. But interestingly, the client who encountered what I regarded as the most pure past life recall was a man who only sought to resolve early childhood conflicts with his mother. A high level VP with a multi-national corporation, he came across as a no-nonsense type, the sort of man who didn't subscribe to any metaphysical beliefs at all. Towards the mid-point of the regression, he suddenly popped back into a vivid scene from the 1870's, and he found himself there, in the scene, with total recall, including the specific date and location. His recalled material went into elaborate details, and he seemed genuinely dumbfounded by the fact that he felt certain the high powered business man in office scene was himself, and the woman who was his secretary back then was his mother today. When I cued him to emerge, he slowly opened his eyes. He took a moment to gather himself, then looked at me and said, "I'm going to have to go home and completely re-evaluate my whole life."

If my hypnosis work had become a more steady business--and certainly, if all of my clients could have been as fascinating as the VP recalling his nineteenth-century life--I might have stayed on in Miami and made a go of it. Maybe even toughed out the lean financial times, and tried to enroll in graduate school for an MSW degree. But things didn't pan out that way, and all signs were pointing to a relocation. Other lessons awaited me elsewhere. The tumbleweed had to tumble again.

It was especially tough leaving Angela and Kimberly behind. Kimberly's uncanny psychic gift had changed my life, forever. In retrospect, I would find it ironic that I had been living in an underground bunker designed to protect people from nuclear explosions on the surface. And yet the material Kim had helped me to pinpoint and release had detonated like a triggered nuke, while I was living underground. My path now took me to a life back on the surface, above ground, following the trail of my surfacing material as it emerged into the light. What had been locked away in darkness had come to light.

Ellen Lyon was right about my life suddenly running as if on parallel train tracks. I really did need to heal, and to integrate my two selves. The amnesiac man was reintroduced to the terrified boy, and I needed to bring them together, onboard the same train, and on one set of tracks. The alienation I felt in the wake of my exploded, regressed material was almost impossible to describe to my friends, even to open minded, metaphysically savvy pals like Angela. It was tough enough for me to understand it. But I took Ellen's words to heart. So it was time to leave Miami, and move on. I would always feel extreme gratitude that I'd been fortunate enough to have her as my therapist during this process. Her therapeutic and hypnotic skills were nonpareil. She was as good as they come, bar none.

I'd lived in South Florida long enough to have tasted all its beauty and indulged in its many colorful distractions--scuba dives from Key Largo to Key West, tripping out on the psychedelic colors at depth; pretty people parties with endless beefcake distractions at pastel villas from the Grove to South Beach; a slew of gyms, from hard core iron to glitz and all variations in between, and an equally varied neon rainbow club scene packed to the rafters with the gym rat crowd. And even the wildly disorienting feeling of showing up to train your spoiled rotten 17 year old high school senior client, only to find his family's waterfront three story villa in the nouveau riche subdivision of *Coco Plum* (complete with home theater, discotheque, home gym, limo and Euro coupes in the motor court, scurrying staff and luxe motor yacht moored in the channel out back just beyond the lagoon pool with the swim-up bar) cordoned off with yellow DEA tape, and a very official looking document attached to the entrance gate asserting newfound government ownership.

I parked in front of the eerily silent palace that day to take a minute to examine the sobering paperwork that had summarily jettisoned my business. Reading the DEA proclamations was surreal. As far as trainer-client finales go, this was an uncharacteristically abrupt, record breaking ending to a training regimen. The sprawling Mediterranean palazzo was destined for a cameo appearance in the last ten minutes of the classic movie showcasing South Florida's 1980's powder-fueled real estate boom, *Cocaine Cowboys*. It was a given that Bolivian marching powder had bankrolled Miami's *Miami Vice*-era resurgence; that simple, stark fact was the big white elephant in the room that nobody ever talked about. Perhaps the relentless sunlight played a role in all the wild goings on, too. Maybe the sun is just too blasted hot and intense in the Sunshine State-- like some medical books say, *tropical neurasthenia*. People going crazy from the sun. Southern fried brainpans. Margaritaville.

No matter. You could say I have a gift for knowing when it's time to go. My gut instincts told me that in order to best deal with the off-world genie I'd uncorked from the bottle of my subconscious, never to be bottled again, I should seek wholeness and healing in a radically different place, way out in the Southwest, a part of the country I hadn't experienced in full yet. I had a hankering to run wild in the wide open spaces, and maybe explore the cowboy way of life on horseback for awhile. So I packed up my things, and Alf and I hit the road and headed for the high desert capital city of alternative realities, Santa Fe, New Mexico.

Tyler, TX: 1960: View from home driveway looking east to the fields. –Family archive photo

Kimberly Lee: Coconut Grove: Nov. 1991

**Alf guarding his Coconut Grove domain
Summer 1990 –** Photos by author

The entrance to the *Space Science Center*, also referred to as Santa Fe's UFO Museum, on a cold February day in 1994. It was here that I would view a classified NASA video with paradigm-shattering footage, confirming a high level Cover-Up to hide from us the reality of UFOs and the alien presence–both here and on the Moon. How the owner, Felipé, got his hands on the video, and subsequently received periodic threatening phone calls about it, make for a real life X-Files story.

— Photo by author

145th YEAR   SATURDAY, AUGUST 27, 1994   50 CENTS

THE SANTA FE
NEW MEXICAN

Space Science Center gives way to UFOs

By ZACK VAN EYCK
The New Mexican

His Space Science Center is now closed, but Felipé Cabeza de Vaca is still open to the idea that humans are not the only intelligent life in the universe.

Cabeza de Vaca, whose nonprofit St. Francis Drive museum served the public for nearly 15 years, said Friday he will focus his future work solely on UFOs.

Eventually, he hopes to open a new facility in Santa Fe dedicated entirely to collecting and spreading information about unidentified aircraft and related phenomena.

"It's the most suppressed field on this planet," the 54-year-old Santa Fe native said of UFO studies. "I intend to continue my research in this field to liberate this knowledge which will serve a cross-section of all inhabitants of the many universes."

He also has tried to make Santa Fe a better place by getting elected into office, but has yet to succeed. He has run for City Council 11 times, including last fall, and has made one run at mayor.

He has never accepted a dime in contributions. Cabeza de Vaca, who grows garlic and onions but makes his living as a plumber, said he will likely run for office

Please see CENTER, Page A-2

CENTER

Continued from Page A-1

again.

His Space Science Center, which closed Aug. 4, housed films, video tapes, books, catalogs and other information from the National Aeronautics and Space Administration.

It also contained a vast collection of materials on UFOs and extraterrestrials, which Cabeza de Vaca now keeps in storage. The center was open to anyone, at just about any time, free of charge.

Suzanne Martinez, a retired special education teacher for Santa Fe Public Schools, said many teachers took their classes to the Space Science Center each year.

"We were interested basically in the science end of it," she said. "It's going to be missed."

When he first started the museum, Cabeza de Vaca said, there was tremendous interest in NASA and the products of its scientific research. In the last six or seven years, however, about 90 percent of the people who visited the center expressed interest primarily in UFOs, he said.

That has been the experience of the founders of the International UFO Museum and Research Center in Roswell, where more than 33,000 have visited since it opened in October 1992.

UFO phenomena "is what is happening the most and what (the public) knows the least about," Cabeza de Vaca said. "When I started the Space Science Center, you couldn't find a picture of the moon here. Now

there's a lot of (space science) material all over Santa Fe, so I don't see a need for it."

He sent an entire room full of materials back to NASA in June — just about everything except a classified film that he says shows footage of UFOs taken by astronauts on NASA missions.

Cabeza de Vaca closed the center later for reasons he is not comfortable talking about, although he hinted that the cost of rent might have been a factor. He is now busy remodeling his home. Soon, he said, he will begin looking for an affordable location for his new venture.

Cabeza de Vaca is convinced that beings from another planet are not only visiting earth but may be planning to stay. "The government, whoever that is," is aware of the visitations but is engaged in a massive cover-up to conceal what it knows, he said.

"These people are coming from a dying star," he said. "It's what we're going to be faced with later on when that happens in this solar system."

He believes the aliens, who he calls "grays," are responsible for a rash of cattle mutilations across the globe, including in New Mexico.

The mutilation mystery is one of many puzzles to which Cabeza de Vaca said he will continue to seek answers.

"There are an average of 70,000 reported UFO sightings yearly" throughout the world, he said. "I've seen takeoffs and I've seen fly-bys. The only thing I really need to see now is when there's a landing or a very close sighting."

The August 27, 1994 *Santa Fe New Mexican*, spliced to show the masthead, the *Space Science Center* article announcing its closing, and the 2nd page. Note that in Column 3, Felipé alludes to the NASA classified UFO video in his possession.

(Find the full abstract of this article in the APPENDIX.)

Nov. 1994: riding "Luna" outside of Nambe, NM, north of Santa Fe. The Peñasco cattle mutilations of September 1996 were about 25 miles towards the mountains on the right. –Author's photo archive

INSIDE

EL DIARIO   B-2
BEST BETS   B-8
MOVIES/TV   B-8

SANTA FE

THE NEW MEXICAN   WEDNESDAY, SEPTEMBER 11, 1996   SECTION B

Chimayó ranchers report cattle mutilation

By RAY RIVERA
The New Mexican

PEÑASCO — An eerie trend that seems to resurface every few years since the 1960s apparently has popped up again.

A mutilated bull — its genitals, tongue, intestines and heart cut out — was discovered Monday in a high-mountain pasture here known as Llano de la Yegua. It was the third reported cattle mutilation in the last two months in Northern New Mexico. The last reported mutilations were in a cluster in 1993-94.

"You talk about aliens and UFOs and don't think much of it, then you see a mutilated animal and it's ... weird," said Chimayó rancher Carlos Trujillo.

Trujillo and the bull's owner, Wilfred Romero, also of Chimayó, graze their 60 head of cattle together through the summer in the high pastures above Peñasco. They

"We got up there around 7 p.m. and it was strange. There were no tracks, no blood, nothing."

CARLOS TRUJILLO
— rancher

viewed the bull after Peñasco resident Albert Ortiz discovered it Monday about 4 p.m. A Taos County Sheriff's deputy and a state livestock inspector also investigated the scene.

"We got up there around 7 p.m. and it was strange," Trujillo said. "There were no tracks, no blood, nothing."

Trujillo returned to the pasture Tuesday to dispose of the 2½-year-old bull — an 1,800-pound, pure-bred Santa Gertrudis. Trujillo said was worth about $2,000.

The bull's heart was removed through a hole, approximately 6 inches in diameter, cut in its chest. The anus, testicles, intestines and tongue also were removed with incisions. Half of the bull's left ear was removed. There was no blood on any of the wounds, the carcass, nor on the grass around the animal. And there were no tire tracks or unusual markings around the carcass.

"Whatever did this has got more power than you or I can think of," Trujillo said. "An animal of this size and weight would have to be tranquilized."

Trujillo said the livestock inspector did not ask to perform an autopsy on the animal. State livestock inspectors could not be

Please see **MUTILATE, Page B-3**

Chimayó rancher Carlos Trujillo stands over a dead bull found Monday in a grazing pasture above Peñasco. The bull's sex organs, tongue and heart were cut out.

MUTILATE

Continued from Page B-1

reached Tuesday.

The livestock board recorded about 30 mutilation incidents in 1993-94 in Northern New Mexico and Colorado. An Eagle Nest rancher alone reported 14 mutilations. The occurrences resembled a rash of mutilations in the

1960s and '70s in several states.

A 1980 investigation by a former FBI agent concluded that most of the mutilations in the '60s and '70s were caused by predators and scavengers.

There were no reports of mutilations in 1995 and the first half of 1996 until two mutilated cattle were discovered in Questa and

Arroyo Hondo near Taos in July. These latest mutilations, along with the 1993-94 incidents, are still being investigated by the state livestock board. The board has concluded that the 1993-94 incidents were not caused by predators or scavengers.

In a 1994 report to U.S. Sen. Pete Domenici, the livestock

board said the investigation into the mutilations revealed "possible involvement of clandestine Satanic groups."

The livestock board also said at the time that the State Police, the state Veterinary Diagnostic Lab and the Albuquerque Police Department were assisting in the investigation.

The September 11, 1996 *Santa Fe New Mexican*, spliced to show the masthead, the Chimayo & Peñasco ranchers cattle mutilation front page story, and the follow-up page.

For a thorough understanding of why I'm convinced the livestock mutilations are on the cutting edge of the UFO enigma (no pun, nor any disrespect to the ranchers intended), go to **page 358** for an eye-opening and mind-expanding interview with **Linda Moulton Howe,** one of the most erudite and insightful UFO journalistic investigators on the planet (start on **page 351** for some fascinating introductory material).

(Find the full abstract of this article in the APPENDIX.)

–reprinted by permission from
The Santa Fe New Mexican

A Space Odyssey

The crux of the UFO problem is that humans can't always be trusted. But astronomer Robert Young's dismissive, "no eyewitness report can be taken at face value," is extremist and reeks of scientific dogma, not the scientific method. Agreed, a large percentage of sightings are easily explained away — anything from mistaken identities to outright hoaxes. But what about the residue that remains once you've separated the signal from the noise?

Years ago I had an extremely intense UFO sighting. I went into shock after the thing — big as a small airplane, brighter than the full moon and making my radio go dead — silently swooped down from thousands of feet up in the dead of night, hovered, zig-zagged and basically scared the hell out of me. At its closest, it was 25 feet away. Time regression hypnosis sessions confirmed all this, and more. If only I could project on a screen for you the tape it indelibly recorded in my memory.

I have a college education, a high I.Q. and a low tolerance for closed-mindedness. So naturally, I found Tom Shroder's intro and some of Dave Barry's comments particularly irritating. To say "they" can't be here because our vast global detection network says so is tantamount to an Indian reporting to his chief leaning against a telephone pole, "No findem smoke signals. Nobody's talkin', Chief."

Wade Vernon
Miami

My letter to the Editor, printed by *The Miami Herald*, Summer, 1987, in response to their *Tropic Magazine* article spoofing and lambasting the UFO phenomenon. This published commentary, submitted to them via snail-mail, is an interesting milestone for me, as it was printed a bit over 4 years in advance of my sessions with Ellen Lyon.

BOOK IV.
THE VIEW FROM THE MOUNTAINTOP

I saw a disk up in the air,
A silver disk that wasn't there.
Two more weren't there again today--
Oh how I wish they'd go away.

–bathroom wall graffiti
White Sands Missile Range
New Mexico, 1967[1]

The notion of Santa Fe as an international alternative healing center and nexus of New Age enlightenment would present itself to me numerous times in the first few weeks after my arrival in April of 1992 in the City Different a/k/a the City Difficult by some locals. It had earned its pejorative moniker by virtue of the troubled job market, scarcity of affordable housing, high cost of living, environmental challenges, and general strife between the multi-cultural natives and all us gringos, whom they perceived as a lumped together faction of invading rich, white elitists. They had a point: the steady influx of jaded Aspen types scrambling to build their getaway wild west cottages in the form of adobe mansions steadily drove up the tax base, so the already near poverty-line natives sometimes had to sell off their homes that had been in their families for generations, just to survive. In some ways, I couldn't blame them for hating the white man's way. Euro-mutts have a decided pattern of discovering the latest paradise, callously exploiting and developing it, and then trashing it, the original inhabitants be damned. But I did find myself wishing that the true Santa Feans understood that not all of us white boys drove Range Rovers or had heated towel racks in our marble master baths. A lot of us struggled to get by day to day too.

Paging through an alternative newspaper, *The Sun*, I was bombarded with ads for every conceivable kind of unconventional therapy. A real standout was the one posted, with a solemn head shot, by the celebrated local healer Navajo Ventana (her conventional identity disguised perhaps at the behest of her sacred lawyer), who described her process thusly: "Utilizing a variety of crystals and sacred Native American shamanic healing modalities, I construct a multi-dimensional matrix around you, simultaneously on the physical, astral and causal planes; healing is spontaneous, occurs at the deepest levels in multiple dimensions, and the results are permanent." Wow, I'm in, where do I sign up, and how much wampum do I bring? I'm not kidding, I really need this.

Though I'd visited the high desert city twice before--once on a long weekend getaway, and again as a stopover on the cross-country road trip Carlos, Alf and I had taken from L.A. to Miami ("leisuring your way across the country like a couple of millionaires!" his father had scolded us via phone from Coral Gables)--I'd always been treated to moderately warm weather. I was in for a rude awakening when the record breaking snows of October through December of 1992 virtually buried us in what some mountain veterans called "that white shit." Soon enough I'd appreciate the comment one of my clients had made to me in half-jest, "Santa Fe has two seasons, July and Winter."

I hadn't deliberately targeted Santa Fe because I needed a place to heal and knew of its reputation, but clearly I was tending my wounds like many of the other recent arrivals rolling into the dusty town like tumbleweeds. Wrestling with the ramifications of my recovered UFO material from Tyler would prove a serious challenge to my equanimity for some time to come. The thing is, we're so thoroughly conditioned by the U.S. government, the Armed Forces, the media, and our peers--hell, even by our family members--to reject outright this kind of material to begin with. The typical knee jerk reaction is that you've been sexually abused, you're lying, or you're just plain bonkers. So in addition to how its mere emergence wears on your psyche, there's the added strain of dealing with your own questions about the integrity of your mental health, along with the ridicule of others. A real double whammy, and once hypnosis has popped the cork on the bottle of your subconscious, there's no going back. The off-world genie is out for good, with zero prospects of rebottling him. There's no choice but to deal with all the things he brings with him when he emerges. Sometimes enlightenment can be incredibly painful and challenging. A gay man who's been taken: I pretty much had the maladjusted outsider gig down cold.

Turning to recreational sports to clear my head became a ritual. Discovering that I was a gifted horseman, I contacted ranchers and began to ride regularly. If you love big dogs, chances are you'll love horses too. Horses, as big and massive as they are, are very sensitive animals, and can tell precisely how you feel about them when you're in the saddle. If you love them, they'll love you back, happy to provide full tilt gallops. My first time galloping, so pure: suddenly I'm the *Enterprise* engaging the warp drive, with the sky and landscape screaming past in blue and green streamers, horse and rider blissing out as one. Warping space is a good thing. I'd found a new heart connection. Man, horse and wilderness: key elements in a primal bliss equation.

I always felt so grateful to the horses for the incredible vistas they took me galloping through. Wilderness riding is incredible therapy. The skies of the high desert are so deeply, azure blue on clear days, and the air so clean you can practically bottle it and sell it for medicinal purposes. I discovered a deep heart connection with both the American Southwest, and the horses that played such an integral part of its heritage. I even lived for a few months on a 1,500 acre horse ranch southwest of Santa Fe, in the astoundingly wide open spaces between the towns of Cerrillos and Madrid, New Mexico. The owner was based in Southern California, so oftentimes if he wasn't in residence with his entourage, the only other man out there was the caretaker a mile away at the gatehouse, giving me full access to the Cook House/Kitchen and the Main House where I showered. My cozy, stone floored cabin boasted a pot belly stove and cot, and my desk, chair and lamp made it a rugged writer's dream retreat; the hinged window overlooking the horse corral suited me just fine. Wonder of wonders, all of the ranch's buildings had full AC electricity, 100% solar powered by an array of solar-tracker panels that automatically tracked the sun all day long like huge rectangular sunflowers. Living off the grid, not bad for the early 1990's.

Alf took to ranch life quickly. His German Shepherd genetics kicked in right away when he saw a slowly drifting group of lazy cows near the horse corral on our first visit out there. As soon as I opened the car door, he bolted out and took off after them and herded them like he'd done it his whole life. Seeing them respond to his zigzagging darts and lunges, jaws snapping at their rapidly galumphing hooves, was an amazing thing to witness. They fell in line, clustered and ambled together with minimal resistance, vocal or otherwise. He was ready for a cattle drive.

One early afternoon day in summer, an Amtrak train's distant whistle announced it would soon be traversing the northern periphery of the ranch on its way westward. I saddled up, shirtless and in jeans and boots, and rode a wild spirited Arab mare out to the distant tracks to gallop alongside the club cars. For awhile, blazing along at a full-tilt gallop, I almost kept pace with the rolling thunder, and waved at the folks behind the black bubble observation windows on top of the cars. Charging ahead on my spirited animal, half deaf from the wind and the train's roaring engine and steady mechanical chugging, gave me a taste for how it must have felt to have been a train robber way back when the wild west was truly wild. Galloping west into the warm sunlight, its rays kissing my skin, no strings, no worries, a man and his horse racing into the wind and the sun and the wide open spaces, a horizon without earth and sky boundaries. The boy and the man running on parallel tracks.

At night, motoring home past 3a.m. from bartending at a new, cutting edge disco near the Plaza--truly a sign that the *Conde Naste Traveler* cover story on Santa Fe had done its job--I'd pull my Chevy up to my humble adobe cabin and say howdy to the horses. Alf, ever faithful, would charge from the front door mat to greet me. Oftentimes I'd kick back on the hood of my car to gaze up at the breathtaking sight of the night sky. The only lights out there were the dull glow on the southern horizon--Albuquerque. Otherwise, sheer blackness. The Milky Way in all its splendor spread over me from east to west like a giant tiara, its stars so bright and close up it seemed the whole galaxy's stellar gems could be plucked from the sky. I used the time to reflect and unwind, and contemplate the abduction material that had exploded out of me back in December. I gazed up there at all those shimmering, glittering lights, and wondered if Kimberly was right, that they were coming back for me before long. In some ways, I guess I was daring them to make their move. You can't get much more vulnerable than being out on a wilderness ranch in the dead of night, miles away from another soul. My raw psyche actually found the near-lunar desolation out there comforting. Like a plug hungry for juice had found its socket.

I enjoyed the night drives home from work, winding down from slinging drinks at the disco by listening to k.d. lang's amazing album *Ingénue* over and over again. Her extraordinary voice serenaded me back to the ranch with *Save Me, Season Of Hollow Soul* and *Constant Craving*; they became my lonesome cowboy standards, custom crafted for a late night drive to an outback cabin. It's funny how songs of loneliness, desolation and soulful hungers can make you more aware of your solitary pains, and yet lift you up at the same time. Lang's haunting vocals were a perfect accompaniment to the kinship I felt with my new, desolate locale. The raw beauty of the ranch's quiet desolation gently soothed my soul.

Because I knew my 2 wheel drive Caprice was no match for the snowed-in roads winter would bring to the ranch, I drifted into town one afternoon and scoped out Santa Fe in search of a new home. I drove north on St. Francis Drive through the ancient, troubled kingdom of tan adobe buildings. Just past the Agua Fria intersection, a billboard-style sign on the west side of the road caught my eye; small wonder, for painted on it was an amply rendered flying saucer, a classic double disc. Doubling back, I slowed down to read, *Space Science Center*, and the sub-heading *UFO Museum*, with a small ground sign pointing to an adobe casita set back a distance from the road. Note to self: check out this place soon. As fate would have it, I would rent a humble barrio casita a few blocks away from it. My inner compass had led me to the right place again. The next installment of my ongoing paranormal learning curve.

A recent newspaper ad in *The Santa Fe New Mexican* had mentioned a weekend seminar on UFOs and government conspiracy theories. The speaker was William Cooper, evidently an ex-Navy man with a lot of stories to tell. I took a look at his book *Behold A Pale Horse* at *The Ark*, the local New Age book store. A lot of the material in it looked interesting enough to buy a ticket to hear him speak, so I found myself queuing up that evening at a local lecture hall on the campus of the New Mexico School For The Deaf on a Friday night in May, 1992. A fascinating cross section of people packed the auditorium to near capacity. With a brief introduction, our speaker entered the spotlight to loud applause.

Cooper was a big bear of a man, and commanded the stage with his bellowing, baritone voice. The brief evening was mostly a teaser to bring us back to the full day's fare on Saturday, a combination of lecture, video presentations and the lobby mini-bazaar hawking his books, pamphlets, audio and video tapes. Although signs forbade any taping of his presentation, I couldn't resist using my micro cassette recorder, hidden in my jacket pocket.

Cooper lived up to his reputation as a conspiracy theorist bar none. His talk that night covered the concept of secret societies, and how he believed the elitist Illuminati and Bilderberg Group

(Google them for an avalanche of information) ruled the world be-hind the scenes, keeping all the ultra high tech toys and technolo-gies to themselves. I didn't have too much trouble with that. I'd already developed a gut sense that the world was run by a cynical cabal I thought of as the Billionaire Boys Club, and all the myriad global governments were merely dog and pony shows to keep the masses dumbed down and feeling sated. A doped up populace is a happy and secure populace. Can we Biggie Size your burger and fries? and so on. But I found it hard to buy into his insistence that all UFOs were from terra firma, high performance flying saucers man-ufactured surreptitiously by the Illuminati with technology they'd scavenged from the Nazi weapons scientists after WWII. None of them were alien spacecraft from points unknown; all of them were simply the favored transportation mode of the secret societies. It also seemed odd when I discovered later that in prior decades, he'd asserted that aliens were visiting, and some UFOs were their flying saucer spacecraft. It seems he'd pulled a one-eighty on his stance. What was up with that, Mr. Cooper?

Interesting side note: Mr. Cooper also proclaimed in his lec-ture that the Bilderbergers had already selected Bill Clinton to be our next president--at a meeting in 1991. He said that should any scandals surrounding Clinton surface and threaten his ascendancy, a third party candidate would be put into place to dilute the vote, and ensure that Mr. Clinton would win the office. Ross Perot was their chosen man. (In fact, a strange presidential candidacy story arc did play out, intertwined with the Bill Clinton/Gennifer Flow-ers he said/she said alleged affair: on February 20, 1992, billionaire businessman H. Ross Perot had announced on CNN's *Larry King Live* that he would enter the presidential race if there were enough signatures to add him to the ballot in all 50 states; then on July 16th, he announced his withdrawal from the race, concerned that he couldn't win and would only cause electoral problems if he stayed in the race; then on September 18th, he qualified for the Arizona ballot, thereby placing his name on all 50 state ballots; on October 1st, he re-entered the presidential race as an independent; and finally on November 3rd, he received one of the largest percentages ever for an independent candidate in a presidential election (18.9 percent). It was argued that the elder Bush may have won the election if Perot hadn't shaved off the victory margin.) This obviously doesn't prove the Bilderberg Group rigged the 1992 presidential race, but the peculiar chain of events in light of Cooper's confident, matter of fact assertions that night in May were certainly interesting. Anybody who successfully ejects a Bush family member from office is fine by me, anyhow. Thanks for the help on that one, Bilderbergers. By all means, keep up the good work.

When Cooper mentioned the name Budd Hopkins, I perked up. I'd read both of the man's bestsellers, *Missing Time* and *Intruders*, and found them to be insightful, thought provoking, entertaining, and sometimes a bit scary. The alien-human hybrid concept introduced in the latter book really floored me. Upon reflection, I thought, why not? I didn't jump on the bandwagon when I read about that scenario, but I was open to a broad spectrum of possible reasons for the Visitor's visits. So, maybe alien-human hybrid breeding was happening, or maybe it was nonsense. Why not keep an open mind? My gut sense: they possess the technology of a star-drive or field propulsion system to get here, so they can pretty much do whatever the hell they want with us. We may already be learning to dance for the aliens without even realizing that we're doing it.

My attention really got zapped when Mr. Cooper shared that Budd Hopkins had met with one of Cooper's associates in Manhattan, and he'd pulled out a wallet ID card identifying himself as a CIA operative. The audience had a mixed response, some people exchanging "Huh?" glances. He rattled on that Hopkins was spreading the "lie" of alien visitations with an intent to invade so as to scare the hell out of the general population to the point of widespread panic, making everyone more sympathetic to the Illuminati's grand plan, the evil machinations of the New World Order. In short, that the Boys believed everybody would accept the consequences of a corporate totalitarian government enforced by a global military police state--one world, one government--so long as the Big Daddies would protect us from the malevolent space invaders. All we had to do was surrender total control to them. That's a pretty high price to pay for bodyguards. I filed this one away in my mental folder labeled, "Hmm." A scrap note to tickle my neurons at a later date.

Bill Cooper started to lose me pretty quickly when his conspiracy theory veered into the JFK assassination. He not only proclaimed that JFK's Secret Service limousine driver, Agent Bill Greer, had shot our president, but that he had a video to prove it to us. On Saturday, the auditorium lights were dimmed and he ran the video clip, projected onto a big screen. Although it did look like Mr. Greer, in the driver's seat, turned around towards Kennedy in the back seat of the Lincoln, I didn't see anything to convince me he had withdrawn a handgun from his jacket with his left hand, turned his head around to the right, locked eyes with Kennedy, stretched his left arm around, aimed the gun over his right shoulder and fired a bullet into the brain of our commander in chief, which is what Mr. Cooper asserted. When the audience balked and grumbled, Cooper replayed the clip over and over, doing a slow-mo freeze frame advance on the big screen, pointing out the exact moment of the deed, instructing us, "See it? Right there! See?!!" Scattered clusters of attendees in the dark auditorium started chuckling, laughing and jeering at him. I couldn't help letting loose a few good horse cackles myself, and a spontaneous chain reaction of laughter arose, blended with random pockets of people mumbling and chatting. Our burly host was not amused at the disruption. He shouted and snarled at the top of his lungs, "DON'T FUCK WITH ME, PEOPLE!!" It sounded remarkably like Broderick Crawford in *Born Yesterday*, as though he was a mobster trying to bully us into submission. I couldn't blame the guy for wanting respect from his attendees, and I'd been impressed with some of his material, but this situation had come to resemble a theater of the absurd. I seriously doubted I'd ever be convinced that the video showed what he claimed it did, even if I'd had access to it with a photographic analysis team. The fact is, for a lot of us in the audience, we simply couldn't see what he was seeing. Sadly, this segment of his presentation only served to tarnish his credibility for me. I strived to keep an open mind for the rest of his seminar.

Mr. Cooper delivered us a mix of truths and possibilities with a few farfetched musings that weekend. When he got to the subject of Area 51, believed by some to be where Black Budget projects (Shadow Government?) are reverse engineering crashed--some say "gifted"--alien flying saucers, he had my full attention again. While I'd heard a lot about the place, and was even open to the possibility that aliens were working alongside our scientists at that mysterious locale northwest of Las Vegas (a/k/a Dreamland), by now I took some of our host's assertions with a grain of salt. I was ready to be wowed by a video he said he'd recently shot, and my gut feeling was, "Ok, Mr. C, show me what you got."

Show us he did. The lights dimmed again. Cooper explained that the video was lensed with a very low lux-rated Canon video camera, a 1992 model, ideal for night filming. He and a few of his associates allegedly had camped out a few days and nights along the perimeter of Area 51 in February of 1992. Apparently that was before the Feds got fed up with intruders snooping with still and video cameras, shooting footage of the top secret and high performance aircraft zipping up, out and around the environs--fed up enough to snatch up and absorb many more square miles of public land into the federal domain in the coming years. This land is your land, this land is THEIR land . . . this land is FED land, and They'll shoot trespassers.

The camera faced east, in the direction of the now familiar mountain rimmed plateau. Patches of dim ground lights and darkness silhouetted aircraft hangars and out buildings scattered amongst the long runways. Presently, something dark and angular rose straight up from somewhere out there in the blackness. There was no jet engine sound. Only silence, save for the occasional murmur of Cooper or one of his companions--a fairly sure sign that the audio was taping in real-time, too. The thing vectored north, then slowly banked in a broad arc and silently turned towards the camera, made evident by the animated voices of Cooper's group gathered on the mountain ridge. As it neared, soaring and darting silently, it revealed a somewhat triangular shape, very large, massive and black except for a line of orange lights glowing along its leading edge. Closing in on Cooper and his comrades, it took on the appearance of a huge black manta ray. The leading edge corresponding to the mouth of the beast glowed with a somewhat sinister looking phosphorescent Halloween orange light--like a luminescent grille, with structural mullions separating the eerily glowing lights. You could clearly hear William Cooper's voice barking out expletives as the graceful behemoth silently dove down at his gathering, looking almost certain to crash into the nosy interlopers. Cooper shouted something like, "Good God, it's going to hit us!" His voice and the frightened voices of his companions were the only sounds you could hear.

Judging by its image on the big screen in front of us, in its closest approach, I'd estimate it was about the size of a B-2 Stealth bomber, perhaps a bit broader at the front, leading edge. At the last second, almost as if it was showing off its performance capabilities to Cooper and his cohorts, the giant craft leveled off and silently streaked past, probably less than thirty feet over their heads. They shouted in unison as the manta ray vehicle cruised on by, then gracefully tilted upwards in a wide open parabolic curve. Its iridescent orange grille leered a malevolent electric grin down at those who dared to view its wicked science up close. There was not the slightest rumble or growl of jet engines--no sign of a conventional propulsion system, no exhaust, no backwash, no humming, no stirring of wind, nothing. Deathly silent. The overtness of its utter silence was enhanced by the cacophony of animated voices beneath it. Then the big black craft silently stole away into the still night. It was disconcerting to witness something so large and massive do a silent flyby.

Cooper showed us other video clips he'd shot at Area 51, mostly far away lights doing artful zigzags across the sky, the sort of footage commonly featured on 1990's TV shows like *Sightings*--and very similar to what I'd seen with Greg Polasek and his family that night in 1968 in Houston's Memorial area. Silently hovering glowing lights, exhibiting sudden speed bursts, then disappearing here and reappearing there. Some of them came a bit closer to the camera, revealing discs and spherical shapes. It all looked very familiar to me. Decidedly non-conventional aircraft. From where, who knows?

I'm no *Industrial Light & Magic* computer animation wizard, but I feel like I have a pretty keen eye when it comes to spotting the real deal on film or still images. The years I've spent stargazing with telescopes, carefully observing planets and stars--even going so far when I was 11 years old as to sketch out the changing patterns Jupiter's moons made transiting and casting their shadows on its surface over the course of a week's worth of night time viewing-- have made me a fairly skilled, dependable observer. The amazing footage of that mysterious manta ray vehicle flying over Area 51 that night in February of 1992 stuck in my mind. Cooper's home video read as authentic to me, without question. That such a massive, exotic looking, obviously highly technically advanced aircraft silently cruised by at such a slow rate of speed led me to believe that we were test flying a vehicle with an antigravity drive propulsion system--or were there alien pilots onboard, too? I regret now that I didn't buy a copy of the video in the lobby that day; it would have been interesting to examine further, and to show to my family and friends. Despite Cooper's extreme, angry outbursts during the viewing of his "Kennedy video," and his mixture that weekend of what seemed like the absolute truth with what seemed utterly fabricated, I do believe the video of that mysterious aircraft in flight, ostensibly filmed at night on the outskirts of Area 51 in early 1992, was a genuine article. It was the flyby of a massive antigravity vehicle, clearly demonstrating its performance capabilities. Silent running.

THE WRITER'S RETREAT

All fantasy should have a solid base in reality.
-Sir Max Beerbohm

    That Fall I moved into a two bedroom adobe house on Ambrosio Street, a few blocks from the Plaza. It was made affordable by its location on the western edge of the barrio, the poorer side of town, nestled amongst a funky blend of small houses, gentrified bunga-lows and a few converted trailers. Though my Latino neighbors didn't exactly roll out the red carpet, and jeers from them and even their kids often greeted me when Alf and I took walks through the neighborhood, it was a good location for me. There was a long nar-row study in the back of the house off of the kitchen, with a small kiva fireplace and glass block windows at the far end. It was a per-fect room for writing--a tranquil space with natural light, but almost zero distractions. A choice writer's retreat.

    The fall of 1992 also brought changes to my personal life in the form of a new boyfriend, Billy Albrecht. Putting this statement to words is a challenge for me, because part of me will always have a streak of the Virginia mountain boy from my father's bloodline--and Virginia mountain boys grow up to be men who are a man's man, the kind of guys who hunt, fish, raise hell, get into fist fights, marry ladies and raise kids. Not the kind who pair off with other guys and call themselves boyfriends. But that's my stuff. I have to deal with the fact that I'm still attracted to a lot of types of women and love to flirt with them, but am much more attracted to masculine men--and have hot sex with them--and don't feel an affiliation with either the straight or gay worlds. My personal interests also don't skew in any particular direction: I love Muay Thai kick boxing, karate and pumping iron, and enjoy shooting my Ruger pistol at target rang-es; I also love cooking clean food, curling up with a good book, oil painting and fine architecture, well designed living spaces and inte-riors, and enjoy the HGTV cable shows about flipping houses. But you're more likely to find me in an old pair of Levi's and a t-shirt than one of Ralph's polo shirts or designer slacks. Like my brother Aaron has told me a few times, "dude, you're a 'tweener'." Which means the gay right's zealots are apt to label me a self-hating homo--which I'm really not. I hate just as many straight, ignorant redneck hicks as I do bitchy, effeminate, prima donna queens. I'm an equal opportunity disliker. Maybe my true feelings could be summed up by a postcard James Dean famously sent to a pal, a photo of himself in a bean field, having inscribed it with, "I hate all human beans."

I met Billy at *EDGE*, the disco I'd been lucky enough to land a bartending gig at shortly after my arrival. He was totally my type, and true to my pattern, another Kraut: Aryan mug, dirty blonde hair, handlebar moustache--a real dead ringer for a young Robert Redford, circa *Butch Cassidy and the Sundance Kid*, with a serious dose of James Dean's smoldering rebel tossed into the mix. When he sidled up to the bar to order a drink, I looked into his sleepy blue eyes and knew I was done for, screwed. I freely admit I have a real thing for bad boys, and we all get what we pay for in love and war--so no complaints there. Flashback to Jessica Lange as Julie Nichols in *Tootsie*: she confides that when she meets a man she's absolutely certain will be especially bad for her, and pretty much ruin her life, "that's when I go in for the kill." Bull's eye, Ms. Lange. Billy was going to be nothing but trouble for me, and I knew in my gut I was gonna take all that he could dish out if it meant I could nail him. (Addiction? Maybe. But really, isn't all cake that way, mostly pleasure and empty calories?) We were destined to be on again/off again for years. At one point, right about the midpoint of our dance of madness, I even contemplated having a custom t-shirt printed with the words: *Sometimes Blondes Are Almost Worth The Trouble*.

By the Spring of '93, I'd settled in and started making a decent living as a body worker, and set up my second bedroom for massage and hypnotherapy. In gearing up to write the novel of my UFO experiences (in its original iteration, I'd disguised my strange and bizarre encounters as fiction, as I hadn't been able to bring myself to own the full range of my flying saucer stuff), I began to frequent *The Ark* to add to my personal library. One book I found there that jumped out at me was *UFOs And The Alien Presence: Six Viewpoints (1991)*. Edited by Michael Lindemann, it's comprised of six in-depth interviews, a few with the biggest names in the field. Although its format allows one to read at will whatever section is desired, I've read it in full, several times. I still return to it.

The first interview was with UFO researcher, nuclear physicist, author and lecturer Stanton Friedman.   When asked why governments are keeping UFOs a secret, Friedman responded:

I'll give you several reasons why all governments are keeping it a secret. You see, it's not just the United States government, it's a worldwide phenomenon. Some people might think I'm saying there's a conspiracy. I'm not saying that at all, although there may be. But I am saying that there are sometimes common interests . . . . Here are several good reasons for all governments to not want to put the UFO data out on the table:

First, they want to figure out how the darn things work. As a top-secret Canadian document said in 1950, 'Modus operandi is as yet unknown.' You've got pieces, you've got wreckage. You set up your secret project, say, a small group working under Dr. Vannevar Bush, the top science administrator during and after World War II in the United States. The first problem is, you want to figure out how they work.

Rule number one is security. You can't tell your friends without telling your enemies. I mean, they read the newspapers too, listen to the radio, watch television.

Second, the other side of the same coin: What happens if somebody else figures out how they work before you do? They'd make wonderful weapons delivery systems and defense systems. You don't want them to know you know they know. Because, you see, if the technology is unusual, there may be ways of countering that, as long as they don't know you know. It's like poison gas. If you've got poison gas and I've got the antidote and you find out I've got the antidote, you're going to change your poison gas. Then I've got to change my antidote, and so forth. This whole sequence has been going on for thousands of years. Stronger shields, bigger spears, swords and all the rest.

Third, and very important: the political problem. Suppose there were to be an announcement tomorrow by highly trusted individuals around the world, George Bush and Mr. Gorbachev, maybe, or the Pope and the Queen--you know, pick your own odd couple--saying that, indeed, some UFOs are alien spacecraft. What would happen? Well, you know darn well the stock market would go down, church attendance and mental hospital admissions would go up. But the big thing that would happen, I believe, is that there would be an immediate push on the part of the younger generation, never alive when there wasn't a space program, for a whole new view of ourselves. Instead of Americans, Russians, Chinese, Paraguayans--just Earthlings. Because obviously, from an alien viewpoint, we are all Earthlings, even though we tend to forget it most of the time.

Gee, you say, that would be great. We could solve all the world's problems--the environmental problems, the political problems--if we all thought of ourselves as Earthlings. But then you realize that there isn't any government on this planet that wants its citizens to owe their primary allegiance to the planet. Nationalism is the only game in town. That's why we spent a trillion dollars last year on things military, in the name of nationalism, protecting ourselves against the other guy, or preparing to attack the other guy, depending on where you sit.

And there's a fourth problem, the religious one. There are a number of fundamentalists who believe that this is the work of the devil, that man is the only intelligent life in the universe. And the rug would be pulled out from under them. And you may recall that on occasion the government had been influenced strongly by fundamentalists. Reagan certainly was.

Then there's the fifth problem: economic discombobulation, if you will. If the public perception, when an announcement was made, was that there would be new means of transportation, new means of energy production, new medical things--a whole new world--there would be a tremendous loss in the stock value of oil companies, power companies, car companies, plane companies. I mean, forget psychological panic; that's a different story. There's still five percent of the people in the United States who don't believe we've been to the Moon. But the economic problems that might arise, should an announcement be made--that's a big difficulty. How do you handle that? And so, even though people might easily accept, as I think they would, the notion of alien spacecraft--most people already do, according to the polls-- that doesn't mean the transition is without its difficulties.

Certainly, the government would have asked psychiatrists and social scientists what would happen, and they'd say, 'Well, it depends on how you broach the announcement. You could make it 'Fear, fear, fear!' Or you could make it 'just' difficulty for religion, economics, politics, medicine, industry. You have a choice.' And I think for many governments, the natural thing to do is to postpone the decision. Let somebody else worry about it. It's a big problem, and I can understand the reluctance. However, I must add that, as a nuclear physicist very much concerned about the proliferation of nuclear weapons amongst countries that I wouldn't trust with a bazooka, much less a nuclear weapon, the only hope I see for a decent future for this planet is an Earthling orientation. By far the easiest way to get that is to recognize that there are aliens coming here.[2]

The first time I read this, it solidified my notion that Stanton Friedman is one of the most insightful voices of reason in this field (and sadly, more than two decades after "Friedman's List" was written up in this book, all of the very same, clearly delineated scenarios are still very much in play on a global scale in the second decade of the 21st century). The "top secret Canadian document" he referred to I'd seen in another book, *Clear Intent: The Government Coverup of the UFO Experience (1984)*. Using the Freedom Of Information Act (FOIA), authors Lawrence Fawcett and Barry J. Greenwood (and other researchers) sued to get access to previously secret and/or classified material concerning UFOs. Among them was this Canadian Department Of Transportation memo dated November 21, 1950, in which Wilbert B. Smith, senior radio engineer, forwarded a proposal to the controller of telecommunications suggesting a formal study of, among other things, using the Earth's magnetic field as a possible energy source. The subject of flying saucers happened to come up in the memo as follows:

> I made discreet inquiries through the Canadian Embassy staff in Washington who were able to obtain for me the following information:
>
> a. The matter is the most highly classified subject in the United States Government, rating even higher than the H-bomb.
> b. Flying saucers exist.
> c. Their modus operandi is unknown but concentrated effort is being made by a small group headed by Doctor Vannevar Bush.
> d. The entire matter is considered by the United States authorities to be of tremendous significance.

The Smith memo was itself classified as "Top Secret."[3]

I wouldn't fully appreciate the significance of the Smith memo until its broader ramifications later became apparent in one of Mr. Friedman's upcoming, exhaustively researched UFO books.

Linda Moulton Howe is another highly regarded UFO researcher/investigator featured in the *UFOs And The Alien Presence: Six Viewpoints* book. Though I'd heard about horse and cattle mutilations over the years, mostly from newspaper articles, I didn't realize until reading her passionately researched, sometimes disturbing material what a strong connection had developed between this grisly phenomenon and the simultaneous presence of UFOs. And I was soon to learn that my new home town of Santa Fe was situated in close proximity to one of the UFO/cattle mutilation hot spots of the country at that time. Yet again, my inner psychic compass had homed in dutifully on a hotbed of the paranormal.

Linda Moulton Howe is an Emmy award-winning TV documentary producer-writer-director-editor and an investigative journalist. Ms. Howe is also a Reporter and Editor of *Earthfiles.com*, her award-winning science, environment and Real X-Files news website (online since 1999). She graduated in 1968 from Stanford University with a Masters Degree in Communication, where she made a documentary for the Stanford Medical Center; her Master's Thesis was about the Stanford Linear Accelerator. She now oversees her *Earthfiles.com* web site, while also lecturing, researching, and writing and producing books and documentary videos. She has been a relentlessly inquisitive investigator throughout her life. *Earthfiles* is a treasure trove of her many inquisitions.

Howe has won numerous Emmy's for her film work, including one for her 1980 documentary *A Strange Harvest*, which chronicled the animal mutilation phenomenon occurring throughout the United States, Canada, Latin America and parts of Europe since 1967.

When asked what the connection was between animal mutilations and the UFO phenomenon, Ms. Howe replied in the *UFOs And The Alien Presence: Six Viewpoints* interview:

It goes back to September of 1967, when an Appaloosa mare called Lady was found dead and stripped of flesh from the neck up on a ranch near Mount Blanca in the San Luis Valley of Colorado. The story of Lady's death made world-wide headlines and was probably the first global public attention to the phenomenon that has come to be known as animal mutilation. Now, why the connection to UFOs?

In that summer of 1967, there had been many reports of strange lights and very odd, zooming little craft in that valley. The Sangre de Cristo mountains are in an area where even the Indians have legends about strange objects and lights flying in and out of the mountains. Besides all these strange lights in the sky, there was also a medical doctor who went to the scene of Lady's death. Unbeknown to anyone, he took some tissue samples from that horse back to Denver, analyzed them under a microscope and concluded that whatever had made the cuts in this horse had done so with high heat. It was as if the cuts were cauterized. But in 1967, we did not have lasers in surgical apparatus, so the doctor was very puzzled by the fact that all of the organs inside the chest area known as the mediasternum had been removed very cleanly. And to his astonishment, there was no sign of blood anywhere in the chest, on the hair, or on the ground. I would say that this lack of blood in the cases that I and others have investigated right from the beginning has been one of the hallmarks of the animal mutilation mystery.

Overall, the strange characteristics of these mutilations suggest that human technology is not involved, and natural causes can't account for it, so that begs the question, How does it occur? Alien activity has to be considered.[4]

Ms. Howe continues that since 1967, the reports included Canada in the late 1960's and early 1970's, with a focus of them in the Great Lakes area of the U.S. and Canada in 1974. Then 1975-76 brought reports of mutilations associated with very large, football field-sized orange glowing craft over pastures; beams of light in others; and eye-witness reports of strange non-human creatures where mutilations were found. Also associated with the occurrences were reports of mysterious dark helicopters; and an upsurge in human abductions during that time, as though something out there had some need, or was anticipating it. "But there has never been a year since 1967 when mutilations have not been reported, so it's not as if they got what they needed and went away," Ms. Howe commented.[5]

It's remarkable how many stranger-than-fiction true life stories of mysterious cattle mutilations are to be found in this interview with Ms. Howe. One standout was the case dated January,1990, in which a director filming a music video on a farm outside of Louisville, Kentucky noticed trees out on the periphery of his production site swaying, oddly so, since the air was still. Advancing towards them with his camera running, he discovered a calf, severed cleanly in half, the lower half missing with an excision resembling the work of a circular saw, with not a trace of blood on the ground nor on the calf's hair, and the exposed eyeball scorched to a blue-white, the poor animal's carcass so recently slaughtered that steam was rising from its remains in the cold air, indicative of a fresh kill. And yet, no object was seen, no sounds attesting to a presence, just the mysteriously swaying trees in the absence of a wind. This tale is told in narrative sequence with other UFO/mutilation cases in this interview, some of them also suggestive that a sort of cloaking device was engaged during these stealth operations, others with mysterious, but silent black helicopters, still others with the sounds of what seemed like heavy duty aircraft overhead, but with no visible aircraft or sources of the sounds in evidence.[6]

Reading this article spurred my interest in her work, so I bought a copy of Ms. Howe's video, *Strange Harvests 1993* with the subtitle: *An investigation of strange aerial lights and unusual animal deaths.* In late 1992 and on through 1993 there was a wave of UFO sightings, or a flap, in northeast Alabama, some captured on video, and many of the sightings coincided with a rash of cattle and live-stock mutilations. Fyffe, Alabama and the surrounding farm communities in that region were such a hotbed of UFO activity in 1989 that international tourists flocked there to watch the skies. (This video became a must-see in my UFO collection; I've watched it and shown it innumerable times, once to an Alabama friend who commented about its revelations, "We're all being played for fools.")

"In 1989 there were people from all over the world coming to Fyffe, Alabama to see a UFO," said Carey Baker, Editor of *The Weekly Post* of Rainsville, AL. "They were curious . . . there was a little fear, but it subsided for most people, almost as quickly as it came. In 1993, we're finding anger, especially from the farmers who have had cattle mutilated."[7]

In fact, Howe's video reports that by the Spring of 1993 more than thirty cattle and goats had been reported dead and mutilated in Marshall and DeKalb Counties of NE Alabama--the largest number of animal mutilations in the region's history. But the majority of law enforcement officials held the attitude that there were no mutilations, only predator attacks. A local newspaper headline countered that, with "Farmer wants answer to calf's mutilation; not satisfied with investigation."

One police officer who disagreed with the predator theory was Ted Oliphant of the Fyffe, Alabama Police Department, whom Howe and her video crew trailed around in their investigations. This was in the midst of numerous nights with strange moving lights, coinciding with cattle and goats found dead with strange bloodless cuts, and no tracks or other evidence found. Locals were nervous about it, and wanted answers.

"The precision of surgery that's been performed on these animals and the lack of evidence that we've got is disturbing and eerie," said Oliphant. "We've got people reporting UFOs, and we've got people reporting mutilated cattle, and people reporting helicopters."[8]

As for the animals in these Alabama towns, their wounds fit the pattern seen in thousands of cases all over the world: the bloodless excision of tissue, usually determined to have been done either with a very sharp instrument or a high heat source. A video montage of newspaper headlines reveals Australia, England, France and

Sweden, each with the same bizarre animal mutilations without any culprits, and usually with UFOs and strange lights in the night skies before the bodies are found the morning after.

Margaret Pope, owner of the Pope Farm in Geraldine, AL, had good reason to take it seriously. She'd lost a cow, it's milk sack cut away, leaving a serrated edge on the cow's belly. She told Ms. Howe, "It looked like somebody had lanced it off . . . it was like a cookie cutter cut around the edges of it, but there was no blood . . . like the edges were sealed; it was an oval shape."[9]

Back in 1990 in Oregon, the cuts on a mutilated steer were examined by the Veterinary Diagnostic Laboratory in the College of Veterinary Medicine at Oregon State University in Corvalis, Oregon. The lab concluded that the notched edge did exhibit a heat induced incision; but it was not possible to tell whether the lesion was caused by a laser. The notched edges of the cuts were similar to a 1975 mutilation in Montana.[10]

The mutilated body of a male calf in Crossville, Alabama found February 6th, 1993, also had notched cuts. They were shown by Officer Ted Oliphant to appear nearly identical to the Montana and Oregon tissue samples. Tissue samples of the Crossville calf and other Alabama mutilation cases were sent to Dr. John Altshuler in Denver, Colorado (formerly of the University of Colorado, and then with his own state of the art pathology and hematology lab); his analysis confirmed that the serrated edges were also cut with high heat.[11]

"In places where the cattle have been mutilated, there haven't been any footprints, any tire marks," said Carey Baker. "The lack of circumstantial evidence here, I think, is a bigger story than the circumstantial evidence. There have been a couple of cattle that have been mutilated after a rainstorm, and there were no footprints and there were no tire marks left. I think it's more the lack of evidence here, than it is the presence of evidence."

When asked for his personal perspective on the mutilations and UFO sightings, seeing as how he'd lived through it all from 1989 to 1993, Mr. Baker replies, "The more open minded people could be concerning whatever phenomenon this is, the more they would begin to understand that maybe we're not alone. And there's a possibility . . . someone speculated that they were doing this, that there are entities from another world who are gradually presenting themselves to us."[12]

Near the end of the video, Ms. Howe looks directly into the camera and says, "Perhaps military and intelligence personnel in the United States and other countries have long known that an alien intelligence is responsible for the highly strange cases of bloodless, trackless animal mutilations. Perhaps officials are afraid of public panic and outrage if that fact were known. Perhaps that is why Air Force Office Of Special Intelligence agent Richard C. Doty told me on April 9, 1983 at Kirtland Air Force Base in Albuquerque, New Mexico, 'That documentary you did, *A Strange Harvest* [1980] upset some people in Washington. They don't want animal mutilations and UFOs connected together in the public's minds,'."[13] Linda Howe's *Strange Harvests 1993* film and video is a must see.

New Mexico cattle mutilations associated with UFOs seen in night skies cropped up in 1993 - 1994, as reported in *The Santa Fe New Mexican* newspaper. One report was from the town of Eagle Nest, New Mexico, northeast of Taos, in which a rancher recorded 14 mutilations, and associated reports of strange lights in the skies. The livestock board recorded about 30 mutilation incidents in Northern New Mexico and Colorado during that period. The board concluded that the 1993-94 incidents were not caused by predators or scavengers.

On September 9, 1996, Carlos Trujillo, a Chimayo, NM rancher found one of his bulls mutilated in a high mountain pasture on his property; its genitals, tongue, intestines and heart were cut out. The heart was removed through a hole cut into its chest, approximately 6 inches in diameter. The tongue, intestines, anus and testicles were also removed, with incision marks. The bull was missing half of its left ear. Not a trace of blood was found, not on any of the wounds, not on the carcass, not on the surrounding grass where the butchered animal lay. And there were no tire tracks or unusual markings around the carcass. It was the third reported cattle mutilation in the previous two months in Northern New Mexico. "You talk about aliens and UFOs and don't think much of it, then you see a mutilated animal and it's . . . weird," he said. "We got up there around 7p.m. and it was strange," Trujillo said. "There were no tracks, no blood, nothing. Whatever did this has got more power than you or I can think of." A virtually identical comment was made by the Eagle Nest rancher after discovering his cattle mutilations.[14]

Watching Ms. Howe's video and then seeing these cattle mutilation reports in New Mexico got me to thinking about my own alien encounter. If you buy into this whole scenario, it feels on the visceral level like a real life monster movie. Scary looking beings from somewhere else using us as lab rats, stealing body parts from animals--and maybe sometimes not even returning their human subjects back home. I guess I was lucky they didn't fully exsanguinate me and dump my lifeless carcass out of the cargo hold while hovering over a field. It would have made some pretty big headlines in Tyler, Texas in 1960.

One of the cases in Ms. Howe's documentary film told of how a rancher had witnessed just such a thing, even that he could hear the terrible noise the dead cows made as their bodies fell from the sky and tumbled through tree branches. They crashed to the earth with such force that their corpses created small impact craters.

For some unknown reason, the ones who took me had also demonstrated an unusual interest in my blood. The casual, matter of fact comment made by my allergy specialist in New Orleans, Dr. Salvaggio, sounds eerie when pondering it, in retrospect, in this context--that it seemed almost as though all the antibodies had been filtered out of my blood. I'm inclined to believe they had been. To what end, that's open for speculation. Age three is an awfully young age for a boy to become a blood donor, especially by force.

My years of miserable allergy shots with their concomitant adverse reactions pretty well sold me that my body had to be challenged by the allergen serums to force it to create a new batch of body-warrior antibodies. Having weekly allergy shots and Jekyll and Hyde reactions in 7th and 8th grade, post doctor's office visits, meant that St. Andrew's School had to keep adrenaline stocked in their nurse's office to zap me out of anaphylactic shock on a routine basis. You think it's tough being the new kid in 7th grade? Try out this routinely repeated shot-shock scenario for even more fun.

The "alien as space monster" theme was explored on the big screen in the 1993 film, *Fire In The Sky*. When I saw the marquee at our local DeVargas Theater, I went to the next showing. D.B. Sweeney did a great job portraying Travis Walton, the forest worker who was purportedly abducted by aliens on November 5th, 1975 in a remote wooded area near Snowflake, Arizona. When Walton and six of his co-workers were driving home from a long day's work in the early evening, they sighted a large glowing disc hovering amidst the trees of the Apache-Sitgreaves National Forest. When they pulled over, Travis foolhardily dashed out of their truck and ran towards the object, stopping beneath it in a clearing. A few moments later, Walton got blasted by a high intensity beam of light from the saucer; it hit him with such force that he flew backwards in the air and landed on his back. His terrified coworkers, fearing him dead, fled the scene with him out cold on the ground. I remembered the story clearly, having read about it in the UPI reports in *The Houston Chronicle* back then. And I'd read the book Walton came out with shortly thereafter in the 1970's, chronicling his five days of missing time, and the fascinating things he'd witnessed onboard. The high points of the encounter were illustrated with color prints of the beautiful paintings Walton's coworker Michael Rogers had painted.

For a boy who grew up enamored with sci-fi, devouring novels and watching as many movies and TV shows of the genre as possible, Walton's onboard time was naturally a dream come true for me: battling diminutive, big headed, white skinned aliens in orange coveralls when he emerged from his blackout on an exam room table, then sleuthing down the curved corridor into an apparent bridge that sported a command chair whose armrest control module gave life and motion to a planetarium quality star field in the overhead dome, and even through the suddenly transparent floor. Now that's the kind of futuristic high-tech display this science-nerd boy loved.

Presently, a door opened. Walton met with another alien, one that fit the occasionally reported description in the UFO/abduction literature of a Nordic type; he was a blonde haired muscular man in a tight fitting blue jumpsuit, about six foot two and two hundred pounds, with a glass bubble type space helmet over his head. He seemed to be completely human, except for the odd appearance of his penetrating golden hazel eyes. Travis was guided by this humanoid out of the ship, to find that he was now in a cavernous hangar deck of what seemed to be a larger vessel. The craft he'd just emerged from looked like the one he'd seen in the woods, only much larger, perhaps sixty feet in diameter and sixteen feet tall; it was shaped like two pie pans connected at the rim, with a dome on top. Two or three other saucers rested at the far end of the room, oval shaped and smaller, with highly polished and reflective surfaces, like chrome.

The Nordic male was soon joined by two other equally muscular men and a female of similar coloring and build, also in blue jumpsuits. The last thing Walton could recall was having them gently push him down on a table and place a sort of breathing mask over his mouth and nose; within seconds, he passed out on the table. His next recollection is of being on the edge of a highway late at night. As he came to, the craft that had returned him was hastily departing up into the starlit sky. He had been gone for five full days.

I felt extremely disappointed to discover that the ending of *Fire In The Sky* in no way resembled Walton's onboard experiences. Instead, the producers had tossed out that material entirely, and replaced it with an outer space monster movie, replete with hostile brownish skinned aliens who performed painful and frightening experiments on a terrified Mr. Walton. (He'd had to break his way out of the ooze filled compartment of an insect hive before he blundered his way to the lair of his sadistic captors. Presumably he'd been sealed away for use later, who knows what for--ask the production company.)

While there's no doubt the film makers had fairly accurately depicted certain elements some people have had in their abductions, it disappointed me that Travis Walton let them do what they did to his story. At the time I felt that he'd sold out, and had given them free reign with his material. I didn't realize until after I'd read the updated account in his new book, released in 1996 (and another in 2010), that he'd had virtually no control over the studio's revision of his onboard scenes. In his new making-of-the-movie book, he came off as a sincere and earnest man who'd been subjected to the unpredictable, serpentine wiles of the film industry. His book is thoughtful and well written, and he clarifies that he fought to preserve the integrity of his truth. By the time he'd learned and accepted that his real story was altered, he came to feel he was lucky that even a modified version of his UFO encounter made it to the big screen.[15] It's actually a very well made movie, with good acting, a great supporting cast and good special effects.

Travis Walton earned my respect, too, for the way he handled the *Larry King Live* TV interview, timed to coincide with the theatrical release of *Fire In The Sky*. Though I only caught the tail end of the show, he scored huge points with me by standing up to Philip J. Klass, the noisy negativist (as Stanton Friedman is wont to say) UFO debunker extraordinaire; I particularly enjoyed his portrayal of Klass as a "disinformation specialist." Not to throw dirt on the old man's grave, but after watching the late Klass maliciously skewer UFO experiencers over and over again through the years with generally baseless accusations of fraud and hoaxes, it made me suspect the dyspeptic and irascible wonk was a government (or Shadow Government) operative, dispatched by The Boys to make mincemeat of any UFO case that might threaten to bring the ETH truth behind the phenomenon out into the open, and deflect any interest in possible hidden, advanced technology that might dramatically improve the quality of life of the common man.

Ever since Walton's encounter, I've sometimes asked myself: Why couldn't I have been greeted onboard by a mute, golden-eyed, blonde muscle god in a tight blue velour spacesuit? If they ever do come back for me, would that be too much to ask? The prospect of not being returned and to stay off-world--maybe not so bad, with Thor as my Star Buddy. I could get a membership at his stellar, zero-gravity gym. Telepathic message to Thor, from Wade: Hey, muscled up starman, I'm game! Let's pump up some time!

After watching the matinee feature of *Fire In The Sky*, it dawned on me that I'd still never visited the UFO museum around the corner from my home, so I decided to make it an outer space-themed day. I was in for an quite an experience.

SANTA FE'S UFO MUSEUM: THE SPACE SCIENCE CENTER

Turning off of St. Francis Drive beneath the eye catching flying saucer sign, I followed the dirt road a few dozen yards to a non-descript rust colored adobe building. Propped up against the wall with vertical pine tree posts was a rustic wooden sign proclaiming *SPACE SCIENCE CENTER* in bold white letters. Against a blue background, vivid painted images of Jupiter and Saturn floated in a sea of stars. The adjacent door, its mullioned windows painted black, seemed an apropos entrance to a building housing the kinds of things our government wants to keep hidden from us. In fact, this humble casita housed a small cache of paradigm-shattering material.

I was warmly greeted by Felipé, the friendly founder and caretaker of Santa Fe's nonprofit UFO museum. His eyes seemed both kind and wise. He happily showed me around the ramshackle space. There were walls of books and shelves of videos. Four rows of movie chair seats salvaged from an old theater sat in front of a screen. I asked him what sort of films he had, and did I rent them. He told me no, I could simply choose a video and he would screen it. I perused his collection, and one caught my eye: NASA UFO Archives, Volume I--how could I not notice a title like that? I asked him about it. He pulled it from the shelf for me and, with a twinkle in his eye, said he thought I'd find it interesting. He told me to take a seat and he'd load it into the VCR/projector. Not knowing what to expect, I sat in the second row. He dimmed the lights, and the video started.

There was no sound. As the screen flickered to life, the title appeared, the same words as on the video box, "NASA UFO Archives." I think it also said, "Volume I." Without preamble, the first scene opens, black and white footage, probably originally Super-8 transferred to video: Cape Canaveral, mid-1960's. The distinctive Gemini capsule atop its booster rocket rests on the launch pad. Plumes of fire and smoke eject from the engine, and the Gemini rocket slowly lifts off. It's barely ten feet from the ground, and suddenly from high up above, out of the blue, a very rapidly moving object appears, coming in towards the Gemini capsule so fast that the eye almost can't process it. As it rushes into view and swells in size, one is able to discern a large, dull metallic disc, somewhat like dark pewter. Its span as it nears looks to be about half the height of the Titan II rocket, so it may have been 40 to 50 feet in diameter, though its speed makes that a rough estimate, and I didn't have the luxury of doing a freeze frame on it. It zips in at full throttle on an intercept course with the rising rocket, and at the last possible second, tilts up on edge like a huge metallic Frisbee, just enough to avoid severing the top of the Gemini capsule. Then it vectors in a huge wide arc back up into the heavens from whence it came, and disappears in a flash. The whole scene plays so fast that you almost have to ask yourself if it really happened. A UFO joy ride! It made me wonder how many people around Cape Canaveral had to be debriefed and sworn to silence about what they'd seen that day.

Fade out. Fade in: again, black and white footage. Immediately apparent is that we're inside one of the Apollo spacecraft; the camera slowly pans across an instrument panel, and makes its way over to a pair of triangular windows, a telltale design instantly identifying the vehicle as the somewhat insect-looking Lunar Excursion Module (LEM). Briefly, the astronaut's space suited hand and arm appear as he maneuvers himself over to the windows. When he reaches one

of them, he places the camera up against the glass, and we find ourselves looking out at the surface of the Moon. The specific quality
of the lighting, both inside the LEM and out on the Moon, is identical to what we've all seen on TV scores of times. There is no doubt
that this is the real thing. One thing that does make this film clip
different than any Apollo footage we've seen on TV? The glowing
disc. A large, pancake shaped, white glowing UFO smoothly swings
into view, hovering roughly ten feet over the lunar surface; it's impossible to gauge its distance with certainty, but my guess would be
about forty yards. As for its size, it definitely appears large enough
to be piloted. The flying saucer holds its distance, and gently drifts
and yaws to and fro, like a sailboat on a lake lazily shifting on its
anchor line. The glow emanating from it looks like captured fire, a
brilliant, high energy plasma glow. There are brief flashes of other
shimmering things hovering nearby, and then the scene fades out
again. Next, fade in as an astronaut in the Apollo Command Module, apparently in lunar orbit, films footage out the window looking
down towards the lunar surface. Presently, the LEM approaches
from below, evidently returning from its mission.

Again, we've seen Apollo mission footage like this before, but this
time the LEM is being escorted by a pair of glowing balls of light,
like illuminated spheres. Again, it's difficult to gauge size, but I'm
guessing they might have been some sort of drones or probes.

A few more minutes of scenes like these, and the video ends.
Temporarily, I'm speechless. I'm dumbfounded that this video
tape isn't in a locked vault somewhere deep inside the Pentagon, or
somewhere else equally secure. Behind me, Felipé switches off the
VHS projector and turns the lights up. I get up and go over to him.
He laughs when he sees the look on my face. The first words out of
my mouth are, "Man! How the hell did you get your hands on a tape
like this?!"

He smiles and says, "A few years ago, I started contacting NASA
for space related information, requesting photos, documents, videos. Then later on, I wrote letters to NASA on official looking business stationery, with my *Space Science Center* letterhead at the top.
I asked them if they had any UFO video footage they might be willing
to share with me. I didn't really have any expectations about hearing back from them. I guess about a month or two later, I received
this tape in the mail, by itself, no letter. It had a NASA stamp on the
outside of the mailer, but no return address. I was very surprised."

"Damn. I can't believe they sent this to you. It seems to me this would be classified," I say.

"I'm sure it's a classified film," he replies softly. "My guess is that they either sent it by mistake--maybe somebody who saw my letterhead thought I was a government research facility, something like that--or else somebody pretty high up at NASA decided to let this information out. Give us a little taste of the things they know. I've gotten quite a few phone calls since then, people claiming to be from NASA, demanding that I return the video. Sometimes they sound very threatening and angry. It sort of bothers me. But I haven't returned it," he says with a quiet chuckle and a twinkle to his eyes.

Without hesitating, I launch into an entreaty for him to let me borrow the video and copy it. I reason with him that people deserve to know this, that this should be widely distributed. His face turns very serious and his eyes go black. He shakes his head and tells me no, that it would be too much of a risk for him to do that. I try to reason with him, plead with him, almost beg him; I tell him that I think it's the right of the American people to see this, to know what is going on behind their backs. But no matter what I say, he stands his ground. It seems there's no way he'll give in, so I relent.

Felipé was probably right to be so stubborn about my idea. I was contemplating making about a hundred copies and mailing them simultaneously to family, friends, and all the media hot shots I could think of, like *Oprah*, *Larry King*, all the major networks, anchors and network news shows, like *60 Minutes*, *CNN*, ad infinitum. Just putting it out there, with the attitude of 'screw it, go ahead and arrest me.' Put me away for thinking that the American people have a right to know, considering our tax dollars bankroll NASA--and the multibillion dollar Black Budget.[16] But Felipé put his foot down, and I respected that.

By its very existence, the NASA UFO Archives (Vol. I, no less!) video spoke volumes about the state of our nation, our world, our global bubble of denial. About the schizophrenic nature of our government, about the stark duality of the world within a world that has been created to house this lie, the biggest lie of all: the lie that says UFOs aren't real, and that aliens aren't visiting us. The biggest story of all time, and they want everyone to believe it's all in the minds of kooks, crackpots, liars, con men and nut jobs. A government of the people, by the people, for the people. But which people is it we're talking about? Who are they? Does anyone know anymore?

* * * * * * *

It was sunny and cool when I got home from viewing the NASA video. I took Alf for a long walk to a nearby park. I had to clear my head and set things right. Even with the intimate relationship I'd had with the UFO experience, the footage I'd just seen shook me up. It certainly wasn't a surprise that astronauts had viewed and filmed UFOs on their missions, that was a given. My lost sense of equanimity today was more about what the footage said about the heartbreaking rift between the American people and our government--ostensibly created by us, to serve us. The sense of betrayal was enormous to me.

When we got back, I fed Alf his usual combo of dry food, canned Kal-Kan and homemade chicken breast with cheese. While he gratefully wolfed it down, I retreated to my study and buried my head in some of my books. When in doubt, all mad scientists happily retreat to the lab, mine being my quiet refuge at the back of the house, its walls lined with books and warmed by a kiva fireplace.

I pulled *Crash At Corona* (1992) from the shelves, Stanton Friedman's seminal work (co-authored with Don Berliner) about the 1947 flying saucer crash outside Roswell, New Mexico. I wanted to review the genesis of the tangled web of lies our government and military began to spin around us way back then. I flipped through and came to the part where Maj. Jesse Marcel, the Roswell Army Air Field intelligence officer, was summoned to the Foster Ranch because Mac Brazel had found mysterious wreckage on his property. Through a serendipitous chain of events, Friedman tracked down Marcel in 1979, shortly before he died. He told Friedman:

> When we arrived at the crash site, it was amazing to see the vast amount of area it covered. It was nothing that hit the ground or exploded [on] the ground. It's something that must have exploded above ground, traveling perhaps at a high rate of speed . . . we don't know. But it scattered over an area of about three quarters of a mile long, I would say, and fairly wide, several hundred feet wide. So we proceeded to pick up all the fragments we could find and load up our Jeep Carry-All. It was quite obvious to me, [from my] familiar[ity] with air activities, that it was not a weather balloon, nor was it an airplane or missile. What was it, we didn't know. We just picked up the fragments . . . this particular piece of metal was, I would say, about two feet

long and perhaps a foot wide . . . that stuff weighs nothing . . . it's so thin, it isn't any thicker than the tinfoil in a pack of cigarettes . . . I tried to bend the stuff, but it wouldn't bend. We even tried making a dent in it with a sixteen-pound sledge hammer. And there was still no dent in it. . . .[17]

Friedman continues the narrative with the revelation that, "A few miles from the debris field the main body of the craft was located, and a mile or two from that several bodies of small humanoids were found. The startling news (possibly including the word that one of the humanoids was still alive) was shot back to headquarters, and the entire nature of the operation changed. Any lingering thoughts that the debris could have come from some advanced Soviet missile or aircraft vanished with the realization that the crew, and therefore the craft, were not from the Earth!"[18]

It was fascinating to me that I'd moved to Santa Fe with the simple objective of exploring the old west and the cowboy way of life, and instead I found all of that, and so much more. Prior to my southwestern sojourn, other than an *Unsolved Mysteries* show I'd seen in 1989 indicating that there might be something to the idea of downed vehicles in New Mexico, I hadn't heard much about flying saucer crashes, and suddenly I was living around the corner from a UFO museum with mind blowing, classified films, and barely a couple hundred miles north of where some alien vessels may have crash landed back in the late 1940's. Myriad UFO books cover the Roswell crash, from all points on the spectrum (unfortunately, at times misleading people to think something crashed in the town itself, but in fact, Roswell was merely the largest city closest to the site, which was more like 75 miles to the northwest, one of them near Corona, NM), but for me *Crash At Corona* does it best. It seemed like a natural progression from this book to Friedman's next, *TOP SECRET/ MAJIC*. It is here that Friedman's tireless sleuthing through documents at the National Archives, the Harvard archives, and countless other records facilities finally paid off in a big way, by delivering key answers to questions behind the UFO cover up. Kudos to Mr. Friedman for his bulldog tenacity and relentlessly hungry mind. The biggest revelations to be found in *TOP SECRET/MAJIC*: that Operation Majestic-12 (a/k/a MAJIC or MJ-12) did exist, and was created in response to the Roswell crash; that Harvard astronomer and tireless UFO debunker Dr. Donald Menzel was one of the twelve; and that Dr. Vannevar Bush was also on the team--as mentioned before in the Smith Memo of November 1950, "A . . . small group headed by Doctor Vannevar Bush."[19]

The Majestic-12 documents first showed up in the form of a roll of undeveloped black and white 35mm film in a brown envelope, left on the Burbank, California doorstep of Jaime Shandera, one of Stanton Friedman's colleagues; the package had no return address and an Albuquerque, New Mexico postmark. Shandera and another UFO investigator, Bill Moore, had been working closely with Friedman on the Roswell crash story; both of them had made many contacts with insiders in the intelligence realm, and they surmised that some of these insiders may have wanted the truth about flying saucers released to the public.

Friedman's excellent book obviously goes into great detail about these documents; for now I'll quote the opening and closing of them, which Shandera and Moore relayed to Mr. Friedman. The film contained duplicate sets of eight pages of documents that were classified TOP SECRET/MAJIC. The title page declared, "Briefing Document: Operation Majestic 12 Prepared for President-Elect Dwight D. Eisenhower: (Eyes Only) 18 November, 1952." The second page listed the Majestic-12 group members, all of whom were dead. The third page declared, "On 24 June, 1947, a civilian pilot flying over the Cascade Mountains in the State of Washington observed nine flying disc-shaped aircraft traveling in formation at a high rate of speed . . . In spite of these efforts, little was learned about the objects until a local rancher reported that one had crashed in a remote region of New Mexico . . . On 07 July, 1947, a secret operation was begun to assure recovery of the wreckage of this object. . . ."

The briefing document stated that the U.S. government was keeping secret that it had recovered the wreckage of a crashed flying saucer about 75 miles northwest of Roswell in early July 1947. Furthermore, four small alien bodies which seemed to have ejected from the vehicle were found two miles east of the main wreckage site. The government took possession of the wreckage and the bodies for in-depth study and evaluation. In September 1947, Operation Majestic-12 was officially established as "a top secret Research and Development/Intelligence operation responsible directly, and only, to the President of the United States."[20]

The full scope and implications of the Majestic-12 document contained in this remarkable book are quite extraordinary. This document deserves to be read in full, and *Top Secret/MAJIC* read and re-read in its entirety to fully grasp its impact. I quote here now from page 4:

> A covert analytical effort organized by Gen. Twining and Dr. Bush acting on the direct orders of the President, resulted in a preliminary consensus (19 September, 1947) that the disc was most likely a short range reconnaissance craft. This conclusion was based for the most part on the craft's size and the apparent lack of any identifiable provisioning. (See Attachment "D".) A similar analysis of the four dead occupants was arranged by Dr. Bronk. It was the tentative conclusion of this group (30 November, 1947) that although these creatures are human-like in appearance, the biological and evolutionary processes responsible for their development has apparently been quite different from those observed or postulated in homo-sapiens. Dr. Bronk's team has suggested the term "Extra-terrestrial Biological Entities,"or "EBEs," be adopted as the standard term of reference for these creatures until such time as a more definitive designation can be agreed upon.
>
> Since it is virtually certain that these craft do not originate in any country on earth, considerable speculation has centered around what their point of origin might be and how they get here. Mars was and remains a possibility, although some scientists, most notably Dr. Menzel, consider it more likely that we are dealing with beings from another solar system entirely.
>
> Numerous examples of what appear to be a form of writing were found in the wreckage. Efforts to decipher these have remained largely unsuccessful. (See Attachment "E".) Equally unsuccessful have been efforts to determine the method of propulsion or the nature or method of transmission of the power source involved. Research along these lines has been complicated by the complete absence of identifiable wings, propellers, jets or other conventional methods of propulsion and guidance, as well as a total lack of metallic wiring, vacuum tubes, or similar recognizable electronic components. (See Attachment "F".) It is assumed that the propulsion unit was completely destroyed by the explosion which caused the crash.[21]

Upon receipt of copies from the developed film Shandera provided him, Friedman went to work with his usual focus and dedication. He began to do research into the twelve names listed as the original MJ-12 members, comprised of six civilians and six military personnel, as follows:

> * Adm. Roscoe Hillenkoetter: Annapolis graduate; active in naval intelligence during World War II; named the first director of the Central Intelligence Agency in 1947.

* Dr. Vannevar Bush: world renowned research scientist; professor at MIT/ dean of MIT between the World Wars; one of the original developers of the analog computer, c. 1930's; during pre-WWII, chair of Office of Scientific Research and Development (OSRD); first chair of Joint Research and Development Board (JRDB), which led to development of the A-bomb, proximity fuse, radar, and 90 or more other high-tech systems with military applications.

* Secy. James V. Forrestal: in August, 1940, President Roosevelt appointed him Undersecretary of the Navy; was instrumental in modernizing and enlarging the Navy; September, 1947, Truman named him the first Secretary of Defense (after the unification of the Army, Navy, and newly formed Air Force that month); the national security implications of Majestic-12 would have made him a natural fit. Upon death on May 22, 1949, was replaced by General Walter B. Smith.

* Gen. Nathan F. Twining: extensive military career, Air Force background; with his experience overseeing new aircraft development, a natural choice for the MJ-12 group.

* Gen. Hoyt S. Vandenberg: West Point graduate; very active in intelligence during WWII; the first vice chief of staff of the Air Force.

* Dr. Detlev Bronk: headed a biology-related research committee under Bush at the OSRD; at one time, President of the National Academy of Sciences; primary field: aviation physiology; probably the best background possible for someone analyzing alien bodies.

* Dr. Jerome Hunsaker: from 1915-1930, instrumental in developing a wide variety of aerodynamic systems; head of aerodynamics at MIT for decades; succeeded Vannevar Bush as the head of the National Advisory Committee on Aeronautics (NACA).

* Mr. Sidney W. Souers: Admiral; very active in Naval intelligence during WWII; named the first head of the Central Intelligence Group by Truman in 1946; in 1947, Truman asked him to become the first head of the NSC.

* Mr. Gordon Gray: active in intelligence work during WWII; chairman of the 5412 Committee, which was in charge of covert operations for the National Security Council; appointed special assistant to President Truman on National Security Affairs, 1950; and in 1951, directed the Psychological Strategy Board of the CIA.

* Dr. Donald Menzel: astronomer; high security consultant to the NSA, CIA.

* Gen. Robert W. Montague: commander of Fort Bliss, near El Paso, and responsible for the White Sands Proving Ground in New Mexico; mathematics expert; solid technology background; chosen to head the Armed Forces Special Weapons Center at Sandia Base (now Sandia National Laboratory) adjacent to Kirtland Air Force Base in Albuquerque; lab's primary mission: development of weapons systems from nuclear weapons designed at Los Alamos.

* Dr. Lloyd V. Berkner: a scientist involved in polar expeditions, solving radio communications problems; served on many national committees; very strong interest in space; head of the International Geophysical Year Program, 1958; served on Robertson Panel, established by then-CIA director Walter Bedell Smith in 1953 to study UFOs.[22]

The fact that Donald Menzel's name was on the list of Majestic 12 members immediately rang true for me; I'd skimmed through his UFO debunking books while in junior high school, finding his explanations for some of the truly unexplainable UFO sightings idiotic--and often transparent lies. And the ridiculous, arrogantly asserted temperature inversion theory he proffered for the July 1952 UFO wave over Washington, D.C.--the idea that such a lauded astronomer would resort to such an unintelligent and easily refuted explanation was an indicator to me that he was deliberately muddying the water and enabling an atmosphere of derision to take hold in the media, which is precisely what the MJ-12 group wanted, obviously. Even as a seventh grader I knew that something smelled really fishy about this whole scenario, it simply didn't add up, and made me wonder about Menzel. So it made perfect sense to me that Dr. Menzel would have been a member of a top secret project like Majestic 12, which sought to keep the UFO/alien truth under tight wraps. What better way to do that than to engage an anti-intellectual smear campaign against UFO sightings and their sighters, delivered by one of our country's exalted high priests of science?

Friedman tracked down a Harvard professor who'd known Menzel quite well for a couple of decades. He asked him, "You knew him very well. How do you feel about the notion that he could have known all about crashed saucers and alien bodies and still have issued loads of disinformation?"

The professor's answer took Friedman by surprise: "He would have loved it. He would have been in on the biggest story of the century and he could show how smart he was by pulling the wool over everybody's eyes!"[23]

At the very tail end of Friedman's remarkable *TOP SECRET/MAJIC* volume is a document whose presence and almost eerie sense of realism I consider to be worth the full purchase price of the book outright (although the entire book is a must-read):

SOM-01: MAJESTIC-12 GROUP SPECIAL OPERATIONS MANUAL.[24] Over the course of its 18 pages, the reader is confronted with what looks to be an excruciatingly well thought out and painstakingly detailed written protocol (can you say: Military?) dictating precisely how Top Secret, undercover MJ-12 ranked officers should conduct a covert recovery of a crashed alien vehicle, specifically, with "biologicals" involved. There is a powerful sense underlying the ordered structure and logical layout of it all, of the regimented discipline and sense of order that the military and its best men would most assuredly have written up and implemented, had this been a reality. To put it bluntly: this whole bloody document looks so real, I had to read it several times to see if I could find anything that looked out of place or "staged." I never have. Its existence, if it is genuine (and I believe that it is), seems to be indicative of another possible hidden reality: that alien vehicles malfunction and crash to Earth (and/or are brought down by various weapons systems of ours) with sufficiently repetitive frequency to mandate that an exhaustively thought out and uniquely stylized document like this exists to begin with, one which spells out a rigidly regimented, step by step protocol for handling every aspect of such a complex and challenging scenario, with the maintenance of secrecy and security its highest priority. We are a planet living in the dark, indeed.

A MEETING OF MINDS: BUDD HOPKINS

Late Spring, 1994. I take advantage of the last snows of the season, and head up the mountain road to the Santa Fe ski basin. A lot of outsiders don't realize that our adobe village is built at the base of a really decent ski mountain, and we're 7,000 feet up. The previous year, I'd gotten clear that I had to find something outdoors to do during the bitterly cold winters--or lose my marbles with cabin fever and pull a Nicholson-esque breakdown à la *The Shining*. My love of the outdoors, coupled with a strong aversion to cold air, dictated that I find something to do out there in it besides wander around, turn blue and cry for Mama. I was learning firsthand the harsh reality of my friend's wisdom--Santa Fe has two seasons: July and Winter. So I took ski lessons at the celebrated Taos Ski School, and found another great outdoor recreational sport for clearing my cluttered head. Whereas the reefs I'd scuba dived in the Florida Keys offered the nonpareil LSD light shows of a silent underwater coral paradise, schussing down a mountain gave me the Zen bliss of head clearing rushes that I'd soon be addicted to. And head clearing rushes are superbly restorative for guys with extra baggage stashed deep down inside their brains.

After a day of speed demon invigoration, I stopped après-ski for a coffee and some newspapers. Thumbing through *The Sun* for kicks, an ad announced an upcoming Budd Hopkins lecture in my humble desert town. Everything I'd ever read of his, and the TV clips I'd seen of him had made me want to meet him, so I bought a ticket pronto. In New Mexico, it was easy to keep UFOs on your mind.

I'd recently seen the excellent Showtime movie, *Roswell*, with one of my favorite actors, Kyle McLachlan starring as Jesse Marcel. Kim Greist did a wonderful job as his wife, and Martin Sheen ate up the screen as a devilishly cryptic Intelligence spook, who taunted Marcel in one of the final scenes in an aircraft hangar with bits and pieces of insider information which he said may or may not be true. It was clear to me that Sheen's character's soliloquy was very much of a mind with a Q & A segment Ralph Steiner conducted with Linda Moulton Howe in Michael Lindemann's book *UFOs And The Alien Presence: Six Viewpoints*. The pertinent passage:

RS: Some people speculate that the human race is poised for initiation into some sort of galactic federation or some greater, increased awareness of our place in the cosmos. In some quarters, you even hear the idea that some event is going to take place in the near future that will reveal the UFO phenomenon on a grand scale. Do you have any thoughts on those scenarios?

LMH: On April 9, 1983, I was taken onto the Kirtland Air Force Base in Albuquerque, into an office where an Air Force Office of Special Investigations agent pulled out an envelope and said his superiors had asked him to show these pages to me. They contained a summary of this government's retrieval of crashed disks and alien bodies, including a live alien from a crash near Roswell in 1949. The paper said that this extraterrestrial had been taken to Los Alamos National Laboratory, where it had been kept until it died of unknown causes on June 18, 1952.

Then the paper summarized some of the information that had been learned from this distinctly alien life form about our planet and its civilization's involvement with this planet. One of the paragraphs said, "All questions and mysteries about the evolution of Homo sapiens on this planet have been answered, and this project is closed."

I remember reading those words, and reading them a second time, and reading them a third time as the implications of such a startling sentence washed over me. Because the paper was implying that this gray alien life form had been able to answer all of the government's questions about the evolution of Homo sapiens. Further, it stated in the paper that these gray extraterrestrials had been personally involved in the genetic manipulation of already evolving primates on this planet, suggesting that Cro-Magnon was the result of genetic manipulation by the gray extraterrestrials. Well, if any or all of that is true, then what we could be coming up to, after decades of animal mutilations and human abductions and UFO flaps and all of the drama associated with the phenomenon over the last four or five decades, might be some kind of an introduction between them and us.[25]

Clearly, the producers of *Roswell* either had studied this interview material, or were privy to an intelligence source with a virtually identical outlook. I'd first been exposed to the theory that the human race may have been genetically manipulated by aliens back in seventh grade, when I read works by Otto O. Binder and Erich Von Daniken, so this wasn't entirely new to me; but the idea of a marooned alien transported from the wreckage of his crashed saucer and kept under wraps in Los Alamos where he was pumped for information was still novel to me, and I was glad that they saw fit to include it as a closing act to a well made UFO film. It also made me wonder if the crash date was correct, and if so, was this unlucky 1949 EBE out scouting for the remains of one of the 1947 Roswell crashes? No matter. Watching it had whet my appetite to get close to insiders again, so the upcoming Hopkins lecture was eagerly anticipated.

Back home, Alf and I went for a jog around the block, and it dawned on me: I had to connect with Mr. Hopkins. Perhaps he'd allow me to interview him like Whitley Strieber had. Worth a shot. When Alf and I returned from our neighborhood jaunt, I tracked down Budd's contact information, and gave him a call in New York City, leaving a message on his machine. To my surprise, he phoned me back a day later.

His telephone persona was every bit as warm as I thought it would be. He questioned me at length about my UFO experiences. Hopkins seemed particularly interested in the issue of my blood being processed during my Tyler encounter, and the ramifications of the trauma to a child and the concomitant amnesia. By now Mr. Hopkins had racked up many years of experience hypnotically regressing UFO abductees to help them, and had become an expert at how the subconscious mind protects us. It felt great to talk to somebody who didn't look on these experiences as an automatic indicator of pathology. To my delight, he consented to an interview, and we scheduled a time for him to come by my house after his lecture.

"UFOs are like flypaper; everything sticks to them," he had said at his talk in a small auditorium. A respected New York artist--and ufologist for the past eighteen years (as of 1994)--Hopkins explained that he meant the phenomenon allows everyone to assign their own meaning, from the far left "saviors from outer space" cultists to Carl "there's no evidence that they're here" Sagan on the right.

Sitting on the sofa in my living room, Budd Hopkins had a kind, avuncular air about him. We quickly established a natural rapport, and it was easy to draw the material out of him. Hopkins told me he was utterly indifferent to UFOs until a daylight sighting in Cape Cod in 1964 spawned decades of personal research. In 1975 a Manhattan neighbor witnessed a UFO landing which Hopkins documented for *The Village Voice*. The response was overwhelming; case material surfaced from people's letters and phone calls, and extensive investigation and analysis led to his first book, *Missing Time*, in 1981. His 1987 bestseller *Intruders* became a CBS miniseries in 1992.

Touching on that topic brought me to something I'd wanted to do for over a year: spring on Hopkins the CIA allegations William Cooper had made in his May 1992 lecture, and see what his off the cuff response would be. Keeping the cadences of casual conversation, I told him that his name had come up in Cooper's UFO & conspiracy seminar; when I got to the part where Budd had supposedly pulled from his wallet a CIA operative's ID card, I paused. For a split second, Budd met my gaze with a blank expression. Then he literally doubled over with laughter, clutched his hands over his belly, and laughed so hard that he slid forward, red faced and cackling, and almost fell off the couch. Once he'd regained his composure, we had a calm discussion about Mr. Cooper's sometimes troubled mind. When I allowed that I'd deliberately set him up for the comment to gauge his reaction, Budd was fine with that. My gut instincts told me that Hopkins's response was genuine--and if it wasn't, he deserved an Oscar for an Academy caliber performance.

Continuing the narrative about his budding ufologist career in the wake of his two books, he said, "I started getting all these cases. And there were these funny things involving missing time, gaps in the stories and pretty agitated people. Using time regression hypnosis we accessed repressed memories." Out came spooky tales of UFO abductions by little gray creatures, "the grays." For the first seven years Hopkins enlisted the aid of psychologists and psychiatrists, sitting in on sessions; then he began conducting the regressions himself.

"After the Travis Walton case in 1975, the UFO community assumed abductions were extremely rare, and took only ten or twelve cases seriously. And in '76 and '77, as I looked at case after case, I began to be aware that it was extremely widespread--which flew in the face of all the perceived wisdom of the time," he said.

In writing *Missing Time*, Hopkins learned that abductions often happened more than once to an individual. "There was this potential of a tracked animal. And then there were the physical scars which resulted from operations--for some reason that had never been noticed or dealt with."

In *Intruders*, Hopkins presented his UFO cases as evidence that the grays are creating a hybrid-human race. Many of his subjects reported meeting with the hybrids onboard. "Obviously the reasoning behind it, as with much of this, has to be speculative. There are two groups never to believe when they talk about this, first the aliens and secondly the United States government--because they both lie, and they both obviously have an agenda. This seems to be a very selfish program . . . and we have no idea why. Are they going to somehow fold themselves slowly into our society and live with us? Are they going to take what they've learned and what they've created and leave? Are we going to be conquered? Nobody really knows. The idea all along has been they're non-malevolent; but I've never seen any sense of benevolence either."

At the time of our interview, Hopkins's latest cases involved children abducted from different locales--even separate continents--brought together onboard and allowed to talk and form friendships. They were then re-abducted every year or so. Sometimes it continued into puberty and the bondings became sexual. Occasionally they would meet in normal life as adults, and recognize each other from their abductions. "Just to make an inference, it would seem the aliens have been following the way we form relationships," Hopkins said.

Several years ago, Hopkins had introduced John Mack, Harvard psychiatrist, to some of his clients. Dr. Mack gradually became convinced these people were actually being spirited away by aliens; the resulting research culminated in his book, *Abductions*. When the news show *48 Hours* did a UFO special, Susan Spencer asked Hopkins why he seemed so saddened by it all, when Dr. Mack seemed very upbeat. He told her, "I've been at this for eighteen years, and John's been at it for four; I've seen a hell of a lot of sadness in eighteen years. Mack feels that this is all very optimistic and they're here to help. But how can you accept the idea of frightening a child? I've seen children just absolutely terrified. I don't think they intend to scare a three year old, but that's the actual effect . . . suicides and nervous breakdowns have occurred. Very few abductees would say, 'I wish this hadn't happened to me'--because they feel it's helped shape them, they've come out on top somehow. But ask an abductee if they'd want their child to go through this; you won't find one."

Hopkins estimated that he was getting about two new cases per day, from all over the country; the fact that he wasn't a licensed mental health professional invited some detractors. Some had accused him of leading his clients while in hypnotic trance to persuade them that they'd been abducted. In fact, Mr. Hopkins had a document he gave his "therapists in training" delineating techniques that steer the client away from the classic abduction scenario to see if they can be led astray from their story. I gave him a lot of credit for that policy.

I mentioned to Hopkins that Jacques Vallee, the celebrated ufologist, has argued that an alien race technologically capable of interstellar travel would only require one human DNA strand to clone as
many copies of "us" as they needed. He therefore dismissed outright
Hopkins's abduction/hybrid breeding concept. Hopkins replied, "I
don't have a theory. I'm reporting what people tell me happens to
them. I don't know what these alien hybrid babies are supposed to
be for, or why it's happening. Vallee's using his theory to refute my
data. It doesn't work that way."

I addressed the issue of the government cover-up, and Hopkins
thought he knew why it was in place. "The government is kind of
stuck. I don't think they know what the bottom line is. All they
can say officially is, 'Yes, they're here and they're flying around and
they can do whatever they damn well please, and they're abducting
thousands of people, and children, and we don't know ultimately if
they're friendly or not, but we'll let you know more when we hear
more about it, thank you and goodnight.' Are they going to say this
enormous thing is going on, and we're impotent to affect it? I mean,
Christ Almighty, what affect would that have on society?"

Looking ahead, Hopkins vowed to maintain his steadfast dili-
gence in offering assistance to people troubled by their strange and
often frightening encounters. "It's extremely important to me to
bring more mental health professionals and scientists into this-to
take this more seriously. And that's what's been happening. We're
expanding our network all the time." His closing words resonated with what Stanton Friedman said about him in an interview: "I
think a great deal of Budd Hopkins's work, partly because he's used
a lot of professional psychologists and psychiatrists with him, and
because I find Budd to be a caring person."[26]

In our phone discussion, Budd and I had touched on the idea
of my possibly becoming one of his hypnotherapists. Since he received letters from all over the country, it followed that not all of the
experiencers could fly to New York and see him for regression. So I
showed him my treatment room with its black leather recliner chair
and ottoman, my diplomas, and copies of some of my hypnosis case
files, with the personal identification information blacked out. He
seemed pleased with what he saw and heard, and we agreed that
I'd be accepting some case work of people he might send my way.
I was touched with how he closed this with, "And try to do some

pro-bono work, Wade. These are people who really need our help, and not all of them can afford a professional hypnotherapist's fee." I agreed, and told him I'd be happy to do that, particularly since I had an insider's viewpoint and was empathic about the challenging life path of an experiencer. Equally impressive to me: Mr. Hopkins stressed that during hypnosis sessions I should always try to steer the client away from the idea of a UFO abduction encounter, if for no other reason than to see if their subconscious mind was actually using screen memories. The whole tabula rasa stance--remain neutral like a blank slate for the client--but with a litmus test twist. Any good therapist knows it's important to let the client find his own way, particularly if both parties are sincerely seeking the truth, and a healing outcome is desired. Budd scored really big points with me on this. When he autographed my copy of *Missing Time*, he wrote, "For Wade, A new colleague, and, I hope, a new friend. Budd Hopkins."

One of Budd's most fascinating comments was that he was intrigued by the fact that, out of the hundreds of male and female abduction cases he'd dealt with, none of them--not one--was a professional athlete. He conjectured that this was because the intensely shocking nature of childhood and early adult abductions was so traumatic that the experiencers had dissociated during the encounter. Ergo, the natural development of mind-body and hand-to-eye coordination had been forcefully severed in youth, and experiencers were naturally less comfortable in their skins. His words rang true for me, and I told him so.

This really hit home to me when I recalled the times my Muay Thai kick boxing guro, Doug Pandorff, had on a few occasions good naturedly shared his bewilderment at the end of our one-on-one sessions about how difficult it was for him to teach me to coordinate the complex sequences of combination punches and kicks. He seemed genuinely perplexed that I couldn't seem to get down and polish and perfect the forms readily and incorporate them into my repertoire of fighting moves. I'd have them down for awhile, and then I'd somehow have a disconnect and lose the pattern, and have to re-learn them. It really frustrated me no end, since I looked up to him as a fierce fighter (his mastery also included Golden Gloves western boxing as well as Tae Kwon Do karate) and a genuinely great guy who sincerely embodied and walked the path of a warrior with a warrior's heart; I wanted to do my best for him, and I wanted it for myself, too. All of his other students mastered the combination moves long before I did--if I did.

Hopkins had wound up our talk with a knockout punch, and provided me with still more food for thought. Not fully comfortable in my skin, mind-body disconnect, and hand-to-eye discordant. Check, check and check.

* * * * * * *

TWO UFO CASE STUDIES:

AMANDA BLACK & DANNY DAWSON

Years passed. I managed to make a temporary peace with the tough high desert winters by skiing every season; in fact making a regular habit of flying downhill on two sticks was among the key elements in my being able to remain in New Mexico for as long as I did, along with horseback riding, iron pumping and kickboxing. Rec-sports therapy.

As promised, Budd Hopkins referred UFO clients my way, all of them female. I can't honestly say that any of the three had me convinced that they'd actually been abducted, which I found a bit perplexing and disheartening. Not to say that I was disappointed that any of the three might actually have missed out on the experience of being scared out of their skins by the whole shocking scenarios involved; but rather, that I may have been dealing with a subculture of people who wanted so desperately to be a part of the whole UFO and alien paradigm that they were willing to fabricate their painful tales out of thin air to become Light Year Club members.

One client in particular, Amanda Black, a Los Alamos housewife, was one I did offer pro-bono hypnotherapy to due to her and her husband's limited finances, as per Budd's request. After our first session, she was phoning me routinely during the week, before or after our Saturday or Sunday hypnosis sessions, commiserating in vivid detail how she'd been taken again during the week--they were coming through her bedroom walls at night to get her on a beam of light, leaving her husband behind in bed (he professed to having no memory of any of these things). One week, between our hypnosis sessions, they had taken her twice. There were episodes of assorted gruesome experiments, and often her captors told her their names. She gave me copies of sketches she'd made of it all.

With Amanda's consent, I found a local clinical psychologist to administer some basic psych tests. According to Dr. Landis, the Locus Of Control test results were not congruent with Amanda's abductions, the unusually low scale score indicative of someone who perceived themselves as possessing a highly developed control of their lives. Dr. Landis said that a person who is being routinely taken against their will, and yanked out of bed in the middle of the night (by alien beings coming through bedroom walls, no less) should have scored much higher, signaling a much more external controlling element dominating their life. She expressed her concern that the low score raised a red flag for the presence of deep seated psychological issues, and a deeply troubled psyche.

Because I perceived that there were some things Amanda was holding back, I gently guided our discussion in our final session toward the concept that perhaps there was another source for a good part of her psychic malaise. She and her husband rather sheepishly revealed that Amanda had been sexually abused by her father from age 8 to 13, and that it had been very traumatic for her. I told her that I believed her mental health issues would best be served by seeking out a licensed clinical counselor to deal with such serious concerns, and help her heal the unhealed wounds from her past. When she asked me, I told Amanda I didn't know what to make of her reported abduction experiences. As with all of the clients Budd referred to me, I admitted to having had a group UFO sighting in Houston, but rarely mentioned my shocking close up solo Houston sighting, and certainly never spoke a word about my Tyler material, in the interest of remaining as neutral as possible. I simply told clients that my mind was open to the full range of UFO experiences, to avoid any contamination of theirs. Tabula rasa.

After a few weeks, I phoned Mr. Hopkins to update him about Amanda. He was sorry to hear the details, but accepted my sincere conviction that I'd be doing more damage by allowing Ms. Black to continue bleeding out more of her troubled tales. I told Budd my gut sense was that it was more comfortable for Amanda's wounded psyche to blame all of her stored up trauma on monsters from outer space routinely tormenting her, when in fact the real monster had been just down the hall in her youth, her father who had repeatedly stolen into her childhood bedroom at night and violated her. I acknowledged that I couldn't prove this, but the evidence was strong enough to warrant cessation of hypnotherapy. I believed Dr. Landis had made a correct assessment, Amanda needed professional help.

I related to Budd that a massage client of mine was a very successful Santa Fe painter who showed his works in galleries all over the country, and was collected widely, both by individuals and corporations. Lars Bloom had shared a UFO story with me I thought Budd should hear about. A gifted abstract expressionist, Mr. Bloom painted many of his wildly colorist oil and acrylic pieces on very large canvasses, sometimes as mammoth as ten plus feet in height by double that or more in width. He therefore employed a driver, Danny Dawson, to shuttle them across the country in a moving truck, sometimes to galleries, sometimes to private collectors or corporations.

On the trip in question--I think it was the Summer of 1992--Danny had gotten a late start leaving Santa Fe. He was driving south on I-25 with a cargo truck full of Lars's latest canvasses, destination: Tucson and Scottsdale art galleries. Sometime after 2am, in a stretch of dark and empty highway between Socorro and Truth or Consequences, he noticed a distant light in the clear dark sky, like a bright star. Its luminosity made it stand out amongst the other stars in the vast blackness that is the New Mexico nighttime sky in the great wide open spaces. But that factor became moot when the "star" began to move in very odd patterns, almost dancing about playfully in the distance.

Danny nervously tried to concentrate on his driving, uneasily endeavoring to ignore the fact that the light he saw through his windshield was growing steadily brighter and larger, and that he was utterly alone on a lonesome stretch of highway in the dead of night. This is where the story gets sketchy; the details fade a bit because Danny didn't like discussing the experience in depth with Lars, or anybody else. His final clear memory was that the light suddenly zoomed much closer, perhaps miles closer in the wink of an eye. It was definitely an unusual, brightly lit aircraft, made no sound at all, and was on a collision course with his big truck. Danny's recollections abruptly blacked out there. The next memory he had was of being on the outskirts of Truth or Consequences--about thirty miles down the road from where he'd been, and over two hours later. He had a very uneasy feeling when he realized where he was, really more of a deep sense of dread. And ever since then he avoided late night drives on lonely highways at all cost. Danny made it clear that investigating his frightening, dead of the night experience any further was out of the question. He'd hang tough with the deep unease, thank you. I still wonder about Danny from time to time today.

I told Budd that I thought Danny's story was a much more likely case of abduction than Amanda's. His resolute unwillingness to explore its fear-inducing elements, the familiar pattern of the strange glowing object's approach, the abrupt blackout, the missing time, and his extreme discomfort with even discussing it in general terms, to me were all hallmarks of a real UFO abduction encounter. And I suspect there are a lot of Danny Dawson's out there in our country: people who'd had extremely close up sightings, and maybe had telltale clues of something more occurring, but didn't want to know more than that. People who chose to live with the nightmares, the fears and uneasiness rather than uncover the truth behind their missing time. For Danny, like many others, hiding the truth from himself with denial may have become his coping mechanism for the rest of his life.

I sometimes shared the general details of my UFO/hypnosis clients' experiences with my other half Billy--but never their names or other identifying factors. He knew my whole background, and had no problem believing anything I'd revealed about Houston and Tyler. He'd had his own sightings before we met. Before his current job with the Department Of Interior, he had been an Air Force officer.

Billy told me stories of when he was stationed at Nellis Air Force Base, the location adjacent to Area 51. Stories about sudden late night emergency drills, times when everything on the base would shut down, and all officers had to remain inside their offices, hangars or bunkers for several hours. The buzz on the base was that they were testing new, classified aircraft. He told me that because of his high I.Q. (165) and his favorable testing aptitudes, they'd given him work assignments that required a high security clearance. He'd been shown schematic diagrams of the Manzano Mountain range, skirting the edge of Albuquerque's Kirtland Air Force Base and Sandia Labs. He told me that many of the diagrams showed that a huge amount of the mountain had been hollowed out, and that a lot of that hidden interior space consisted of multi-tiered hangers used to house experimental and unconventional aircraft. With all the tales circulating about Kirtland and Sandia Labs, who could argue with that?

A short time after he'd left the service he had a close up UFO sighting one night with a friend up in the Manzano Mountains. They had pulled over onto one of the overlook spots of the mountain road, killing time before heading to a concert, enjoying the night sky. A light, like a star, had risen up from the distant mountain range, one of the Manzano peaks. It zigzagged and bobbed around, and about three seconds later was suddenly a huge disk hovering overhead in front of their car. As they stared up at it through the windshield, Billy and his friend exchanged verbal descriptions of the object with each other, talking out loud to be sure they were seeing the same thing. It made no sound, and there was no exhaust. Billy said it looked metallic, was brightly lit up with a white glow on top, and had a dull crimson glow emanating down below. The underside looked embedded with an unusual cross-hatched network of lights, in an odd pattern like the folds of a brain hemisphere. The huge disk didn't stay long, it silently hovered for less than a minute, then darted back to the mountain, once again becoming a moving star in the distant sky. Several miles silently traversed in a few seconds, thousands of miles per hour in the atmosphere, silently darting away, perhaps on its way back to an aircraft hangar in the hollowed out Manzano Mountain complex Billy had seen in diagrams.

BOOK V.
INTEGRATION:
ASSEMBLING THE PUZZLE PIECES

By the late Spring of 1999, I'd already had my fill of zero degree temperatures during Santa Fe's long winter nights. Even my heavy-duty parka couldn't always prevent the brutal wind gusts from feeling like I was being stabbed by icicles on nighttime hikes. The last straw was when the temperature of our high desert refuge was chilly enough to warrant wearing a long sleeved sweater at noon in early July that year (time to update the bi-seasonal, July and Winter myth). A sweater needed in July, midday. Surreal. The cold, coupled with an ever increasing elitist disposition imported by the jet-set clique of interlopers (I'd by then renamed our one horse town *Mayberry With An Attitude Problem*) rekindled my wanderlust. So finally in September of 1999, I packed up my belongings, completed my sign of the cross over America, and hightailed it back to Austin. Goodbye, high desert; hello again, Texas heat. New location, same drill: use the trainer, body worker, hypnotherapist gigs to buy writing time.

It was at this time that I gradually warmed to the idea of re-writing my UFO book in its true life format, owning all of my personal and difficult to accept Tyler material and getting it in print. My newfound resolve meant that I pored through my research notes accrued over the years.

Part of my UFO questionings naturally meant that I was curious about the physics behind flying saucers, and so way back in the Summer of 1979, I sought out an in person interview with Dr. John A. Wheeler, the former colleague of Einstein. I used to sit in on some theoretical physics classes at U.T. just out of curiosity (as long as you were an enrolled student, U.T.'s policy was you could monitor any class you liked). Upon retiring from Princeton in 1976, Wheeler had joined the faculty of The University of Texas at Austin, and founded the Center for Theoretical Physics. I jumped at the chance to meet with one of the great minds of the day. Happily, Dr. Wheeler's secretary obliged me with an appointment. It would give me the opportunity to query about the nature of the invisible forces operating in our Universe. I was clear enough already in my own mind that the off-world technology of alien saucers possessed a mastery of gravity, antigravity and field propulsion, but I'd always wanted to ask a top notch physicist a question probing into what I thought might be one of the most subtle forces of all.

On that day, Dr. Wheeler, a distinguished looking gray haired man with a kind face, gave me a warm greeting and had me sit down in his office. I told him I was writing a book, and only had one question for him. That question was: did he think it was possible that the primary force in the Universe--the force that made particles group together to form atoms, and atoms to group together to form molecules, and eventually sufficient numbers of atoms and molecules to coalesce into stars and planets--could it be that this powerful, pervasive, magnetic and electromagnetic-gravitational force underlying everything, was *love*?

Dr. Wheeler smiled, and silently stared off into space for a few moments. Then he answered me with a quiet, measured reply, "I can't help but think that we scientists are missing out on tremendous insights in our failure to look at the Universe in a more poetic fashion." Because a significant part of the UFO phenomenon was related to subtle energies, I found this concept relevant, becoming more so with time.

Dr. Wheeler's words not only revealed how open minded he was, they also showed that he acknowledged the importance of thinking on multiple levels and perhaps utilizing unconventional tools as a means to get at the truth. Ever since reading Fritjof Capra's *The Tao Of Physics* (1975) I'd been attuned to the concept of physics and metaphysics being somehow interwoven, and part of a greater whole--intertwined elements of a multidimensional reality.

As mentioned, back in the Fall of 1991 while working as a tarot reader for *Psychic Central,* Ryan the manager had told me that my Third Eye would open up dramatically from giving so many psychic readings on a daily and weekly basis. It finally did do just that, quite noticeably in one session.

I'd been there about a month and a half, and late one evening I had a man on the phone, Josh, who was seeking a general reading. A few minutes into it, I paused in my talk to contemplate things while staring down at the spread of cards on my desk top. With no warning, the cards blurred, and I suddenly had a full color scene appear, floating in space as if in a small virtual bubble in front of my forehead (a/k/a my Third Eye), much like a holographic image hovering before me (the phone room was candlelit, no other lights were used, so it glowed in the dark, which was really far out). I told Josh, "I'm seeing a truck, it's a bright reddish-orange pickup truck, with very shiny metallic paint . . . it looks like an old Ford . . . hmmm, maybe around mid 1930's, early 1940's," I said. I could see it like it was on a dais, with spotlights illuminating it, in the center of a dark showroom, with the edges of the image fuzzy and faded out to darkness, not unlike the typical visual movie trope of a psychic vision.

Josh started laughing. "Uhhh, yeah," he chuckled. "That's pretty amazing," he drifted off.

"And I'm getting that you restored it, is that right?" He laughed again, and told me I'd nailed it. It was a 1935 Ford pickup with red-orange metallic paint; he'd just finished overhauling and refurbishing it, and was contemplating selling it. I had just experienced *bilocation*. There's no way I can prove Josh was being truthful with me, he could have been stroking my ego, and made it up that I was so accurate. But at $2.99 a minute per reading, I'm inclined to think callers wouldn't very likely give out false positives for such a pricy session (an hour would be a high end shrink in 1991 dollars, clocking in at $180.00+ when you figure in the taxes, etc.). My "visual bubble" experience with him brought to mind something I'd read before. Back in the early 1970's, I recalled reading a newspaper UPI article about a New York artist/psychic named Ingo Swann. Swann had known of his psychic gift since childhood, and had begun to develop it as an adult. Much later, the excellent Jim Marrs book *Alien Agenda: The Untold Story Of The Extraterrestrials Among Us* (1997) would connect the realm of psychic powers with research into the UFO enigma (I've read it twice, with more readings to come; it's one of my favorites in the genre).

By late 1971, Swann had teamed up with Dr. Karlis Osis and his assistant Janet Mitchell at the American Society for Psychical Research (ASPR) in New York to conduct psychic experiments, some of them involving getting a visual lock on a distant object or target. The work was done in conjunction with Dr. Gertrude Schmeidler who was with the Department of Psychology at the City College of the City University of New York (and also a board member of ASPR).

The experiments generally went well, with more "hits" than "misses." It was during this work with Dr. Schmeidler and Dr. Osis that Swann was approached by Dr. Harold E. Puthoff, a physicist at

Stanford Research Institute in Menlo Park, California, with a proposal to test a psychic tool or technology known as Remote Viewing, but only under the strictest scientific protocols. Thus began nearly twenty years of remote viewing (RV) study at SRI. One investigator referred to it as being "the most severely monitored scientific experiment in history."

While at SRI, Swann developed a method of assigning coordinates to determine remote viewing targets; that way, if the target was the Lincoln Memorial, for example, the viewer wouldn't be drawing upon memories of what he already knew about it. Swann quickly honed his skills at coordinate remote viewing (CRV).

Dr. Puthoff alleged that two CIA men had approached him at SRI in 1972. They knew of Puthoff's previous experiences as a Naval Intelligence Officer, and then as a civilian employee of the National Security Agency. The CIA agents told him that there was increasing concern in the intelligence community that the Soviet Union was taking parapsychology and the development of psychic skills very seriously. It was thought they might try to harness its potential for use in psychic warfare--Psi Ops. Because most working Western scientists considered the field nonsense, the CIA saw the SRI remote viewing experiments as a perfect fit for their need to have a low profile classified investigation. The CIA negotiated with Dr. Puthoff, and a working relationship was thus forged, with Ingo Swann as their first guinea pig.

The remote viewing study, named Project SCANATE (SCANning by coordinATE) proved successful enough to warrant funding until 1976, when the program was taken over by the U.S. Army's Intelligence and Security Command (INSCOM). More than a dozen military intelligence officers were trained using protocols developed by SRI.[1]

The early 1970's UPI article about the SRI program was impressive. What really grabbed me was when Ingo Swann started using remote viewing to observe various bodies in our solar system. His visions weren't taken seriously at the time, but were later confirmed by various NASA deep space missions.

Swann was hungry for diversions while undergoing seemingly endless tests with the SRI scientists. In March 1973, he noted that NASA's Pioneer 10 space probe was headed for Jupiter, and would begin transmitting data about the planet in early December, nine months away. Swann realized it would be a fascinating test of his powers if he conducted psychic probes of Jupiter to compare with upcoming Pioneer 10 feedback. Results: Swann's visions were in such detail that they could not be fully evaluated by Pioneer 10's data. But full confirmation of the accuracy of his descriptions came six years later, with the Voyager 1 and 2 probes in 1979. Among his accurate hits: that Jupiter had a ring--though much thinner than Saturn's, so not visible with our telescopes--as well as high winds, towering mountain ranges, and thick cloud cover.

In March of 1974, Swann decided to further his space pursuits and conducted remote viewing sessions of Mercury, in advance of the upcoming Mariner 10 flyby. Again, he correctly identified Mercury's thin atmosphere, magnetic field, and helium trail streaming out from the planet in the opposite direction of the Sun. These observations were completely contrary to the scientific thinking of the time, but were corroborated with our instruments.

Jim Marrs's engrossing book *Alien Agenda* goes into much greater detail about these case studies. The book is in fact a veritable treasure trove of cutting edge UFO/alien material. Suffice to say that Ingo Swann's visions, coupled with scientific feedback, established the credibility of remote viewing as a tool for exploring outer space and other dimensions. Swann laid the groundwork for the off-world experiences of other remote viewers. What might sound like sheer fantasy to much of the general public would become the norm for these government trained psi-spies. Knowing that their superiors probably didn't want to hear the bizarre reports about UFOs and alien races, the viewers created "Enigma Files" in which they recorded some of their most extraordinary cases. Their real life X-Files cases involve remote viewings to strange, wondrous and sometimes ineffably beautiful places far beyond Earth.[2]

Ingo Swann's remarkably accurate details about our solar system and its planets may never be acceptable to hard core, closed minded skeptics, particularly religious fundamentalists who would likely regard remote viewing as something evil or a work of the Devil. But my life path of openness to alternate and paranormal realities has given me a different perspective. My one brief glimpse into a similar state, when I bilocated and saw Josh's pickup truck that night at *Psychic Central*, convinced me of its reality. I see remote viewing as a powerful exploration tool, ideally suited to help us discern the ultimate truth about UFOs and alien life forms.

Alien Agenda zeroes in on the CIA's Stargate program, in particular the RV experiences of three members, David Morehouse, Mel Riley and Lyn Buchanan. The three had many unexpected encounters with UFOs in their psychic travels--known by them as "aerospace anomalies." This encouraged them to further probe Enigma targets. They began to realize that they were finding answers to the UFO mystery where others only had theories--because their remote viewing brought them direct knowledge.

Their focused quest in this realm began with NORAD (North American Aerospace Defense Command), and some of its satellites called DSP's (Deep Space Platform), which orbited more than 500 miles in space. The DSP's used advanced radar technology to monitor missile launches. But the DSP's were not programmed to monitor some of the high-performance craft they detected streaming into earth orbit from deep space. These fast moving anomalous objects were referred to as "fast walkers." When given a photo of a glowing object taken by satellite to use in RV sessions, the team detected that there were humanoid life forms inside the craft, and it was hovering above a nuclear storage facility. Their impression was that the Visitors were taking an inventory of the number of armed warheads at the site; ultimately the psi spies realized that these fast walkers were utilizing off-world technology, and when they traced them back to their points of origin, determined that some of them came from subsurface locations on the Moon, and from Mars, and that they sometimes came to rest in underground locations on Earth. It was also noted that the remote viewers weren't always able to study all of the vehicles they found in their sessions--there were simply too many of them.[3]

Adding further depth and color to the off-world RV sessions: multiple reports of airless planets scattered about our galaxy harboring what was described as relay towers which seemed to catapult spaceships at super-luminal or trans-light speeds from one galactic coordinate to another far distant one, allowing them to bypass the limits of space and time and get places in the wink of an eye. A network of Trans-galactic slingshots--sounds like an advanced alien culture's answer to simplifying and streamlining interstellar galactic travel.[4]

The RV adepts also described a hierarchy of aliens of multiple species and provenance in operations about the Earth, different alien confederations each with myriad agendas; some of them were described as humanoid survivors of a Martian cataclysmic episode far back in time, who lived in underground Martian colonies, and also had hidden underground and undersea bases on Earth. They harbored resentment towards earthlings for so callously raping our planet's biosphere, as they themselves would like to live here amongst us, but seem for now simply to be using the planet for replenishing their stores of raw materials and some food staples, using their spacecraft as shuttles back to their Martian home bases.

Others in the hierarchy were the classic "greys," perceived as highly technologically advanced bio-engineered worker drones from another star system; on the one hand, they altruistically assisted the Martians, perhaps even helped save them, but nevertheless were also the ones who rather systematically abducted humans for medical experiments. Still others were perceived to be non-physical entities, referred to as "Transcendentals" by some of the RV adepts, beings who could be described as angels. Transcendental energy beings were characterized as the ones who were in charge, overseeing and monitoring the operations of the physical aliens, and yet they, too, answered to a higher authority located at a distant point outside our galaxy. In some ways this harkens back to Arthur C. Clarke's sci-fi classic *Childhood's End*, with the alien Overlords physically interacting with the human race, while being beholden to and operating under the aegis of the non-physical Overmind. In the face of the utterly mind blowing enormity of our multidimensional universe and multiverse, and when realizing how bewildering the scale and scope of the Creator's handiwork truly is, I take comfort in embracing this concept. Somehow, it gives me hope that highly evolved Beings of Light might actually be helping the Creator run things. I'm thinking He could probably use the assistance.[5]

* * * * * * *

THE SOURCE:
NASA, THE ASTRONAUTS & UFOS

Moving back to Austin in 1999 made me reflect about growing up in Texas, much of it in Space City, USA. Back then, my biggest heroes were astronauts. They started out as hot shot pilots, then moved on to Houston's Manned Spacecraft Center for full astronaut training with NASA (I was extremely envious).

In 1965, Gordon Cooper was the last American astronaut to brave the perils of space alone, orbiting the earth twenty-two times. Cooper would become one of the most vocal of the astronauts when it came to the extraterrestrials. During a 1978 United Nations meeting to discuss UFOs, a letter Cooper had penned was read; it included a statement testifying to his belief that extraterrestrial craft and their occupants are visiting us from other planets, and are clearly a bit more advanced than we are.[6]

The groundbreaking UFO documentary film *Out Of The Blue* (2002) features a segment wherein Gordon Cooper relates a UFO experience he'd had while stationed at Edward's Air Force Base on May 2, 1957. He was supervising the filming of the installation of a system for precision landings being installed on the dry lake bed for F-86 fighter jets. His team had both still and movie cameras on hand. Suddenly a flying saucer flew directly over the cameramen. Three landing gear apparatus extended, and the craft touched down about 50 yards away. Cooper described it as the typical saucer configuration, double lens shape and metallic; his team picked up their cameras and moved on out toward the landed craft, filming. The saucer then lifted off, put its gear back in the well, and climbed out rapidly and disappeared equally fast. He knew that the U.S. didn't have any vehicles like that. And he was virtually certain the Russians didn't have any craft to compare with it either. He had to look up the regulations on who to call to report it, which he did, and they ordered him to develop the film immediately, put it in a pouch, and send it to them by commanding general's plane to Washington; he followed their protocol, and it was the last he'd ever heard of the film. When asked if he watched the film, Cooper replied, "We didn't have the chance to run it. I had a chance to hold it up to the window and look at it. It was certainly good film." When asked if Cooper ever kept in touch with anybody about the film or discuss it, he replied, "How would I keep in touch with anybody about it? There's no way within the military or within the government to keep track of something that is classified--unless you're directly involved with it, and I was not."[7, 8]

For many years, stories have circulated that our manned space-flight program has persistently experienced UFOs in space, from Gemini to the Apollo and Space Shuttle missions. Perhaps many of the reports were covered up because nearly all the astronauts have been military officers who obey their strict orders to remain silent. But according to transcripts of the technical debriefing after the Apollo 11 mission, astronauts Armstrong, Aldrin and Collins reported an encounter with a large cylindrical UFO even before reaching the Moon. Said Aldrin, "The first unusual thing we saw . . . was pretty close to the Moon. It had a sizable dimension to it."

Otto Binder, ex-NASA space program member, gives a closer look at how the astronauts were muzzled; he accused his former employer of censoring Apollo 11 transmissions (and others) by using radio channels unknown to the public. He attests that unnamed ham radio operators monitored Apollo 11 transmissions after the astronauts touched down in the Sea Of Tranquility and overheard one of the astronauts exclaim, "These babies are huge, sir . . . Enormous. Oh, God, you wouldn't believe it! I'm telling you there are other spacecraft out there . . . lined up on the far side of the crater edge. They're on the Moon watching us." The NASA UFO Archives film I saw at the *Space Science Center* in Santa Fe in the early 1990's certainly makes this conversation completely and utterly credible.[9]

Maurice Chatelain, formerly in charge of designing and building the Apollo communication and data processing system for NASA, attests to a story similar to Binder's. In his 1978 book, *Our Ancestors Came From Outer Space*, he wrote, "When Apollo made the first landing on the Sea Of Tranquility, only moments before Armstrong stepped down the ladder to set foot on the Moon, two UFOs hovered overhead." Additionally, he stated, "The astronauts were not limited to equipment troubles. They saw things during their missions that could not be discussed with anybody outside NASA. It is very difficult to obtain any specific information from NASA, which still exercises a very strict control over any disclosure of these events." Chatelain stated that NASA keeps a tight lid on UFO/alien related information, and even has a list of instructions for agency administrators explaining how to avoid disclosure of sensitive information.[10]

Chatelain further stated in his book that all Apollo and Gemini flights were followed, both at a distance and sometimes very close up, by extraterrestrial spacecraft. The astronauts informed Mission Control about each UFO occurrence, and were instructed to keep the encounters to themselves. Chatelain even maintained that rumors circulated within NASA alleging that Apollo 13 carried a small nuclear device to be set off on the Moon for seismic testing--but barely made it home after being disabled by a UFO ostensibly protecting "some Moon base established by extraterrestrials."[11]

In the late 60's and early 70's, I avidly followed the Apollo missions, buying books about them, and watched the black and white transmissions of our astronauts on the Moon. Who could forget the footage of our NASA boys hitting golf balls, or gallivanting around the dusty lunar surface in the lunar rover? I vividly recall all the buzz in the scientific community that the Moon was going to be the next big frontier, destined to become a base housing low-gravity laboratories for scientific research--even proposing that we'd have breakthroughs in the development of integrated chips in such a low-g environment. So why did the Moon program suddenly come to such a screeching halt after Apollo 17?

Don Ecker, Research Director and columnist for *UFO Magazine*, has stated, "For years, rumors abounded that we were asked to leave early on--by the much speculated "somebody else on the Moon"-- but that would have caused massive questioning and perhaps panic if we suddenly stopped, so we completed the program and then went into hiatus . . . this idea is pure speculation and rumor."[12]

Timothy Good, author of the popular UFO book *Above Top Secret* (1989)--a really great read--contributed to this speculation when he published an unattributed report of a conversation involving astronaut Neil Armstrong. According to Good, an unnamed personal friend overheard Armstrong at a NASA symposium say, "It was incredible . . . the fact is, we were warned off. There was never any question then of a space station or a moon city . . . I can't go into details, except to say that their ships were far superior to ours both in size and technology--boy, were they big! . . . and menacing." Armstrong then allegedly explained that the final Apollo missions were completed because, "NASA was committed at that time and couldn't risk a panic on Earth."[13]   (Again, the classified video I'd seen--several times, several visits--at the *Space Science Center* makes this ring spot-on accurate and true for me. Indubitably.)

As Moon researcher George Leonard noted, "We put enough billions into the U.S. space program to put all the major cities of America out of debt, and then some. And after the successful Ranger and Surveyor and Orbiter and Apollo flights, we dropped lunar exploration like a hot potato."[14]

Throughout the Gemini and Apollo missions, NASA received fairly routine sighting reports from astronauts. On September 13, 1966, Gemini 11 astronaut Richard Gordon and Charles "Pete" Conrad saw something strange on their sixteenth orbit of the Earth. Conrad contacted Ground Control. He reported that he had just seen an unidentified object about 50 miles downrange. It was metallic, and revolving at more than one revolution per second. He pulled out a camera and took three photographs. The UFO dropped down in front of them before it quickly disappeared. As shown in a *UFO FILES (History Channel)* television broadcast, it is quite evident that the clear color photos Conrad took of these glowing objects/vehicles are not our spacecraft. The Gemini 11 encounter is the only UFO case NASA acknowledges to be unidentified. The images and background story can be viewed online at:

mercuryrapids.co.uk/articles/UFOFILESBlackBoxUFOSecrets.htm

As the Gemini program led into the Apollo, astronauts reported more and more UFO sightings. And when NASA initiated the Space Shuttle program in 1981, an unprecedented wave of UFO sightings ensued, including extraordinary audio and video evidence. One controversial recording came from STS-29, the space shuttle Discovery flight of March 13, 1989. Reportedly intercepted by a ham radio operator in Maryland, one can hear the Mission commander, astronaut (and USAF Colonel) John E. Blaha, state, "Houston, this is Discovery. We still have the alien spacecraft under observance." According to Bill Birnes, the publisher of *UFO Magazine*, "This was one of the NASA transmissions that wasn't successfully sanitized back in 1989. 'We're still looking at the alien.' It was plain as day." And yet, NASA denies the very existence of UFOs to the general public. They've been covering up their UFO material from the very beginning.

In fact, two U.S. Presidents were stonewalled by NASA when they approached the agency for UFO information. In October of 1969, then Governor of Georgia Jimmy Carter had a sighting which he filed with NICAP--a sharply outlined self-luminous object, about the size of the Moon, which maneuvered silently before ten witnesses. During Carter's presidential campaign, he promised to make public heretofore hidden government information. In an effort to make good on his promise, President Carter had his scientific advisor, Frank Press, write a letter to NASA on July 21st, 1977, requesting that the space agency begin an investigation into the UFO phenomenon. Apparently, when word of this pending project reached the Air Force, Colonel Charles Send wrote NASA. In the closing of his letter, he stated, "I sincerely hope you are successful in preventing a reopening of UFO investigations."

In December 1977, in a letter addressed to Frank Press, NASA officially declined President Carter's request. It states, "I would therefore propose that NASA take no steps to establish a research activity in this area."

President Clinton didn't fare any better. When he requested that his Associate Attorney General Webster "Webb" Hubbell look into the existence of UFOs, he was unable to make headway. When the *Out Of The Blue* producers contacted Mr. Hubbell in a 1999 phone interview, he told them as much, and suggested that they contact Mr. Clinton. They did contact President Clinton, and by way of reply received a letter from NASA dated March 19, 1999, which stated that, "No government agency is responsible for investigating UFOs, because there is no factual evidence that alien life exists on other planets, or that UFOs are related to aliens."[15]

Having watched the NASA UFO Archives, Volume I multiple times, I see their denial for what it is--a bald faced lie. You might even say that NASA is completely full of crap about this. We're talking Texas-sized bullshit--in a really, really huge way.

And in Austin, Texas, I was finally able to take my conviction about this to the next level.

* * * * * * *

THE NASA INSIDER: CONFIRMATION

And so it was that in the Summer of 2004 I gave an in-house massage to a guest at a local upscale hotel, the *Renaissance Arboretum*. The concierge who contacted me knew of my interest in physics, UFOs and the paranormal, and told me I'd probably find the guy he had booked for me interesting. He was a NASA engineer.

Since I don't recall his real name--and because of what he revealed to me, I wouldn't be able to divulge it anyway--let's call him Dr. Jones. At the appointed time, I knocked on his door, he greeted me, and I set up my table next to his writing desk; it was fairly groaning with a laptop computer and stacks of bound reports, notebooks, and journals. My curiosity being what it is, I couldn't resist glancing at his paperwork while he was face down on the table. A report on the top of the stack bore an official NASA emblem masthead, some sort of retrospective paper about the Gemini space program--wildly coincidental given what we were about to discuss. He was the author of the piece, and there was a Ph.D. after his name. He was in Austin for a symposium.

Our ninety minute massage went well. Dr. Jones remained silent throughout, which is my preference, since silence allows me to get into the flow of the work. At the end, I told him to take a few minutes to relax while I washed my hands in the bathroom. When I returned to the sitting area, he was already up from the table and dressed. He smiled, his eyes a bit glazed--usually a good sign after a long massage. He complimented me, saying that our session had been one of the best deep tissue massages he'd ever had. I thanked him and started packing up my table and gear.

Sensing he was an easy going type, I said, "Say, I hope I'm not being too nosy, but the concierge told me you're an engineer with NASA, and . . . well, I'm interested in space travel, things like that. Would you mind if I asked you a few questions?"

"No, not at all," he said. "I don't have anything going for a few hours. Fire away."

"Well . . . I couldn't help noticing your symposium paper there, with the NASA logo on it. What sort of engineer are you?"

"I have a Ph.D. in Aerospace Engineering and a Master's degree in Electrical Engineering," he said.

"Wow, great combo. So would it be okay with you if we talked for a few minutes about UFOs?" I asked as I folded up my sheets and towels.

He smiled. "Sure," he said. I plunged into a brief overview of my UFO experiences in Houston (but not Tyler), and all the research I'd done, all the people I'd met and interviewed, and all that I'd read in the Jim Marrs book *Alien Agenda* about how we aborted the Apollo Moon program because we couldn't guarantee our astronaut's safety on missions--because of aliens who already had a base there. And to wind up my wide ranging soliloquy, I told him, "And to put a cherry on top of the whole thing, when I was living in Santa Fe, I went to a non-profit UFO museum called the *Space Science Center*. The owner showed me a video that was titled, NASA UFO ARCHIVES, Volume I." Dr. Jones focused a bit more attention on me after I'd said that. "And I found the footage, ummm, illuminating, to say the least. It opens up with a black and white video clip of the Gemini rocket on the launch pad. Smoke and fire erupt from the exhaust nozzles, and the Titan booster lifts off. And when the rocket is barely ten feet off the ground, suddenly a huge dull metallic disc, almost like a giant Frisbee made of pewter, zooms in from out of the blue, and comes zipping down towards the Gemini capsule, and--"

"You saw the flyby," he calmly interrupted with his deadpan statement. He laughed out loud when he saw my shocked reaction. I laughed nervously as well. It took me a moment to center myself and recover my focus. His words were still reverberating in my head when I resumed.

"Well then," I said. "So then we both know of this video footage. Then you probably know that the next scene shows another black and white film clip of what is clearly the LEM on the Moon, and an astronaut moves the camera over to one of the triangular windows. We look out, and we see the lunar surface--and a large glowing pancake shaped disk, glowing white, hovering over a crater's edge, maybe fifty yards away."

"Ahhh, interesting; I wasn't aware that we had any saucer footage on the Moon," he casually commented. His blasé attitude about this floored me. I figured I had a captive audience for a few more minutes, so I ran with it. "So then, what you're telling me is that the whole tamale is true--that Roswell happened, aliens have crashed their vehicles here, and multiple alien races from different species, different star systems--maybe sometimes even different dimensions--are coming here, they've been here for a long time, and NASA and a whole slew of other super secret agencies have known the whole truth about UFOs for fifty plus years--and they're keeping it all from us. And we really did abort the Apollo program because aliens didn't want us interfering with their Moon base. And from that point on, we couldn't guarantee our astronaut's safety if we had continued our lunar explorations. And we've been trying to reverse engineer flying saucers at Area 51." I felt like I was bombarding him with a Gatling gun full of questions and concepts, but I knew I'd never see the man again.

He smiled and paused to gather his thoughts. He put his glasses back on, and scratched his short salt and pepper beard pensively. It struck me that he looked a bit like a young Einstein, before the crazy hair. Presently, he said, "I have a high level security clearance, so you'll have to understand, I'm not at liberty to discuss any specifics with you. But what I can do is . . . let me put it to you like this: everything you've just said here is completely in keeping with what I know to be the truth. And nothing you've said is at all off target or off base. Rather, you've been completely on target with every statement that you just now made."

I shook my head, but not in surprise--nothing my government and its myriad acronym agencies do to us could possibly surprise me anymore, no matter how dark or nefarious the deeds. "But why all the subterfuge? Why are we hiding something so important from the public?" I asked.

"It's the nature of the game," he replied. "All of those whom you might call the Big Players are hiding things from each other--meaning the Army, Navy, Air Force, NSA, DIA, CIA, and yes, even NASA. And obviously private industries and high tech corporations are included too. We're all hiding what we know about this sort of super technology from each other. It's a race. Everybody is trying to develop it ahead of everybody else. There's no sharing of material or of experimental breakthroughs when it comes to this kind of stuff. Look, one thing I can tell you is, right now at NASA we're working on experimental propulsion technologies that would absolutely blow your mind--things that are almost beyond your imagination." Dr. Jones looked at me with a plaintive expression, like he wanted to tell me more, but instead, he stopped there, smiled at me and shrugged. It was sort of like the inscrutable facial expression worn by the tall skinny gray alien at the end of *Close Encounters Of The Third Kind*, as he gazed into the eyes of his human counterparts, standing beneath the behemoth Mother ship just prior to its departure.

I thanked Dr. Jones for his frankness, shook his hand, gathered my gear and left. Driving home, I felt a bit dazed by our exchange. It brought to mind how supposedly even J. Edgar Hoover couldn't get the Army to share the spoils of the crashed saucer at Roswell. An FBI memo penned by Hoover which seems to allude to the crash, salvaged via use of FOIA, and dated 7-15-47, includes the passage, "we must insist upon full access to discs recovered. For instance in the [Sw or La] case the Army grabbed it and would not let us have it for cursory examination . . . ."

A copy of this can be found online at:

www.cufon.org/cufon/foia_001.htm

If not there, simply Google it--Googlebots can find anything. While there's no way I can prove Hoover was talking about Roswell (though I think the 'Sw' was a code for it via the geographic locale, or via 2 adjacent letters in the town's name), it does clarify for me that as far back as the very beginning of the cover up, one can find reference to fiercely guarded insider information, even between high level government and military sources. Like Dr. Jones said, "We're all hiding what we know about this . . . from each other."

* * * * * * * * * *

When I got home from my eye-opening crash course in UFO Cover-Ups 101 with Dr. Jones, the things we'd discussed stayed with me--in particular, that he'd essentially green-lighted everything I threw at him; paramount among them were some of the allegations made in the groundbreaking Jim Marrs *Alien Agenda* book, so I pulled it from my home library shelf and thumbed through its well worn pages.

One of the more fascinating parts of the book relates how, in mid-1996, under the auspices of the American Association Of Remote Viewers, Inc., seven experienced remote viewers with proven performance capabilities were tasked to examine the UFO issue.

Equally fascinating: the UFO related queries were presented to the viewers with numerical coordinates representing the questions.

Regarding the craft, they were perceived to be hundreds, in some cases thousands of years ahead of our technology, and their propulsion systems involved electromagnetic energy and radiant energy (no big surprises there), calling to mind Billy Meier's beautiful Pleiadian mentor Semjase's term for her interstellar saucer: beamship.

Some viewers perceived that our government possesses saucers stored in hidden bases. One viewer even went onboard an alien craft piloted by three greys and two humans who seemed to be soldiers, perceived that they were a crew, and that the humans gradually were learning how to operate the saucer--an alien/human driver's-ed class. During one flight, the viewer went "inside" the grey pilot, allowing him to experience the steering of the vehicle. One of the human soldiers apparently detected the viewer's presence, and told the grey to "let him go." The soldier then entered some input into a device on his lap, and the viewer felt like he was on fire, as though being bitten and stung by a pile of fire ants. The viewer then lost the connection.

An even more telling account is in the form of a supposed 1950's meeting between aliens and President Dwight Eisenhower. The scene: Daytime. A hot, desert scenario is perceived, and an airstrip is present, along with aircraft that sound like jet planes. It is a restricted location, an Air Force base with guard towers. On a blacktop highway comes a convoy of black cars, accompanied by jeeps. Three flying saucers approach from the open sky, and they land, one after the other. The first one opens up, and one of the black cars and a couple of jeeps approach it. Three greys emerge from the craft, a tall one and two smaller ones. From the black sedan emerges a tall, older man, two military men and a civilian. The viewer is startled to realize the older man is President Eisenhower. The men and the greys get into a large black Cadillac and are shuttled to a building nearby (presumably a proud first for the luxury automaker). They are escorted quickly inside by big men in military police attire.

The viewer perceives that this meeting was preplanned. The group convenes in a conference room. The greys express an interest in opening and maintaining communication with us. The civilian asks how this would be done, and the tall grey responds that they'll be given an apparatus for communicating both locally and long distance. The tall one also asks that humanity not fear them. When Eisenhower asks why they're here, the tall grey responds that they have come to assist us with anticipated upcoming changes, and will help speed up our development. One of the smaller greys adds that they're not here to wage war. When Eisenhower asks how long they've been around, the short one replies, for a time longer than our recorded history, and that our forefathers, who were our makers, abandoned us. (Talk about serious childhood abandonment issues!) The tall grey says that they want an understanding between our cultures, but that they weren't seeking an alliance with us. Eisenhower asserts that he wants a friendship forged, and a sense of cooperation with them, and the tall grey confirms that they want the same thing. He adds that they will continue to observe us and watch what we do, how we interact with them, but their intent is for there to be a continuation of ongoing activity here.

Abruptly, the tall grey turns "towards" the remote viewer; apparently he knows the viewer is present. One of the military men perks up and wants to know what's wrong; nothing, says the tall grey, just that they have a visitor. The military man asks how that can be possible, and the small grey says that it's not yet time to explain this. The viewer states that his RV session ended abruptly at this point, as though his psychic presence was being blocked, because there were still things the aliens wished to talk about in private, without the intrusion of a psi spy.[16]

To someone not versed in the ways of the psychic realm, this whole scenario may sound like a sci-fi movie, but it rings true for me. Metaphysical doctrines have long referred to the "akashic record," a so called river of information which exists on a higher vibrational plane of existence, one which contains the entire history of all that has happened in the past, present and future. In the akashic record, all events in this river are as one, and exist in the *eternal now*. Those possessing psychic gifts (such as gifted clairvoyants or remote viewers) can tap into this river to access this information. Therefore, a viewer conducting a remote viewing scan from the mid-1990's (or from any present date) back to the early 1950's is not such a stretch. But what an extraordinarily twisted up pretzel of events ensues, as doubtless other viewers have gone back to this Eisenhower meeting, and still others will continue to go back there. So how many times will the tall grey alien notice that "they have a visitor?" The UFO mystery is inextricably woven into the larger mystery of time.

257

This account wasn't the first time I'd heard allegations that Eisenhower met with aliens, and it also dovetails with the timing of the "saucer invasion" of Washington in July of 1952. Their blatant intrusion into restricted airspace over our nation's capital seems like the aliens were saying, *We mean business. Deal with us.* In fact, the Washington UFO wave seems to mark the beginning of the end with regards to the concept of full disclosure on the part of our government concerning what they knew about UFOs and aliens. Donald Keyhoe had alluded many times (up until that fateful summer) of how he'd been tipped off by the Air Force and various intelligence sources to an approaching time when the lid to the UFO cover-up might be lifted, and people would be told we're not alone. But in fact the CIA (created by the stroke of Truman's pen when he signed into law the National Security Act of 1947, weeks after the Roswell crash) clamped down the lid more firmly than ever in the aftermath of Washington's UFO invasion. The idea of full public disclosure, to the extent that it was ever a possibility, seems to have vanished from that point on. Perhaps in the aftermath of the airspace intrusions that July, Truman wasn't ready to trust our invaders enough to deal; or the timing wasn't right for a meeting. But apparently it was when Eisenhower became president in 1953. Therefore this remotely viewed scene of Eisenhower forging a top security alliance between the aliens and our government makes perfect sense to me. The aliens had forced their hand, we called them, and everybody showed their cards. But who really won the chips in that pot?

The remote viewers in this UFO targeted study from *Alien Agenda* also seemed to agree that there were ongoing "turf wars" over our planet. Some sort of secret conflict was going on around the world in the present day, in which two groups were battling for global dominance, one viewer referring to them as two factions of the same party. They don't see eye to eye about things, with one side feeling responsible for our planet, while the other wants to use, dominate and rule Earth. A larger group of numerous kinds of aliens don't like what's going on, but steer clear of direct involvement. They simply try to keep the peace while preventing these factions from harming us or any other beings. And in one rather eerie and ominous remotely viewed scenario, one of the RV adepts saw a veritable armada of very large ships on the outer fringes of our solar system, poised as though waiting for something or someone to trigger an initiation sequence; the adept referred to the tableau as being a bit reminiscent of "Luke Skywalker and the bad guys."[17]

This last RV session calls to mind "Semjase's Warning to Mankind" to me. Every time I've read it, I've gotten a powerful sense that it was a valid remote viewing session, and that the viewer was hitting the target accurately. And each time I reflect back on what my NASA aerospace engineer, Dr. Jones, said about there being a technology race, I tie that in with a vivid visual of this edge-of-solar-system armada, and it tells me one thing clearly: perhaps the race is not only about each "team" wanting to create the most high performance antigravity craft they can, as though they're in a competition; maybe it's also about protecting the Earth from hostile factions that would like to snatch our planet away from us, and "beat us into bondage." Factions that fit Semjase's eloquent description of those who "live in a barbarism even worse than yours," and that, "more and more, the time approaches, when a conflict with these becomes unavoidable." So perhaps we're beefing up an ultra high tech team of starship troopers to defend us from the barbarians at the gate?

UFOs AND THE SOVEREIGN STATE OF YOUTUBE

It's a given that the internet and the resultant worldwide web have transformed utterly the nature and flow rate of information we share with one another on an unprecedented, global scale. The domain name for the site YouTube was registered on Feb. 14th, 2005-- to me quite fitting, as this remarkable web presence, founded by former PayPal employees Chad Hurley, Steve Chen and Jawed Karim, has become a Valentine's Day gift for all of humanity that gives over and over and over again to all of us Earthlings, every day, 24/7, in a relentless flood of information, sound and imagery (much of it enlightening, informative, entertaining and enjoyable, and some of it, well, not so much). YouTube's growth statistics are staggering, going from a reasonably successful upstart in 2005 to a multi-billion dollar global powerhouse today. On December 21, 2012, *Gangnam Style* became the first, and so far the only, YouTube video to surpass one billion views (don't ask me why), and Lady GaGa's music video *Bad Romance* surpassed the 531 million mark in views as of September 2013 (that, I can understand--she's wonderful, I love her, more power to her, and God bless Ms. Germanotta for her *Born This Way* foundation and its efforts to curtail and end the bullying of gay youth). As of March 21, 2013, the number of unique users visiting YouTube every month reached the 1 billion mark. I therefore refer to this internet powerhouse of information transfer as a sovereign state, by virtue of the definition: a state which administers its own government, and is not dependent upon, or subject to, another power.

In that vein, YouTube is proving itself to be the great equalizer and democratizer when it comes to bringing to light information heretofore hidden and/or classified. With regards to UFOs, a primary case in point can be found at the following web address, where a YouTube video, *Aliens In the NASA Archives--More Stunning NASA UFO Anomalies Captured On Film*, uploaded on June 1, 2009 by the YouTube user LunaCognita, shares some UFO footage filmed by various NASA missions:

http://youtu.be/X-RPWhigpQg

Even more striking than the video images themselves (some of which are most assuredly alien spacecraft in Earth orbit, in my opinion, one of which is at digital marker 7:47 - 7:54), is an embedded audio file of a July 23rd, 2008 radio interview conducted with former Apollo astronaut Dr. Edgar Mitchell, Lunar Module pilot of Apollo 14, to be found at approximately the five minute and thirty second mark, transcribed here:

Interviewer: Do you believe in life on other planets?

Edgar Mitchell: Oh yes, there's not much question at all that there's life throughout the Universe, we're not alone in the Universe.

I: You're convinced that we're not alone in the Universe?

EM: Oh, I know for sure we're not alone in the Universe. And I happen to be privileged enough to have been in on the fact that we have been visited on this planet, and the UFO phenomenon is real, although it's been covered up by governments for quite a long time. It is a real phenomenon. And there's quite a few of us--it's been well covered up by all of our governments for the last sixty years or so, but slowly it's leaked out, and some of us have been privileged to have been briefed on some of it.

But I've also been in military circles and intelligence circles that know beneath the surface of what has been public knowledge, that yes, we have been visited. I have been deeply involved in certain committees and certain research programs with very credible scientists and intelligence people that do know the real inside story, and I am not hesitant to talk about it.

There's quite a bit of contact going on. We have been visited. The Roswell crash was *real.* And a number of other contacts have been real and ongoing. It's pretty well known for those of us who have been briefed and have been close to the subject matter. There are UFOs in the skies all the time that are very likely alien craft--now not all of them are. I suspect some of them are home-grown; I suspect that in the last 60 years or so that there has been some back engineering, and the creation of this type of equipment, that is not nearly as sophisticated, yet, as what the apparent Visitors have.

Back in the late 1980's and early 1990's, when I first started hearing more and more increasingly credible reports about the Roswell crash (which may now include 2 crashed alien vehicles, perhaps having collided with each other in the air on July 4, 1947, before tumbling to the earth roughly 75 miles northwest of Roswell; and yet another crashed vehicle in '48 or '49, also in New Mexico, and at least one of these involving a live alien that survived and was spirited off to Los Alamos where it was kept alive in isolation for a couple more years of information retrieval), one simple fact dawned on me: if Roswell was real (as Dr. Mitchell states in an utterly sober and factual manner in this interview), then the whole game changes in the wink of an eye. I liken this concept mentally to a giant garage floor hosting a spread of upright dominoes, where the Roswell crash(es) are at the central source of the layout--the core dominoes. Spiraling outwards in multiple concentric circles for many feet are other dominoes representing a myriad of other UFO and alien-related encounters. So, for me, Dr. Mitchell's statements have heralded the intitation of the trigger mechanism of the core dominoes, and all of the others in sequence have tumbled in the ultimate chain reaction of information. There's no going back.

As if this audio interview wasn't enough, another, full video interview with Dr. Mitchell was posted on YouTube on February 14, 2013--interestingly, on the 8th anniversary of the groundbreaking web site's founding:

http://www.youtube.com/watch?v=7AAJ34_NMcI

In this video, posted by the YouTube user StephenHannardAD-GUK, Dr. Mitchell further elaborates on the Roswell crash and ensuing cover up, as manufactured during the Truman administration. Here is a partial transcript referring to the time immediately following the Roswell crash of July 4, 1947:

> Because of my openness to these things, I did have many of the old-timers in the military and in the intelligence community over the years wanting to get it off of their chests before they passed away, and they allowed me to interview them and talk to them about it. And so my ideas became fairly well solidified about the fact that we have been visited. We have to remember that, right after World War II, the Army Air Force got separated, and then there was the Air Force, a separate branch of service. And that the

OSS, which was the Office of Special Services, was disbanded, and eventually became the CIA, so that, here was a major military organization, and a major intelligence organization, totally in disarray, newly founded, and didn't know what they were doing after World War II, and weren't really reorganized yet.

And as a result of that, President Truman convened a very high level committee to examine the alien or UFO phenomenon. They did come to the conclusion that it was alien, and the military rightly came to the conclusion that if they're hostile, there's nothing we can do about it. Therefore, their choice was to deny it and to hush it up, and to create the National Security Act of 1947, which validated that deception, and covered it up, and allowed the group to go underground, as it were. And we've been living with that now for 50 years. It was really the beginning of the whole cover up, the entire denial of this phenomenon. And the addition of dismissal, disinformation, misinformation to cloak and to discourage investigation, to misinform--it's just been continuous for many, many years now.

Eventually it came from the fear, I believe, of not being able to protect and do their duty, to the notion of power and control--controlling the knowledge and the technology. And the group involved is still doing it. We have created our reality here, and we have created it right now rather badly, it's not a sustainable reality. We have created it with our science and technology, instead of using it for the greater good, it's been captured by greed and self service, which is rife, and instead of using all of our technology and our brilliance and genius for the greater good, we use it for self service. And that's not gonna work. It's important that we look at our civilization, our place in history, use our tools of science for greater understanding to promote the greater good, and that's what it's all about.

High technology captured by greed and used by the few, rather than shared with the many. Kudos to Dr. Mitchell for his concise assessment: the reality of what the Billionaire Boys Club has done with back engineered, high-tech saucer technology. Bull's eye.

THE PHOENIX LIGHTS: LARGER THAN LIFE

*They have no pattern for contacting people. It is by pure
chance so the government cannot determine any patterns
about them . . . to a certain extent, they want to puzzle
people . . . and they are trying to confuse the public's mind.
He is telling me they want everyone to believe some in them so
we will be open to their invasion. . . .*

**--Herbert Schirmer, Nebraska police officer
from his 1967 hypnotic regression[18]**

A prime example of the ongoing UFO cover up and utilization of blanket denials (to the extent of utterly ignoring the incredibly widespread and abundantly documented extremely close up sightings) by the U.S. government, the Air Force, local officials and the media surfaced in the late 1990's. On the night of March 13th, 1997, strange lights and mysterious objects were seen by hundreds, and possibly thousands of witnesses throughout the Phoenix, Arizona area, perhaps in part because so many people were outside to observe the Hale-Bopp Comet. The mysterious sightings, ultimately grouped under the rubric The Phoenix Lights, weren't widely covered by the national network news shows. One eyewitness described the object as appearing like, "five lights in a 'V', one in front and two on each side . . . a perfect triangle."

Months later, the June 18th, 1997 *USA Today* front page feature story of the mysterious Phoenix incident brought national coverage, and some TV news reports; I can still hear Tom Brokaw's NBC broadcast, his folksy baritone voice proclaiming, "Something strange in the night skies over Phoenix. . . ." with the now famous image of the boomerang shaped lights affixed in the black Phoenix sky as a backdrop to his introduction. (These, in fact, were almost certainly the signal flares the National Guard reportedly had dropped in an aerial training exercise around 10pm that night, conveniently not explaining the giant, illuminated boomerang sightings that had already come and gone long before the flares were dropped--but certainly helping to muddy the waters when the video footage analysis of the 10pm lights indicated they were signal flares. Can you say, obfuscation?) For me, the solid reports of the real thing evoked the same feeling of disorientation as the black manta ray vehicle in William Cooper's video--times ten. Something so huge isn't supposed to float weightlessly, or cruise by silently at barely 30 miles per hour, or silently dart off in the wink of an eye. An ominous aura pervaded here; a sense that unchallenged technological superiority was biding its time, quietly scouting around. No need for hasty exits when nobody is telling you to go, let alone high ranking authorities even acknowledging that you're there. This hovering, drifting enigma evidently looked like it could house scores of Cooper's manta ray gliders in its cargo hold. Its black silence in the dark sky seemed to pose the question, *is this the next phase of an invasion?*

As reported in the Sci-Fi Channel's UFO special, *Out Of The Blue*[19], UFO researcher Michael Tanner received about 800 reports regarding the Phoenix Lights that night. Between 8:15pm and 8:45pm, reports came in of enormous craft at varied altitudes, silently making their way through the center of the valley of Phoenix, led by a formation of glowing balls of light. This scenario was clearly visible through much of the state. Witnesses reported seeing an enormous triangular or boomerang shaped aircraft; one estimated the size of the craft from the nose to the end of the left wing where it passed him as being over 5,000 feet long; as it moved across the sky, it would block out and unblock stars; it was reported being a bluish-black color, gun metal black, and completely silent the entire time. One witness reported that when looking up "into it" as it passed overhead, that it had "distortion waves," like the heat waves you'd see rising from the streets on a hot day in Phoenix. Another said, "the object we saw, if we opened up a newspaper (holding hands up overhead) you could not block out the object that we saw." Said witness Mike Fortsan, "People say, 'Mike, no, you saw a B-2 bomber;' my response was we could land all 40 of our B-2 bombers on the wing of that craft."

Googling the topic, a wealth of information emerges. The *USA Today* newspaper article dated June 18, 1997 states that witnesses of the event of March 13, 1997 generally agreed on three points:

1) The object was enormous. The most conservative estimate was 1,000 feet wide; computer analysis of videotape footage taken that night places the figure closer to 6,000 feet--more than one mile in size.

2) The object made no sound.
3) By the time the object arrived in Phoenix, it was moving slowly--no more than 30 mph, and sometimes it even stopped.

The article continued that air traffic controllers at Phoenix's Sky Harbor International Airport saw the lights with their naked eyes, but nothing appeared on their radar screens. "Weird, inexplicable," said Bill Grava, one of the controllers on duty that night. "I still don't know what to think, and I have no idea what it was."

Dana Valentine of Phoenix and his father, an aeronautics engineer, saw the object fly directly overhead at an estimated altitude of 500 feet. "We could see the outline of a mass behind the lights, a gray distortion of the night sky." Tim Ley, a neighbor down the road added, "It was so big and so strange. You couldn't actually see the object. All you could see was the outline, as though something was blocking the stars."

Michael Tanner and Jim Dilettoso, owners of Village Labs, a computer firm in Tempe that has designed special effects for Hollywood, ran an analysis of videos shot by amateur observers. The lights in the sky were perfectly uniform, with no variation from one edge to the other and no glow. The two men ruled out lasers, flares, holograms and aircraft lights as sources, but had no idea what they actually were. Tanner interviewed witnesses to determine a chronology of events. His present view: There were four objects, including the V formation. All came from the north at the same time, and after being seen by thousands of people for nearly two full hours, left the way they came.[20]

A reporter on the local Channel 15 news cast on March 14th stated, "Explanations have been tossed about, that they were flares, that they were planes flying in formation. But we checked with the FAA today, we checked with Sky Harbor, and we checked with Luke Air Force Base, and there's been no official explanation of those strange bright lights last night."

A truck driver named Bill Greiner was driving south on I-17 towards Luke Air Force Base; he witnessed two fighter jets scrambled to intercept two UFOs--glowing orbs of light--directly overhead. He'd seen the two huge orbs silently hovering over the base, and suddenly heard the roar of the fighter jets, their afterburners on, flames coming out the back. The jets shot straight up towards the orbs, and the UFOs disappeared in a flash.

Peter Davenport is the Director of the *National UFO Reporting*

Center, based in Seattle, Washington. He was featured in the late Peter Jennings's ABC Special on February 24th, 2005, *UFOs: Seeing Is Believing*. Mr. Davenport oversees his organization's web site (*nuforc.org*), an impressive forum dedicated to the collection and dissemination of objective UFO data. Regarding The Phoenix Lights, Mr. Davenport sounds off on the *ufoevidence.org* web site, and here is a summary of his findings, titled "2nd Anniversary of 'Phoenix Lights' Incident:"

> Possibly thousands, or tens of thousands, of witnesses on the ground saw at least one object pass and/or hover overhead; they described it as being huge, gigantic, or unimaginably large; most witnesses said the object was triangular in shape, with anywhere from five, to "many, innumerable," lights on the leading edge of the object; it was capable of very rapid flight, probably even supersonic flight, although few witnesses reported any sound emanating from it. Coming in from the vicinity of Henderson, Nevada, the object then appeared over Phoenix, where it is reported to have hovered for 4-5 minutes in the vicinity of the intersection of Indian School Road and 7th Avenue.

> Approximately seven hours after the incident, a person identifying himself as an airman stationed at Luke Air Force Base (20 miles west of Phoenix), phoned the *National UFO Reporting Center*. He stated that the U. S. Air Force had launched two F 15c fighter aircraft from Luke AFB, and that one of the aircraft had "intercepted" a gigantic object over the intersection of Indian School Road and 7th Avenue. It was also reported by this individual that the onboard radar of the intercepting fighter had suddenly gone to a condition of "white noise," and that the lights on the anomalous object simultaneously had suddenly dimmed in unison and disappeared from the pilot's sight. (It should be noted that nuforc.org was unable to corroborate all the details of this "intercept" report.)

According to Davenport, on March 14, 1997, senior officers from Luke Air Force Base stated they knew nothing about the incident, and the base had received no reports from the public regarding the event. Long distance telephone bills, which indicate calls to the base, contradict the statements made by these officers. Furthermore, a witness who had phoned in his report to the *nuforc.org* hotline stated that he had called both Prescott Airport and Luke AFB to report the sighting. The female operator at Luke AFB volunteered to him that their switchboard had been deluged with reports about the strange object. Later statements from Luke AFB maintained that they had received no calls about it.

Some military personnel and UFO investigators claimed that the entire event was caused by "military flares" that had been released at approximately 2130 hrs. (Mountain time) by a flight of Air Force A 10 aircraft in the general vicinity of the Gila Bend Bombing Range. The bombing facility is located approximately 60-80 miles to the southwest of Phoenix. This event, if in fact it did take place, occurred at approximately 2130 or 2200 hrs., some 45 minutes after the UFO sighting had already occurred and moved on, over northern Arizona, Phoenix, and Tucson.[21]

UFO researchers quickly countered that the flares explanation did nothing to explain the sightings all over the Nevada and Arizona areas. How could flares--in a fixed formation, no less--have been seen traveling across the state by so many people who steadfastly maintained that the gigantic object, which blocked out the stars, had glided directly over their heads?

Mike Fortsan of Chandler, Arizona disputed the Air Force's flares explanation. Fortsan stated that he saw the object fly directly overhead and that he plainly saw a "solid object." He claimed that a 737 disappeared as it passed above the object. Mike noted that the object was over a mile wide and flew about 30-35mph at an approximate altitude of 1,500 to 2,000 feet.

According to the *Out Of The Blue* special, a reporter complained to Phoenix city councilwoman Frances Barwood about the lack of response from the local government. Barwood states, "What this reporter said was that they had gone to every level of government including the city of Phoenix, and nobody would talk to them . . . that the object went from north of Prescott and all the way to Tucson . . . and nobody would give them any answer; they either told them they were not going to talk about it at all, or that there was nothing, that nothing happened."

Responding to pressure from his constituents, Governor Fife Symington held a press conference on June 19, 1997. He stated, "I'm going to order a full investigation of this through the DPS. We're going to make all the necessary inquiries, and we're going to get to the bottom of it. We're going to find out if it was a UFO. . . ." Later that day, the governor held an unscheduled press conference to announce that he'd discovered the culprit of The Phoenix Lights. Standing at the podium before the cameras, he stated, "And now I'll ask Officer Stein and his colleagues to escort the accused into the room so that we may all look upon the guilty party." Pan to the doorway, where two men in suits and sunglasses escorted a tall, silver space suited "alien" into the room, its over-sized head and large black almond-shaped eyes facing the media. As the gathered crowd laughed, the "alien" clutched its hands to its face, in an "oh my!" pose. Symington chuckled and said, "Now this goes to show that you guys are entirely too serious."

Not everyone was amused. Geraldine Greiner stated, "I think pretty much everybody laughed at it. I thought it was really disgusting. He was just dismissing everything that these people had seen, and said, just like, we're all looney tunes. I just thought that was a really improper attitude for somebody in his position."

Another witness to the UFO incident said, "Nobody in state, local, Federal [government] or military wanted to talk about it. They didn't want to interview witnesses, they said it didn't happen. They said it didn't concern them. There was one councilwoman, Frances Barwood--she's the one who spoke up and said why don't we investigate? And then when it came time for her election she lost. The press, the media, everybody else was really trashing her badly."

Barwood closed the segment with, "Why is everybody so afraid? And the only thing I can think of is, one, they're afraid for their careers--which comes to a point of what's more important, your soul or your job? And I guess to them it's their job. And the second thing is the scorn factor."[22]

Peter Davenport's assessment was that the witnesses who reported sightings of the object(s) over northern Arizona and Phoenix that night included architects, physicians, law enforcement officers, educators, attorneys, airline pilots, scientists, real estate brokers, and other seemingly reliable citizens. He affirmed that many of them were qualified observers who reported their sightings to *nuforc.org* in eloquent, written form.

The citizens of Arizona were justifiably upset about how blatantly they'd been blindsided by the media, the government--and governor--and the Air Force. Just more of the same denial and derision m.o. from the fifties, sixties, seventies and eighties. Deny or ignore everything, proffer feeble, inadequate cover stories (signal flares that hover in perfectly stationary position, maintaining a sharp boomerang outline even while slowly cruising over the city, blocking out a large expanse of stars--almost too rich for words) and ride it out.

A full decade later, in March of 2007, former Governor Symington came forward and spoke with CNN reporter Steve Tuchman, acknowledging that he'd seen the huge object also. He took him to a Phoenix park, and pointing out at the mountainside, said, "If you'd been here ten years ago and standing out here and looking out there at the lights and the view, you would have been astounded. You would have been amazed." Symington was alluding to the gargantuan object whose string of bright lights defined a boomerang or arrowhead shape, the silent and larger-than-multiple-football-fields, mile wide behemoth referred to as the Phoenix Lights. When asked why he didn't come forward and say something back then, he explained that he did tell his family, friends and staff about what he saw early on, but kept mum to avoid widespread panic. "I think as a public figure you have to be very careful about what you say, because people can have pretty emotional reactions, and I said my goal wasn't to try to stir the pot."

Symington's sighting was in Sumida Park, where a crowd of people were out enjoying the Phoenix evening. When the giant object appeared out of the northwest, he described it as "this great big massive craft . . . it took out a whole chunk of the sky. And you could see other aircraft in the distance. But airplanes looked like little toothpicks compared to the size of this craft. And it came right at us . . . not a sound. And it just moved, sort of gently and smoothly, not that fast--it just glided over Squaw Peak. And we had a magnificent view of it silhouetted." Governor Symington observed the object for nearly a minute before it sped away across the Valley. "It just all of a sudden disappeared," he said. "I've served in the U.S. Air Force, and I've been flying all my adult life, and I've never seen anything like that before. It obviously had unbelievable aeronautical capabilities, something way beyond anything we have in our inventory, that's for sure . . . the lights were really brilliant. And it was just fascinating. I mean it was enormous. It just felt otherworldly. In your gut you could tell it was otherworldly." As for his assessment of the sighting, Symington commented, "I suspect that unless the Defense Department proves us otherwise, that it was probably some form of alien spacecraft."

UFO STANDARD BEARERS:
RICHARD DOLAN & LESLIE KEAN

It is here that the amazing works of two UFO book authors, Leslie Kean and Richard M. Dolan, become quite germane to the topic of cover-ups and denials, the Phoenix Lights being a late 1990's prime example of our government and military applying the Triple-D Modus Operandi of Disinformation, Derision and Denial.

It was a pleasure to discover and read Richard M. Dolan's seminal work, *UFOs and the National Security State, Vol. I* (2002)--and the subsequent Vol. II is equally impressive. In this exhaustively researched book, Mr. Dolan deftly intertwines the UFO phenomenon, the military-industrial complex, and the machinations of the National Security apparatus. From the Foreword, penned by legendary UFO researcher Jacques Vallee:

> In the last fifty years the various branches of the military and intelligence community in the United States have so clouded the reports of the UFO phenomenon that the citizenry has been left not just uninformed but indeed misinformed. This may not have been the intent, but it is indeed the result. Those who truly care about democracy are justified in asking that the government come clean about what it knows and--most importantly perhaps-- what it doesn't know about a phenomenon of such far reaching consequences for our science and our society.[23]

Vallee veers off into his "some might be real spacecraft, while others might be multidimensional phenomena" argument here, but the thrust of his foreword is precisely what I perceive the crux of the problem in our society to be: that The Big Lie has metastasized, and not only are we kept out of the loop about UFOs--to say nothing of the no-doubt miraculous back engineered alien technology and its myriad quality of life enhancing wonders--but along the way, the terminal cynicism of those who invincibly keep and guard the saucer secrets has poisoned all aspects of our "democratic" government, down to its very core and roots. (Certainly one can harbor a healthy skepticism that America even has a democracy anymore: pre-emptive military strikes against an oil rich foreign land, missing

WMD's, a corporate war of aggression facilitated by a borderline mobster vice president whose "former" company cleaned up to the tune of tens of billions of dollars in profit via his delivery to them of the Iraqi war theater based upon falsified information, CIA extraordinary rendition flights to foreign torture centers, a military-industrial complex that has gone completely mad with power, and an NSA that every day is revealed to be keeping an ever more watchful eye on virtually all of its own citizens, never mind persons of interest--the list is legion, but that's a good start.)

As Dolan's book points out, 1947 was a banner year for UFOs: flying saucer sightings were popping up all over the country, with the Roswell crash aftermath bringing home the gravity of the situation. Amidst this frenzy of activity, our government passed the National Security Act on July 26, 1947, creating a unified National Military Establishment (NME), a National Security Council (NSC) and a Central Intelligence Agency (CIA). The Act represented the most comprehensive overhaul of our nation's national security apparatus imaginable.

The Act established the CIA, but prohibited it from any spying operations in the United States. But a future Director Of Central Intelligence, Allen Dulles, wrote a loophole into it, enabling the CIA to "perform other functions and duties related to intelligence affecting the national security as the National Security Council may from time to time direct." Bottom line: the agency soon had a virtual blank check to carry out any mission it desired, wherever and whenever, without Congressional oversight, and even, when necessary, without the consent of the president. Funding came from Congress, contributions from wealthy citizens, and myriad private ventures, many of which were highly nefarious, to say nothing of immoral or illegal.[24]

As Dolan states in the Sci-Fi channel special *Out Of The Blue*, "The National Security Act of 1947 allowed for the creation of a state within a state." He asserted that the Act created an intelligence infrastructure capable of maintaining a very high level of secrecy, "one with an enormous amount of power, an enormous amount of wealth, an enormous amount of secrecy and latitude of action. . . ."

Barely a dozen years after the enactment of the National Security Act, it seems the private cabal that had seized control of the UFO issue in its entirety in the late 1940's--regardless of whether you believe in Majestic-12 or not--had by then fully secured a stranglehold on all things UFO, quietly, behind the scenes, unfettered. On January 17th, 1961, Dwight Eisenhower's farewell address to the nation now seems like more of a eulogy for our democracy than a warning; he soberly warns us about the serpentine relationship between the U.S. military and the industrial sector which gladly supplies the military with weaponry and equipment (and which today uses skilled lobbyists with extremely deep pockets to purchase the best congressmen and senators corporate money can buy). Eisenhower worried that the military industrial complex would become too powerful if its influence went unchecked:

> *In the councils of government, we must guard against the acquisition of unwarranted influence, whether sought or unsought, by the military industrial complex. The potential for the disastrous rise of misplaced power exists, and will persist. We must never let the weight of this combination endanger our liberties, or democratic process.*[25]

Brigadier General Stephen Lovekin (now reserve) served under President Eisenhower in the White House Army Signaling Agency. He recalls Eisenhower's concern with this complex issue: "What I think he was telling us is that the military industrial complex will stick you in the back if you are not totally vigilant. And I think he felt like he had not been vigilant; I think he felt like he trusted too many people. He realized that he was losing control. He realized that this [UFO] phenomenon, whatever it was we were faced with, was not going to be in the best hands. As far as I can remember, that was the expression that was used--it's not going to be in the best hands."

This sentiment is echoed by other men well aware of the inside information. As Edgar Mitchell, Ph.D, Apollo 14 astronaut, comments to an interviewer, "Whatever activity is going on, to the extent that it is a clandestine group, a quasi-government group, a quasi-private group, it is without any type--as far as I can tell--of high level government oversight. And that is a great concern."

And Stanton Friedman relates, "The thing that you have to remember is, that absolute power corrupts absolutely. When you've got the knowledge, and the budget, and nobody looking over your shoulder, there's a great temptation to think that you know best-- to heck with the public, they don't know enough to have a sensible opinion. That's unfortunate."[26]

In 1989, Whitley Strieber published the novel *Majestic*, proffering a fictionalized tale of the aftermath of the Roswell crash of 1947, the retrieval of the wreckage and the alien bodies. Typical of his work, it is a beautifully written and sometimes poetic account of some of the things that may have happened that fateful year, of decisions made by our president and his MJ-12 advisors, and how their actions forever changed the structure of our government and the nature of our democracy--assuming you still believe a democracy exists anymore.

In the Foreword of the book, we are introduced to the central fictional character, Wilfred (Will) Stone, one of the core members who took part in the UFO cover up. He reflects upon what he and his cohorts have done:

> I was among the architects of one of the worst mistakes that has ever been made. . . . We who fought World War Two and the Communist menace have only one legacy beyond the armed and furious world we have given you.
>
> In 1947 somebody from outside this world attempted to form a relationship with mankind. First contact fell to the United States government. Fresh from victory and full of pride, our generation failed the test. We made a horrible mess of it. We did not understand the subtle and terrifying--the magnificent--thing that they were . . . they thus represent absolute and total change, the collapse of economic civilization and the end of days. They are freedom; the soul in the open sky. Because they stand for such radical change we in the government saw them as a threat to the United States.
>
> Instead of proclaiming their arrival up and down the land as we should have, I participated with a group of men who hid it behind a curtain of denial and ridicule. We posted guardians at the gates and spread a net of rumors and lies to protect our secret knowledge.

Near the end of the narrative, we see inside a White House cabinet meeting attended by Will Stone, in which an angry and exasperated Truman, surrounded by his MJ-12 advisors, is heatedly demanding feedback and answers to determine what course of action to take. Truman fears there could be an alien invasion coming, and is inclined to go on the offensive. Some of his advisors suggest he wait for further developments, while one of the hawks counsels that we have to show that we're in control.

Truman makes his decision, that every U.S. military base be alerted that they are to intercept and shoot down the saucers on sight. The meeting ends, and one of the attendees concludes that everything would be handled militarily for the time being, until control of our airspace is regained. Will Stone reflects that regaining control of our skies was never going to happen--and even that of our minds; that "we had just made a catastrophic mistake, and were lost."[27] The bittersweet ending of this beautiful book (I've read it multiple times, like a lot of Mr. Strieber's other works) basically sums it up with the simple concept: *We blew it.* No arguments from me.

Richard Dolan's *UFOs and the National Security State, Vol. I* hammers home a compelling argument in the same vein as Strieber's Majestic. He also pays homage to Donald Keyhoe's invaluable, tireless efforts to pry open the closed doors in the field. Keyhoe's ties to military and intelligence sources were so strong that Capt. Edward J. Ruppelt (as aforementioned, formerly in charge of Project Blue Book) confided in him that the CIA was behind the Air Force's habit of debunking UFO sightings. He told Keyhoe, "We're ordered to hide sightings whenever possible, and ridicule the witness. We even have to discredit our own pilots."

About Keyhoe, Dolan writes:

A few words about Donald Keyhoe are in order. There is no question that Keyhoe was the most important UFO researcher, ever. Only James McDonald came close, but even McDonald's impact fell short when compared to that of the Major. It is not simply that Keyhoe wrote five books, along with various articles, about flying saucer cover-ups. It is not simply that he was the driving force behind the world's most important civilian organization [NICAP]. It is that Keyhoe, nearly by himself, opened up the field of UFO research and made it possible for others to follow. . . . In 1952, the year that UFOs seemingly engulfed America, Keyhoe was there, writing about events and forcing out information.

Despite Keyhoe, of course, it is within the national security apparatus, and not among civilians, where most of the pieces of the puzzle exist. UFOs have national security implications, if for no other reason than that they have involved military personnel of many nations. The subject is therefore subject to secrecy protocols, a situation that has existed for over fifty years, and is unlikely to end any time soon. During the period under review in this volume [1941 - 1973], the military struggled to submerge its involvement with the UFO problem. The existence of Project Blue Book until 1969, however, had prevented this from being complete, and NICAP used

Blue Book, with some success, as a kind of wedge to obtain more information. That wedge was gone after 1969, but for a period of about ten years
(1975 to 1985), the Freedom Of Information Act also provided an effective
tool to get at the UFO problem. Unfortunately, changes during the Reagan
era have since limited its usefulness, and the military dimension to the UFO
problem remains locked away within the classified world.

Some believe this is as it ought to be. Can the public really handle the truth
about aliens? If the presence of others constitutes a threat to humanity, for
example, what could the average person even do about it? There are those
who believe that secrecy about UFOs is in the public's best interest. Whatever the value of this sentiment--which I do not share--the 'public interest'
has never been the main concern of those making the decisions. Ultimately, *a national security apparatus exists not to protect the public, but itself.*
The attachment of Americans to the fiction of a representative government,
or--God forbid--a democracy, has clouded their ability to see their society
for what it is: *an oligarchy that uses the forms of democracy to appease
and distract the public*. [all italics mine--wv] Whether or not this is the best
solution to organizing millions of people into a body politic, it remains folly
to imagine that an oligarchy is not concerned with maintaining its position,
to the exclusion of all else.

If we accept the reality of an alien presence, as the UFO evidence suggests, we must be willing to consider that presence as a threat. The record
of military encounters with UFOs suggests that this is the case. Since the
public is completely unprepared to meet this threat, one can only hope that
those groups which have been dealing with it will act in the public interest.
During the period under review in this study, those groups did not always
work in the public interest when it came to other matters. There is little
reason to believe it was, or is, any different regarding an alien presence.

Since the 1970's, the subject of UFOs has become more complex. Encounters are as widespread as ever, and even more plentiful than in the early
years. At the same time, UFOs have received a thoroughly schizophrenic
cultural treatment. Within popular culture, UFOs and aliens possess a
cache they never had during the cold war. Yet, the bastions of 'official
culture'--academia, mainstream media, government, and the elements of
national security--continue to ignore the subject or treat it as a joke. One
can plainly see that neither *ABC Nightly News* nor the *American Historical
Review* deems the subject worthy of analysis.

Among organizations studying UFOs, the situation is one of extreme
division, far more so than in the simple days of NICAP, APRO, and Blue

Book. Throughout most of the century's last three decades, very little effort was expended by the larger organizations either to end government UFO secrecy or, it appears, to reach the public with a coherent message. Instead, they have spent their efforts squirreling away huge amounts of data for . . . who knows what end?

In addition to these, there now exists organizations that serve in a kind of professional debunking capacity. The military and intelligence community continue to show myriad connections with UFO organizations, and several instances of UFO disinformation planted by intelligence personnel are known. The result has been three decades of fragmentation and perennial wheel-spinning. How can one make sense out of such confusion?[28]

How, indeed? In fact it seems today that most people, especially in America, are too busy being consumerist zombies, upstanding citizens of Hypnos who are too zonked out on the hypnotronic bubble we live under and inside of to bother wondering about the biggest issue of all time, UFOs and the alien presence. As Dolan reasons,

The easy thing to do with the UFO evidence is to ignore it. Much harder is to confront it, study it, and ask, 'Just what does this mean?' If we look at the evidence with no prior positions, no expectations of what are the limits of the possible--if we are purely empirical about this matter--then we can easily conclude that alien visitation is the most probable explanation. Others have found us, or our world, and continue to find this world of interest.

Why should this be so difficult to understand? In fact, it is not. No one doubts our own ability, perhaps soon, to find another planet somewhere that supports life. Nor do most scientists doubt the existence of intelligent life elsewhere. Despite the supposedly impenetrable nature of interstellar space, there are people currently working on breakthroughs in propulsion technology, and even a few qualified scientists who believe that the speed of light may not be the ultimate barrier.

Could it be that others have found us? I believe they have. How they arrived, I do not know, but I can speculate that they may not be able simply to walk about on our planet's surface and that they may have good reasons for making themselves scarce. I can speculate that others might find the Earth's immense resources and biodiversity to be of great value, something that, despite the possibilities of life in this universe, may yet be special in important ways. These others, whatever they are, could well be genetically engineered, or even to some extent artificially intelligent. They may not, after all, be natural biological organisms. Looking into the next fifty years of our own future, it is not difficult to imagine things in this way.[29]

Dolan's words, "The easy thing to do with the UFO evidence is to ignore it. Much harder is to confront it, study it, and ask, 'Just what does this mean?',", form a natural segue to one of the most remarkable books in the genre of UFO research and exposition I've seen emerge in many years, Leslie Kean's *UFOs: Generals, Pilots And Government Officials Go On The Record* (2010) . The culmination of ten years of research, Ms. Kean's marvelous opus definitely lives up to the front cover blurb from noted theoretical physicist Dr. Michio Kaku, in part, "a treasure trove of insightful and eye-opening information." I didn't simply read this book, I devoured it. It's a fantastic piece of work, meriting multiple readings.

Confronting the UFO phenomenon head on, pondering its conundrums and mysteries, and relentlessly posing the question, *just what does this mean?*, would be a good way to describe the process that seems to be operating throughout Kean's lucid and incisively reasoned narrative. I first learned of her and her book through a YouTube video in which she expounded on the implications and importance of Fyffe Symington's UFO "coming out" event in 2007, on the tenth anniversary of the Phoenix Lights encounters. Just in those few minutes of screen time, I realized that the way her mind works meant that I'd probably enjoy reading what she was able to make of the perplexing UFO subject. When her book debuted, I found I was correct, it represents investigative journalism at its finest, and Michio Kaku spoke the truth in singing its praises. It now rests on one of the shelves of my home UFO library, alongside all the other works cited here.

Leslie Kean's book zeroes in on the UFO phenomenon as though she had patiently, systematically dissected it into discrete slides, placed each section under a powerful microscope, observed closely and dispassionately, took notes, and reported her findings without bias. Of particular interest and impact to me was Chapter 11, *The Roots Of Debunking In America*, detailing how the UFO cover up escalated following the July 1952 Washington, D.C. saucer invasion, essentially giving birth to the intelligence community's debunking modus operandi which I've coined The Triple D's: Disinformation, Derision and Denial.

In the wake of that historical press conference in which General

Samford spoon fed the public the idiotic explanation of temperature inversion--in spite of seasoned pilots and radar men disputing it--to dismissively negate the presence of the incredible high performance glowing objects buzzing the capital and registering solid radar returns on properly functioning scopes (disinformation & denial), I was unaware until reading Ms. Kean's book that later that year, H. Marshall Chadwell, assistant director of scientific intelligence for the CIA, sent 2 memos addressing their concerns regarding ongoing aerial intrusions, to the Director Of Central Intelligence (DCI). One stated:

> Sightings of unexplained objects at great altitudes and traveling at high speeds in the vicinity of major U.S. defense installations are of such nature that they are not attributable to natural phenomena or known types of aerial vehicles.

The second Chadwell memo that year, titled "Flying Saucers," addressed the issue that the DCI must be given the power to begin conducting whatever research was necessary "to solve the problem of instant positive identification of unidentified flying objects." The CIA understood that they needed a "national policy" regarding "what should be told the public regarding the phenomenon, in order to minimize the risk of panic," as per government documents. It was then determined that the DCI would "enlist the services of selected scientists to review and appraise available evidence." As a result of this decision, the CIA set up a crucial meeting that permanently changed the nature of media coverage and the official attitude toward the subject of UFOs. The results of this meeting help explain the ever present disengagement of American officials during the ensuing decades.[30]

Thus came the inception of the Robertson Panel in January of 1953, chaired by H.P. Robertson, a Caltech physics and weapons specialist, who oversaw a handpicked advisory panel. Their primary concern was that the proliferation of UFO reports could clog communications channels to the point that false alarms might dangerously confuse the signals, and defense agencies might not be able to distinguish a bona fide attack, should the Russians decide to use a faked UFO Wave as a smokescreen in advance of a first strike.

The Panel then decided to use the debunking Triple-D's as a means to dull public interest in UFOs by utilizing the mass media tools of movies, television, and newspaper and magazine articles. The CIA took it a step further, partly by having their intelligence operatives keep the real facts from respected researchers through disinformation, and by actively infiltrating civilian UFO interest groups to further engage the Triple-D technique. Between the two, the CIA could invisibly spin and guide public perception, drastically reducing public interest; simply put, who really wants to be ridiculed or have their intelligence quotient questioned when it comes to something as silly as flying saucers?

By then already having been with the Air Force's beleaguered Project Blue Book for a few years, astronomer J. Allen Hynek sat in on most of the Panel meetings, discovering that their apparently pre-set game plan was for the best UFO evidence to be virtually ignored. Said he, "The implication in the Panel Report was that UFOs were a nonsense (non-science) matter, to be debunked at all costs. It made the subject of UFOs scientifically unrespectable."[31]

Hynek was certainly no stranger to this debunking Triple-D mindset, as he would eventually oversee for nearly twenty years (1951 - 1969) the sorrowful sham that was Project Blue Book, himself partaking in the CIA's very same underhanded tactics, with its revolving door staff of low-level officers and yokels and their ever increasingly ridiculous explanations for sightings, even the extremely solid ones that deserved much better. The Michigan sightings of 1966 and Hynek's shameful swamp gas theory come to mind. (It should be noted that in his 1972 book, *The UFO Experience: A Scientific Inquiry*, he does somewhat redeem himself by owning up to his role in the Triple-D debunking, but didn't want to fight with the CIA and Air Force all those years, lest he lose access to the Blue Book files altogether. I found it easy to forgive him; he was in over his head and swimming with the intelligence sharks. He obviously had made a full 180 degree turnabout on his ETH stance--as witnessed by his comments on the 1973 Hickson/Parker abduction in Pascagoula--and founded the Center For UFO Studies, long before his death in 1986.)

In the backwash of public outrage about how the Michigan UFO wave and its many solid sightings were handled, a series of congressional hearings on UFOs were held, and it was decided that the Air Force would find a university willing to conduct an independent scientific study of the phenomenon beyond the purview of Blue Book. The University Of Colorado agreed to the task, and physicist

Edward U. Condon would head it. He immediately came across as a closed minded yahoo to me, and in fact, in October 1968, a full month before the study even began, Condon was widely quoted as saying that it was "highly improbable" that UFOs existed. "The view that UFOs are hallucinatory will be a subject of our investigation, to discover what it is that makes people imagine they see things."[32] So much for the scientific method.

The real cherry on top of the botched banana split the Condon report ultimately became was the infamous leaked memo of project coordinator Robert Low, ruminating that the "trick" of the project would be to describe it "so that, to the public, it would appear a totally objective study, but to the scientific community would present the image of a group of nonbelievers trying their best to be objective but having an almost zero expectation of finding a saucer."[33]

In stark contrast to Low's assessment, the August 4, 1967 *Time* magazine essay, *A Fresh Look At Flying Saucers*, noted that "discussions of UFOs have begun to appear in the pages of such respected journals as *Bulletin of the Atomic Scientists* and *Science*. A few responsible scientists now put their reputations on the line by suggesting that saucers may be vehicles from outer space." One such scientist was University Of Arizona atmospheric physicist James E. McDonald, who commented after extensively studying the Project Blue Book files, "I think that UFOs are the No. 1 problem of world science. I'm afraid that the evidence points to no other acceptable hypothesis than the extraterrestrial. The amount of evidence is overwhelmingly real."[34]

The final act of the Condon-Low festival of incompetence and prevarication was their nearly 1,000 page study, completed in 1968, *The Scientific Study of Unidentified Flying Objects*. Condon's conclusion, that "nothing has come from the study of UFOs in the past twenty years that has added to scientific knowledge," flew in the face of the fact that nearly 30% of the cases remained unidentified, which might have registered as an invitation to engage in further investigation--had he even bothered to actually analyze any of the cases himself, much less read the finished product (reportedly, he never read a single page of his eponymous study). Willfully or not, Condon had done a bang up job acting as the Air Force's stooge.

In 1969, our local public library in New Orleans had a copy of the Condon Report (the full "*TSSUFO*"), and I marveled at it, because of its heft and size as well as for how blatant it was that such a sizeable chunk of the UFO sightings featured and studied remained "un-identified." Paramount amongst them to me was the daylight disc photograph taken by farmer Paul Trent of McMinnville, Oregon, on May 11, 1950. I've always thought it could be described best as a metallic double-decker Frisbee with a radio tower centered on the top. Full details and images of it are available online at the excellent UFO site, ufoevidence.org, at the following URL:

ufoevidence.org/photographs/section/1950s/Photo301.htm

William K. Hartmann, an astronomer from the University of Arizona, meticulously investigated the original negatives, and con-cluded, as per *ufocasebook.com*, that "an extraordinary flying ob-ject, silvery, metallic, disc-shaped, tens of meters in diameter, and evidently artificial, flew within sight of [the] two witnesses." (*UFO Casebook*/ufocasebook.com/B.J. Booth)

Like a collapsing house of cards, the Air Force used Condon's sham report to justify shutting down Project Blue Book on Decem-ber 17, 1969. This gave scientists and all persons in authority from then on a rubber stamp to deny and/or ignore any further incoming reports, based on how Condon's conclusions (never mind how in-correct they were) established that UFOs had no basis in reality, and serious analysis of them would be a waste of time. It also granted the media and television's ersatz-pundits who love to breezily dis-play their faux expertise the right to deride with extreme prejudice UFO experiencers sight unseen, in perpetuity. In essence, the CIA's potent psychic poison, the Triple-D protocol, had done its job, and all further serious studies of UFOs and the alien craft behind some of them could be ongoing, out of the public eye. Funny thing, even though Project Blue Book folded in December of 1969, the blasted mysterious flying things have refused to go away to this day.

What was actually going on behind the scenes, as revealed in a FOIA-accessed document known as "the Bolender memo" (named after Air Force Brigadier General Carroll H. Bolender and drafted in October 1969), was the general's directive that Blue Book be closed, with the understanding that there were already regulations in place channeling serious UFO reports elsewhere, within the hidden sys-tem, and out of public sight. The memo also makes clear that UFOs were hereafter to be handled on the inside, that they can affect national security, and in some cases a "defense function" may be necessary to respond to them.[35] One marvels at government duplicity in full bloom.

So officially today, in the second decade of the 21st century, UFOs (and alien spacecraft) don't exist--unless they do. In which case, the proper official response protocol, now in default setting for decades, is to engage the Triple-D (or simply ignore the events altogether, as in the Phoenix Lights incidents), and study the case behind the scenes. They don't exist--unless they do. And aliens aren't visiting--unless they are. But not in the real world. Except in umpteen often amazing and compelling YouTube videos and images, virtually clogging the Triple-W--as in world wide web. Welcome to the modern world of UFOs in America--our king-sized version of a Twenty-first Century Mindfuck. It's truly a wonder to behold.

Leslie Kean rightfully devotes chapters of her book to The Hudson Valley UFO Wave (early-to-mid 1980's, giant boomerang and triangular objects) The Phoenix Lights (1997), as well as the giant triangular (and sometimes rectangular) UFOs over Belgium in 1989 - 1991 (Google that for extraordinary color photos, as well as in Kean's book). But for me, the case that best illustrates the current tactic of outright ignoring a solid UFO sighting, going way beyond the realm of the Triple-D protocol, is in her chapter, *Incursion At O'Hare Airport, 2006*. It's absolutely extraordinary that such an event took months to filter into wide media coverage (reminiscent of the Phoenix Lights); even more so when you realize how many pilots and qualified observers at Chicago O'Hare Airport witnessed in broad daylight for five full minutes (some say as long as 15 minutes) a disc hovering above the United Airlines terminal, prior to its wink of an eye departure, leaving behind a perfectly circular, cookie cutter hole in the clouds, through which ground observers could gaze upwards to view the blue skies peeking down through the overcast cloudbank blanketing the area at about 2,000 feet.

The gist of the event: evidently scores of people saw this daylight disc, including ground crews, an aircraft mechanic, United workers, civilian witnesses, and pilots, one of whom announced its presence via ground radio for all personnel to be informed. Some pilots in line for takeoff even opened their cockpit windows to lean out and have a look. The consensus was unanimous: it was a silent, metallic disc; the only reported variations were that some thought it was spinning, and the size estimates varied from 22 to 88 feet in diameter.

To fully appreciate the extent to which starkly real, in-your-face UFO sightings are now, in the 21st century, met with a tactic of simply ignoring and denying their existence, even by FAA officials, and how pilots fear losing their jobs if they come forward and speak on the record because of airline UFO-report squelch policies, read this engrossing chapter in full. Suffice to say that the FAA's lame explanation of "hole punch cloud" holds about as much validity from the standpoint of atmospheric physics as Thomas Mantell's 1948 fatal plane flight, nearly 60 years prior, being the result of his chasing after the planet Venus--also in broad daylight. Personally? This makes me long for the days of Hynek's swamp gas. Marsh gases would have been an equally sound explanation for the solid, hovering daylight disc that dared to show itself at O'Hare's United Terminal C-17 on November 7, 2006. As Ms. Keane so aptly concludes:

> Why is our government uninterested in a strange, highly technological object hovering over a major airport, as reported by competent airline personnel? What about passenger safety? Or national security after 9/11? Or just plain scientific curiosity about an unexplained phenomenon? Official distaste for dealing with the UFO phenomenon is entrenched to the point of being not only counterproductive, but possibly dangerous.[36]

In my view, an exceptional UFO book of the investigative journalism genre will feature multiple cutting edge reports, and in this regard Kean's *UFOs: Generals, Pilots And Government Officials Go On The Record* does not disappoint. Chapter 7's *Gigantic UFOs over the English Channel, 2007* offers an absolutely amazing pilot-and-passengers-in-flight tale that includes multiple airborne and ground visuals of the two enormous, brilliantly illuminated objects in question, coordinated with solid radar returns. It also offers a blunt appraisal of the stark differences between how Great Britain's CAA (Civil Aviation Authority) handles larger than life, in your face UFO sightings like this one was (openly inviting UFO case filings, and offering state of the art investigative assistance) vs. our almost heartbreakingly pathetic FAA (arrogantly dismissing and/or outright ignoring solid UFO reports, and routinely resorting to simply lying to explain away that which they know they cannot explain). It's almost worth buying Kean's book for this one chapter alone.

The FAA as Keystone Cops is a scenario replayed, sadly, in Chapter 22 of Kean's book, *The FAA Investigates a UFO Event "That Never Happened,"* regarding the Japan Air Lines 747 cargo jet flying just north of Anchorage, Alaska, that was repeatedly buzzed, almost in an aerial dogfight fashion, by a gigantic, aircraft carrier sized, roughly Saturn-shaped UFO, on November 7, 1986 (interesting coincidence, it fell exactly 20 years prior to the Chicago O'Hare cloud-puncturing hovering disc).

True to the Triple-D protocol default setting we live with, I vividly recall a TV news report in Miami in which the local newscaster chuckled his way through his comments that perhaps the captain and flight crew of the JAL 747 were partaking of some of the wine housed in the cargo hold. It's really gotten to the point in America where these talking head faux pundits seem compelled to ridicule and deride cutting edge UFO reports and those who report them, lest they risk having their pseudo-intellectual credentials revoked. I suspect some of our alien visitors regard Sol-3 as one of our galaxy's premier Remedial Reading planets.

In Richard Dolan's and Bryce Zabel's excellent book, *A.D. After Disclosure: When The Government Finally Reveals the Truth About Alien Contact* (2012), we find that in fact, the JAL 747 pilot of this 1986 Alaska event, upon landing, filed a formal report of his aircraft's intense UFO encounter, only to be punished by his airline, which barred him from flight status for over three years. (Moral of the story for commercial pilots: reporting UFOs can sometimes damage both your bank account and your employability.) Also revealed here is that in 2010, commercial pilots were experiencing such a high volume of UFO encounters that pilots based out of London's Heathrow Airport began keeping each other apprised of troublesome "UFO hot spots" privately, out of the loop of their respective airlines. It sounds like an intelligent way for seasoned pilots to avoid paying the same high price the JAL 747 captain had, all the while endeavoring to keep their aircraft and passengers safe from troublesome unidentified aerial interlopers hailing from destinations unknown. Fly the friendly skies.[37]

UFOs: STANDOUT CASES

In all my years of UFO research and readings, there have been numerous cases which stood out from the others. But two which I'll share here now, I've chosen because they illustrate some important points: one, that the military world, due to its preset, rigorous protocol of fiercely guarding compartmentalized information, is likely far richer in unusual UFO cases than we'll ever know, by virtue of the very insular world it exists in, and the officers who by and large remain silent when told to; and two, there may well be a myriad of cases similar to these which we'll never know about, which offer just as potent a dose of alien craft and alien encounters as these two. Both of these come from a book published in 1984, *Clear Intent: The Government Coverup Of The UFO Experience*, by Lawrence Fawcett and Barry J. Greenwood.

Late summer through the fall of 1973, the U.S. was experiencing a UFO wave. Television, radio and the press were saturated with a full spectrum of reports, from lights in the sky seen from a distance to actual UFO landings with occupants emerging (as in the Hickson/Parker October 1973 sighting and abduction in Pascagoula, MS). What was not common knowledge was the corresponding uptick in military UFO encounters just prior to this public wave.

Fawcett and Greenwood, both well regarded UFO investigators for over 20 years at the time their book appeared in 1984, including multiple memberships in professional UFO organizations, were approached by a man claiming to have been a former Navy officer. Their background check on "Ed Sims" (pseudonym) did reveal a naval service record with a duration that jibed with his story. When they interviewed him, Sims held a position with a Connecticut-based technical firm, and asked that he remain anonymous. The authors complied. In return, they got a fascinating inside story of how it's possible to have to pay a price in the military if you know more than is good for you.

Ed Sims was a member of the U.S. Navy in 1973, and served on assignment onboard a nuclear submarine, the USS Abraham Lincoln. His story begins at night time; his sub had just passed through the Panama Canal the previous day. Outside on the conning tower that night, Sims and three fellow officers were filming algae in the water, and night lights. His ship's photographer was using both a 35mm still camera, as well as a 16mm movie camera. Without warning, a 100 foot wide circular crimson-colored disc (presumably glowing) swooped down out of the dark sky and circled in a wide arc around their submarine about ten feet over the water's surface. Their sub immediately lost all navigation and sonar, and essentially floated dead in the water. Their photographer captured the entire encounter on both cameras. The huge crimson disc circled their sub two or three times, then shot off at a high speed, quickly disappearing. As soon as it was gone, navigation and sonar resumed their normal functioning.

Sims and the other three sailors with him went below to report to their executive officer what they'd just witnessed from the conning tower. As per orders, the photographer developed all the films immediately, and then presented everything to his commander. The four were warned by the executive officer to say nothing about their encounter to anyone else onboard. Apparently the photographer later told Sims that everything they'd seen had come out clearly on film. Sims told Fawcett and Greenwood that the incident wasn't mentioned again until their sub docked in California.

Upon docking, Sims was about to take advantage of his liberty, when the skipper summoned him and his fellow UFO experiencers to his quarters. Once there, two civilians and one Navy officer questioned them about their UFO encounter for about an hour. They were told that their liberty was cancelled, and military police escorted all four men off ship and took them to an unknown location on the base. Their final destination was a small eight by ten room with just a desk and a chair, where they were left alone and confined for about an hour, with no visitors. One of the civilians with an intelligence background (Sims thought it could have been CIA, FBI, OSI or Naval Intelligence) accompanied by a naval officer, entered the room. Interrogations then began.

In a style seemingly lifted from a film noir detective movie, the interrogators aggressively tried to convince Sims that he hadn't seen anything during his UFO encounter. When Sims countered that even the ship's photographer successfully recorded photographic evidence of the object, they told him that he and his comrades were lying and had made up the whole thing. Sims stated that this grilling interrogation lasted for hours; when the intel men couldn't break Sims's account, they left him there, saying that they'd be back to see him tomorrow. About forty-five minutes later, Sims was escorted by guard to a base location where he would spend the night; they said they'd come for him in the morning. The next day, two MP's returned him to the same cramped room with the desk and chair, where he was interrogated for eight full hours by new civilian and Navy personnel. The same brainwashing rhetoric ensued: he hadn't seen anything, and he and his fellow crewmen were lying.

On the third day, the same protocol was played out, but afterward, three civilians with an attaché case entered the room. They told Sims they believed his story, and proceeded to show him photos of different UFO types, asking that he look through them to try to identify the craft he'd seen. Sims alleged that all the photographs they showed him were eight-by-ten glossies of a variety of saucer-like objects. Some were shaped like cigars, some like elongated footballs, some like ice cream cones; still others appeared like two car headlights. He was able to find a photo that closely matched the object he'd witnessed from the sub's conning tower that night. He then had to sign a secrecy document stipulating that if he revealed anything about the sighting or the photos they'd shown him, he'd be court-martialed, fined, and placed in solitary confinement for an extended period of time. He signed. Then Sims was told to report back to the USS Abraham Lincoln, and was escorted there by military police.

Once back onboard, he was summoned by the commanding officer, who informed him that they were transferring him; he was to pack all of his belongings and report back when ready. Once back, he was instructed to sign another secrecy document, and was reminded of the dire consequences should he reveal anything to anyone about the whole ordeal. He was then taken off the sub, and put on an airplane to Hawaii, the location of his next duty station. Through acquaintances Sims heard that his three unfortunate fellow crew members were similarly taken off board and flown to other duty stations in other scattered locales; the four men never saw nor heard from each other again. Fawcett and Greenwood attested that Sims showed no signs of lying, and sought no publicity for his encounter. An Air Force source the authors considered reliable confirmed the account, that a huge disc-shaped UFO had in fact circled the USS Abraham Lincoln that night.[38]

I chose one other case from *Clear Intent: The Government Coverup Of The UFO Experience* because it illustrates the possibility, in a similar vein to the Roswell crash recovery, that operatives behind the scenes may have worked on other Special Access Projects handling recovered vehicles, under the protective canopy of National Security. This UFO case, also from 1973, was provided to the authors by a former Naval Intelligence officer. While based at the Great Lakes Naval Air Base in Chicago, thanks to his high I.Q. and mechanical aptitude scores, he'd risen in rank quickly and had a high level security clearance. One night on guard duty, he was Officer of the Guard and had been assigned to a quonset hut that he'd been told had highly top secret material inside. He was to keep all unauthorized personnel 100 feet away from it, and not to look into any windows, and not allow anyone in or out who didn't have proper identification. He was given a package to be delivered to and signed by the OD (Officer on Duty) in the hut. The OD was busy inside, so it was decided that the guard would be allowed into the hut and into his office. He was met by three burly MP's who escorted him down numerous hallways. He was instructed that once he reached the inner warehouse area, he was to go directly to the OD's office and not pay attention to what was going on. In the warehouse en route to the office, he saw an extremely unusual craft to his left. It was possibly thirty to thirty-five feet long, and twelve to fifteen feet across at its thickest part, and tapered off in front to a teardrop shape. The whole craft tapered in the back to a very high edge, which looked to be razor sharp; the bottom of it extended about three quarters the length of the craft and then sharply angled upwards. His quick glance revealed it didn't appear to have any seams, and had a bluish tint if you looked at it, but when you looked away you perceived white lights. It rested on a pedestal of four by four wooden blocks, held up by crossbeams underneath it.

A few months later, the Naval Intelligence officer was in San Diego to help install missiles in a sub. He met another Naval officer in a bar, a man who'd been on a destroyer, and the sailor had a story of an unusual craft his vessel and crew had tangled with. They'd shot it down in the Pacific in about 350 feet of water with a surface to air missile. It was extracted from the ocean floor by the *Glomar Explorer*. No one was even sure it was an aircraft because it didn't look like anything they'd seen before. Although they had downed the object with a direct hit, they hadn't put a dent in it, but the impact had sent a concussion through the craft. The sailor said that they'd been able to pull some sort of life form from out of it, but it was dead. The two were drinking beer, so the Chicago officer took everything with a grain of salt, until the sailor sketched what the strange aircraft looked like. It was identical to the craft the Intel officer had seen up close in the high security warehouse that night at the Great Lakes Naval Air Base in Chicago. The sailor told him it had been shipped by rail from San Diego to Chicago to be worked on; they had to get it out of the San Diego area because word of what they'd shot down started leaking out.[39]

THE WEB:
A GLOBAL TREASURE CHEST OF UFOs

There is no place in this new kind of Physics both for
the field and matter, for the field is the only reality.

--Albert Einstein

The advent of the internet and the World Wide Web in the early to mid 1990's (originally conceived on the desktop computer of
Sir Tim Berners-Lee on March 12, 1989 in his office at the CERN
research center in Switzerland) would dramatically change the intelligence playing field, insofar as our global culture of instant information has made keeping secrets much more challenging than
back in those quaint, good old 20th century days, before we lived
immersed in the cacophony of a 24/7 streaming-info world of sensory overload available at the tap of keys on a keyboard, the mouse-
over of a hyperlink or gentle strokes on our iPhone or iPad screens.
On the down side, I've witnessed friends, relatives and "Friended"
strangers alike get entrapped in Facebook's hypnotic trance dance
of relentless narcissistic selfies, to the extent that some are compelled to post life-changing photo updates at roughly 15 minute daily intervals, chronicling in Technicolor detail every single step of
their amazing, not to be missed out on lives (is this what Andy really
meant?). Consequently I'm having a love-hate affair with F-Book
or the Face-BooBoo. Especially when you factor in the creepy Big
Brother NSA relentless and "totally innocent" info-hoovering. On
the up side, the web as a resource for useful, enlightening and sometimes amazing information is not to be underestimated, which obviously holds doubly true for UFO cases with streaming video footage
of what appear to be miraculous machines cavorting hither and yon.
In Leslie Kean's *UFOs: Generals, Pilots And Government Officials Go On The Record*, more than twenty pages are dedicated
to the remarkable UFO wave over Belgium that lasted nearly two
years, from 1989-1990. Often encountered were the gigantic triangular craft, some of them over 100 feet on a side, which invariably
seemed to be on a leisurely cruise through various smaller Belgian

towns and communities, sometimes slowing to a halt and hovering for many minutes, as if posing for photos. It was regrettable that these hundreds of sightings--some of them coming into close proximity to their viewers or even flying slowly over them (nearly 2,000 were officially reported and logged)--occurred before the soon to be ubiquitous cell phone cameras could have snapped up scores of still photos and mpeg movie clips. A few rather amazing photos were snapped nonetheless, featured in Kean's book, as well as on the web. Some of them, when subjected to analysis and image adjusting, revealed a somewhat dusty halo pattern of photons scattered around the triangle's periphery, not unlike photos of iron filings around a magnet, suggesting that the craft emitted a strong magnetic field.

A few of the current internet sites featuring them matter of factly refer to them as the Black Triangle, the TR-3B "Astra," alleging that they're part of a USAF experimental Black project known as Aurora. One site, **hidden-truth.org**, goes into some detail regarding its supposed background, how and where it was tested, the technology behind it (reverse engineered alien technology), and offers a few compelling schematic diagrams of it. The TR-3B specs on this site are generally credited to a man named Edgar Fouche, who ostensibly worked on the development of this exotic aircraft at Area 51. (During their wave, Belgian officials asked the U.S. if we had any experimental craft doing flyovers over there, and were assured we didn't; and we queried them back, as we'd had our own giant black triangles darting about in various U.S. locales, behaving in very similar ways, often casual flybys with wink of an eye departures.)

The online schematic diagram of the vehicle's underside reveals a large central circular device referred to as a Magnetic Field Disruptor (MFD), the location of which precisely coincides with the brilliantly illuminated central and circular light source in many of the triangular UFO sightings. The diagram further elaborates that this is a Mercury Plasma Accelerator Ring, rotating a high energy metal-infused plasma up to a blinding 60,000 rpm, pressurized up to 250,000 atmospheres, supposedly reducing the gravity and mass of the craft by 89%, making it exceptionally light and easy to propel by the three hydrogen-oxygen rocket engines, each of the three placed near the triangle's corners.

Here's the link, have a look for yourself:

hidden-truth.org/secrecy/unknown-crafts/black-triangle-ufo-tr-3b-qastraq

Even more in-depth and fascinating information about the TR-3B can be found on the Dark Government web site. There is a nearly 10 minute video there which specifies the principles theoretically operating behind the craft's revolutionary engine, officially referred to as the Plasma Torus Anti-gravity Centrifuge Engine (PTACE). The theory proposes that by cooling (super-cooling) the liquid metal substance down to the point where all of its atoms coalesce into a single wave function (150 degrees Kelvin), referred to as Bose-Einstein Condensation (Google/Wikipedia for more info), the atoms behave as a single atom, and this results in two effects which play an important part of its ultra-high performance: super conductivity and super fluidity.

View the video to explore some theories behind this advanced technology via this link:

www.darkgovernment.com/news/tr-3b/#idc-cover

If you watch the video you'll see and hear that the core of the PTACE is a hollow donut-shaped metallic container filled with a specially mixed ferro-fluid or magnetic fluid. As per Edgar Fouche's explanation, it would contain an ionized mercury plasma gas, infused with countless tiny iron particles to act as super conducting super fluid accelerators when you have pulsing electromagnets firing at specific intervals around the perimeter of the big torus container, revving up the rpm's, faster and faster, up to "relativistic speeds." Once the fluid attains these relativistic speeds, theoretically the magnetic field created can warp the space-time continuum, producing the effect of anti-gravity, while simultaneously dramatically reducing the craft's mass.

The concept that we already have high performance antigravity vehicles of our own is bolstered by the fact that there was a document (which was uncovered in the 1980's, declassified from the Wright Patterson Technical Library, of the Wright Patterson Air Force Base) which revealed that beginning in the 1950's, for a solid ten or twenty years, all of the big player aerospace companies in the U.S. were conducting research on electro-gravitic technology, also known as an electromagnetic form of antigravity propulsion. In 1958, after a special conference set up by the Air Force that hosted all the myriad research teams, all reference to this technology in the so-called "open literature" disappeared. So who knows what these guys could have come up with from decades of Black Budget

funded projects? An answer of sorts appeared in the May 2001 issue of *Popular Mechanics*, in the article titled *When UFOs Land*. It
revealed that (then) recently declassified documents explained that
during the 1950's and 1960's, the U.S. Air Force was experimenting
with electrostatic drives. It was based upon the concept that lift
and propulsion could be created by inducing into an airframe an
electric charge that matches, and thereby repels, the air surrounding it. An aircraft of this type would consume an enormous amount
of electrical power, and it seemed that the Air Force knew how to
create that: other declassified documents indicated that they had
built compact-sized nuclear reactors that would fit onto an aircraft.
Evidently they were also experimenting with what was known as a
magneto hydrodynamic generator (MHD) to extract huge amounts
of electricity from fast moving streams of molten metal (which obviously echoes the TR-3B's magnetoplasmadynamic centrifuge engine). According to the *Popular Mechanics* article, engineers knowledgeable with these systems asserted that if MHD units somehow
became unstable, some of the liquid metal circulating in the unit
would have to be ejected. (This could be one possible explanation
for sightings of low altitude hovering UFOs squirting out liquid metal onto the ground; whether alien craft or reverse engineered flying
saucers, cutting edge technology can get a little messy sometimes.)

* * * * * * *

Dr. Steven M. Greer:
The Disclosure Project

Further bolstering this concept of an antigravity engine utilizing liquid metal can be found in Dr. Steven M. Greer's remarkable book *DISCLOSURE: Military And Government Witnesses Reveal The Greatest Secrets In Modern History.* This extraordinary and groundbreaking book was released in 2001, the same year that Dr. Greer, a former North Carolina emergency room physician who quit his lucrative medical career to investigate UFOs full time, hosted *The Disclosure Project Press Conference.* Presented at the Washington, D.C. National Press Club on May 9, 2001 as a means to publicly "out" the subject of UFOs, the alien presence on and around Earth, and the hidden alien technologies we've recovered and back engineered from their vehicles, it was and still is an extraordinarily ambitious and awe inspiring endeavor to behold. You can view all eight parts of it, beginning with Part 1, on YouTube, at this link:

http://www.youtube.com/watch?v=nlLrxYIVPmg

In Dr. Greer's *DISCLOSURE* book, a procession of military, government and research scientists he interviews offer insightful and sometimes shocking revelations. Among the testimonials: confirmation of Eisenhower's Air Force base meeting with aliens in 1954; the fact that we did have a live alien from one of the Roswell era crashes kept in captivity at Los Alamos labs for a few years; the concept that the entire human species was created by an extraterrestrial race, via genetic engineering and DNA modification over thousands of years. And that when a CNN reporter asked Mikhail Gorbachov during his second visit to the U.S. if we should eliminate all of our nuclear weapons, Mrs. Gorbachov interjected an off the cuff comment, "No, I don't think we should get rid of all our nuclear weapons, because of alien spacecraft."

In Greer's book one also finds case after case of encounters, many with what seem to be alien craft, often disclosed by ex-military officers whose backgrounds as solid armed forces members beyond reproach speak for their integrity and reliability. I'll look at a few of them here, the first one for its remarkable coincidental details regarding a freshly crashed craft of alien origin.

Lance Corporal Jonathan Weygandt joined the Marine Corps in 1994, and was ultimately stationed in Peru in 1997; his job: providing perimeter security to a radar installation. One night in March that year, he and two fellow officers (both Sergeants), along with others, were detailed on a potential rescue mission, and packed into several hummers, heading out into the deep forest to secure a crash site that involved an aircraft that was "possibly friendly." When they arrived at the scene, dawn was breaking. Weygandt noted that the wooded area leading to the craft looked scorched, as though a high heat source, like a laser, had cleanly burned a swath through the trees, leading to a cliff ridge overlooking a ravine. Following the path, they came upon an unusual dull metallic craft, a roughly 60 foot long egg or teardrop-shaped vehicle buried at a 45 degree angle in the side of the cliff. The exotic craft was leaking a syrup-like fluid, and upon closer inspection Weygandt noticed that its greenish-purple color appeared to fluctuate and change its appearance a bit each time you looked at it, as though it was alive, exhibiting a different shade of greenish-purple on each viewing. The dazed officer made his way down into the ravine and around the craft. The single flashing light on its surface, and its deep, rhythmically oscillating bass hum seemed to indicate it was still operational, but presently the humming sound steadily faded, then grew silent. In the process of scoping out the vehicle, he got some of the syrup-like liquid on his cammies, which discolored and ate away at the fabric; later he realized it had eaten away some of the hair on his arms. Entranced by what he'd found, the officer noted three hatches, one which seemed half open; and he perceived that there were life forms onboard which he believed were transmitting thoughts to him, even though he was not one to believe in that sort of thing.

Weygandt was jolted from his trance by his angry fellow officers yelling at him to back away from the vehicle. They were soon joined by DOE people, then others in hazmat suits, as well as men in black cammies without nametags who forcefully arrested him, cuffed his hands, tied his legs together, secured him to a cot, and flew him away in "a huge 47" (presumably he meant a twin-rotor Chinook helicopter). He was taken to a secure facility, a base where he saw a lot of people of multiple nationalities. His first one on one encounter was with an unidentified Air Force Lieutenant Colonel whose threat echoed what Glenn Dennis said he'd been told in the wake of his Roswell crash experience back in 1947, that "they would never find you out there if we took you out in the jungle." In an interrogation room, he had a bright light shone in his face for roughly fifteen hours; he recognized one of his interrogators, in black fatigues, as being one of those in black cammies at the crash site. He growled and yelled at Weygandt, asking him what he saw, if he was a patriot, if he liked the Constitution. When he replied in the affirmative, the gruff man in black shouted, "We are on our own program. We don't obey. We just do what we want!" His fellow interrogators chimed in with their own shouts and curses. Weygandt was then threatened by them, "We will do you and your whole goddamn family!"[40]

It seems like a fair assumption that the man in black fatigues was speaking on behalf of a UFO-related quasi-government (or Shadow Government) that operates above the law, one in which they have their own program, and they just do "whatever the hell they want." Of equal significance here is Weygandt's description of the syrup-like liquid dripping out of the damaged craft as being a greenish-purple hue, changing shades each time you'd look at it. As per the TR-3B information video (**darkgovernment.com** web site), the ideal liquid metal fluid for the plasma torus centrifuge engine is described as having tiny flakes of iron to provide the required magnetic properties; and in modern day metallic-flake paint jobs--for automobiles for example--the duplicolor paint changes appearance, depending on how it's viewed, because ground up flakes of shiny metal are thoroughly blended into the mixture, and they exhibit multi-hued colorations from different angles and in different light.

Another eye opening interview from *DISCLOSURE: Military And Government Witnesses Reveal The Greatest Secrets In Modern History,* is dated December, 2000. It's the almost shocking account related by Mark McCandlish, a well regarded aerospace illustrator who has worked for numerous top U.S. aerospace corporations. His testimony regards his colleague, Brad Sorensen, who alleges that he was taken inside a hangar at Norton Air Force Base, where he was shown a few remarkably advanced and fully functional aircraft that were reputedly alien back engineered vehicles. His narrative further claimed that the U.S. government has manufactured these sophisticated antigravity propulsion vehicles, and we've had them for quite awhile, and that they've been developed in part via the study of alien vehicles over the past 50 plus years. The interview and article go into considerable detail regarding the inner workings of these antigravity craft, which makes for fascinating reading.

Before we explore this amazing *DISCLOSURE* interview further, due to the unique propulsion methodology of these exotic, reverse engineered vehicles--which purportedly make use of what's known as the Zero Point energy--a bit of background information regarding this technology is in order.

In 1982, I came across a fascinating book, *The Holographic Paradigm And Other Paradoxes: Exploring The Leading Edge Of Science* (1982), edited by Ken Wilber. The book consists of multiple interviews regarding what was then (as now) cutting edge concepts in physics and metaphysics, and how the realms overlap. In Chapter 5, Renee Weber interviews celebrated physicist David Bohm, a colleague of Einstein's. In a nutshell, Bohm posited that the human mind--as well as the Universe--operates like a hologram, or holographic image; it is referred to as the holographic model of consciousness. One factor about holographic images is that when you have a holographic photographic plate (using the laser-based photographic method), if you cut away even a small square inch section of any part of that plate, and shine a light through it, that small piece of the original still has the full contents of the entire photograph emerging from it. Ergo, part of the holographic paradigm is the principle that all parts have access to the whole. In that same way, Bohm believed that the human mind operates holographically, not only within and by itself, but that the human mind and human consciousness also have full access to the totality of the Universe. All parts have access to the whole. The humanitarian implications of this concept are evident if you consider this implies that we human beings, and indeed all life, are all interconnected on a higher level of energy. All is one, invisibly interconnected energetically.

This holographic model also applies in a rather staggering way for what would have to be the ultimate form of propulsion technology, via tapping into the Zero Point energy. In Weber's discussion with Bohm, the physicist states that if you're pondering the electromagnetic field in empty space, every wave possesses what's known as a zero point energy, below which it cannot go. So that if you were to total up all the waves in any region of empty space, it would be found that they contain an infinite amount of energy, simply because an infinite number of waves are possible. And therefore if you considered that the gravitational field was composed of waves in this way, you'd find that there was a certain length beyond or below which the gravitational field would have no definition because of this zero point movement, and length would not be possible to define. The place where it fades out, according to Bohm, would be at 10^{-33} centimeters, a very short distance for measurement of the wave of an electromagnetic or gravitational field; and if you did calculate the amount of energy that would be available in one cubic centimeter of open space, with that shortest possible wave length, the energy in that one cubic centimeter would be immensely beyond the sum total energy of all the known matter in the entire universe (I re-read that statement a lot for the sheer fun of it). The two then say,

Weber: In one cubic centimeter of space?
Bohm: Yes, and therefore, how is one to understand that?
Weber: How do you understand that?

How, indeed. Bohm then continues that empty space has all of this energy, with matter that we perceive in our daily lives representing a slight increase of energy, and that therefore matter can be likened to a small ripple on the surface of this tremendous ocean of energy, with some relative stability, and being manifest all around us. Matter is just a ripple against this background of energy, a ripple on this ocean of energy. And this ocean of energy is not primarily in space or time at all.[41]

Over time, Dr. Bohm's references to the Zero Point movement and the virtually infinite supply of energy accessible in any single point of open space in the Universe became more and more prevalent in subsequently published works. In the Prologue of Lynn McTaggart's fascinating book, *The Field: The Quest For The Secret Force Of The Universe* (2001), she shares that for decades, scattered around the world, a variety of scientists respected in their various fields have conducted well-designed experiments whose results have been utterly astonishing: that all human beings, as well as all living things, are an amalgamation of energy fully enmeshed in an enormous energy field which interconnects everything in the world (may the force be with you). And that this pulsating energy field is the core engine of our consciousness and our being. And further, that if the equations representing the Zero Point Field energies--an ocean of infinitesimally small energy vibrations oscillating in the micro-spaces between things--are factored into our most basic concepts about the nature of matter and energy, then the underlying reality is that our Universe is a vibrant, oscillating, vibrating ocean of energy, and everything is connected to everything else through a vast underlying energy network. Finally, that all living things, human beings included, are essentially packets of quantum energy relentlessly exchanging energy and information with this infinite ocean of energy. And, as in Dr. Bohm's holographic postulations, our minds operate on a level implying a constant interconnection with the Universe; so the quantum information vibrating in our bodies and brains--human perception--is the result of an ongoing interplay between the subatomic particles of our brains and minds and the infinite ocean of energy in which we're swimming.[42]

In the Mark McCandlish *DISCLOSURE* book interview, we open with his friend Brad Sorensen having invited McCandlish to a private air show at Norton Air Force Base (fully closed down as of 1995). Sorensen had done quite well for himself by designing and securing the patent rights to exotic aircraft he would dream up, and collected royalties from clients who wanted exclusive licensing rights to them; he'd become a millionaire while still in his twenties. At the last minute, McCandlish had to back out because of a project deadline, so Sorensen attended the air show with one of his clients, a by then retired government Big Wheel (BW).

On the field, the Air Force Thunderbirds were about to begin their show; the BW breaks away and tells Brad to follow him. They proceed to the other end of the airfield, leaving the crowd behind, and come to The Big Hangar (its actual moniker on the base). Sorensen describes it as looking like they'd merged four giant Quonset huts together, with work areas and shops around the perimeter, and more sensitive intelligence areas deep within the vast space.

The BW dispatches a guard at the entrance to retrieve the man in charge from inside. He returns shortly with a distinguished looking, well dressed gentleman in a suit, who warmly welcomes the BW as a known associate, and accepts Brad as his assistant. The suit green-lights them into what is apparently an exhibit meant strictly for military officials and local politicos with high security clearances. As they enter, the BW instructs Brad that they'll be shown things he wasn't originally aware they'd be displaying, "stuff you probably shouldn't be seeing." He is told to keep quiet, not to ask any questions, and just to follow along silently and enjoy it.

Their distinguished host is very accommodating to the BW and his cohort Brad, making sure that they see everything. Included among the aircraft on display are the losing competitor to the B2 Stealth bomber, as well as the Lockheed Pulsar, also known as the Aurora.

After taking in all the exotic aircraft, Sorensen relates that they came upon a big black curtain that apparently divided the immense hangar in half. They are escorted through the parting curtains into another cavernous area, which at this time was in the dark. Presently someone hit the lights, and Brad immediately realizes what the BW meant about things he shouldn't be seeing. There are three flying saucers hovering silently above the floor of the vast space; no cables from above suspend them, and no landing gear support them down below. They float silently above the floor, as though it was their natural state. Massive craft floating peacefully in silent, hover mode, occasionally listing and gently drifting in place like boats moored at dock, but in this case, wafting on an invisible sea of energy.

Brad notes that the three saucers are staggered in sizes small, medium and very large; the smallest of the three he estimated to be 24 feet wide at the base, the middle one had a full 60 foot base diameter, and the largest one spanned a mammoth 120 - 130 feet in diameter at the base.

The performance of the smallest vehicle is featured in a video tape playing on monitors in multiple displays around them. It begins by resting on the ground of what looks to be a desert scene, which Sorensen assumes is Area 51, or someplace similar. As the video footage progresses, the saucer makes three quick hopping motions (probably not unlike the classic pogo stick jumps I'd seen the UFO doing in my Houston group sighting), and then zips straight up vertically, silently vanishing into the open sky in a scant couple of seconds. No exhaust rush, no sonic boom, no sound at all (it's unclear if the volume on the video was turned off).

Sorensen's narrative then elaborates about how there was a cutaway illustration showing the vehicle's inner components, and in fact the smallest saucer hovering before them had some panels removed to reveal hidden parts, which included an extendable robotic arm. The dome at the top of the craft was actually a large spherical crew capsule, and wrapped around it and below it were 15 or 20 copper coils in stacked layers.

He notes that the craft's lower segment, presumably the heart of the drive system, consisted of 48 sections radiating out from the center, like pizza slices fanned out on-edge in a circle in precisely symmetrical spacings, with each of the 48 metal slices loaded up with half-inch thick copper plates. Sorensen surmised that this was a series of plate capacitors, that someone had figured out how to harness the Biefield-Brown effect--a process whereby you charge up a capacitor to provide lift towards a positive plate (more about the Biefield-Brown effect later).

The crew compartment had a central column, around which were fastened four ejector seats. The crew capsule had no windows; he conjectures this might have been because the vehicle's propulsion system likely harnessed a half million or a million or more volts of electricity while in flight [perhaps making conventional windows impractical; or perhaps it was necessary to have uninterrupted metal surfaces in order for the ultra high voltage craft to function properly--wv]. Sorensen was told that the pilots wore glasses enabling 3-D viewing outside the craft.

Brad was told that these flying saucers (presumably at least partially reverse engineered from alien technology) were known as ARVs--Alien Reproduction Vehicles--and nicknamed the Flux Liner (because they run on ultra high voltage electricity). Looking them over, he realized that what he saw hovering before him were essentially open-air transformers, or giant Tesla coils; when a powerful jolt of electricity is sent through the big coils, it surrounds the craft with an electromagnetic field. The logic behind having all of the pizza slice plate capacitors was so that the craft could be precisely controlled and vectored in any direction; current directed to one single thin slice of capacitors would result in the craft engaging a specific trajectory. Forty-eight of them enabled finely tuned, precision course changes, no matter if in vertical jumps, sharp right angle turns, zigzags, rolls, pitches, yaws, or any combination thereof.

Reading this *DISCLOSURE* interview in full is time well spent, as it plunges into much greater depth about how the Flux Liner operates, with schematic diagrams; and how the ARVs circumvent Einstein's light speed barrier (the argument always having been that the mass of any object accelerated up to and beyond the speed of light would become infinite, ergo an impossible scenario), by virtue of the fact that the craft absorbs the Zero-Point energy, while keeping all of that energy from interacting with the vehicle's atomic structure, thereby preventing any alteration of mass in the positive direction. Perhaps the true genius of the ARV is that when you activate the propulsion system with that initial jolt of electricity, every part of it begins to become mass-canceled. Even on the sub-atomic level, the electrons furiously dancing around their nuclei become mass-canceled. So as this stardrive engine (and all of its structure, and all of the electrons streaming through the enormous Tesla coil inside of it) achieves virtually total mass cancellation, the vehicle becomes the ultimate zero-mass super-conductor, and consequently its propulsion efficiency goes flying off the charts. The ARV's mass gets bled away more and more as it accelerates, so essentially, the faster it goes, the faster it wants to go. In no time, you warp out of space time, leaving the speed of light barrier in the dust (sorry about that, Albert--although your equation did allow for loopholes).

Apparently one of the three star generals admiring the exhibit casually commented that the ARV vehicles on display could travel at the speed of light, or better. Super-luminal vehicles. Warp drive, without all the fuss of matter-antimatter engines. Introducing the Flux Liner. This baby runs on--are you ready for this--electricity. (View the fascinating schematic diagram in the *DISCLOSURE* book.) And this "insider's only" air show was to have taken place on November 12, 1988.

Mark McCandlish referenced yet another insider, this one named Kent, who told him that he'd seen the Flux Liner for himself when he was working as an aircraft crew chief at Edwards Air Force Base in 1973. He basically blundered past a slightly opened Quonset hut hangar door one night on his way to a work site, and took a peek inside, where he beheld the ARV hovering in place silently, exactly the way Brad Sorensen had described, and matching its physical description. This dovetailed with another account shared with McCandlish, when the man in question (sometime before 1982) took a look into a hangar when he shouldn't have, and also saw the Flux Liner ("flat at the bottom, with sides that sloped up to the top"). In seconds the fellow had a machine gun barrel at his throat, was blindfolded and hooded by guards, and subjected to 18 hours of grilling.

The narrative also stated that these ARVs were exclusively used by Air Force Intelligence, the CIA and the NSA, and conjectured, no wonder NASA funding had dried up dramatically (and the shuttle program eventually got mothballed in 2011). What's the point of it, when an ARV can get you to the edge of the solar system in hours? At the speed of light, 186, 282 miles per second--or about 670 million miles per hour--once you've left Earth you'd be orbiting Pluto in 4 - 7 hours (it's a trip of 2.66 billion to 4.67 billion miles, depending upon the planet's--or demoted planetoid's--specific elliptical orbital position around the Sun relative to ours).

There are a few other choice "accidentally seen back engineered saucers at military bases" accounts in the McCandlish narrative. The overall message in this particular *DISCLOSURE* interview is the presentation of evidence that seems to establish a simple fact: super advanced propulsion technology exists, and it is being deployed, and has been deployed for decades. But alas, with an unfortunate kicker thrown in: it's only for a very select few, in very high places, and behind very high security clearances. "We, the people" are decidedly left out of the loop.[43]

I have come to trust in Richard M. Dolan's integrity as much as I do that of Stanton Friedman and Linda Moulton Howe. His background as a historical scholar definitely reveals itself in his uncanny ability to ferret out information, as witnessed by his remarkable books. So it was no surprise to me to come across in his *UFOs And The National Security, State Vol. II* (2009), added details about Brad Sorensen's experience with the ARV. Mr. Dolan's supplemental findings: that Sorensen and his companions actually were flown on an Air Force passenger jet from the Norton Air Force Base air show (same date, November 1988), and shuttled 50 miles northwest to the Lockheed Skunk Works in Palmdale [which made a lot more sense to me--wv], to the real location of the Big Hangar (apparently Sorensen deliberately withheld this part of the story originally); that the event was staged for insiders in hopes of gaining increased financial support for projects classified 'black,' or SAR (Special Access Required) programs; and that the ARV, aside from being capable of superluminal travel, also possessed incredible instant acceleration capabilities, that it could go from a stationary, near-ground hovering position to an altitude of 80,000 feet (a bit over 15 miles) in 2.5 seconds, meaning it could perform an acceleration burst from a standing still "0" to nearly 22,000 miles per hour (6 miles per second) in the wink of an eye. Most tellingly, Sorenson reported that the ARV "looked ancient," as though extensively used, and had already conducted a general survey of all the planets in our solar system in search of life, but none was found. My best guess is that the increased financial support the insiders sought that day turned out to be a great big Two Thumbs Up.[44]

I consider the McCandlish interview to be one of the most striking testimonies in Dr. Greer's book, full of compelling semi-technical passages, particularly regarding the "how things work" disclosures about the ARV propulsion system. Because of this, my curiosity was stoked, so I Googled the name Brad Sorensen on Friday, October 11, 2013. At the press of a key and one mouse click, I discovered a web presence that fully establishes his background, history and credentials, precisely as represented in *DISCLOSURE*, i.e., he is a California-based, very high level, highly respected, award winning product designer with a stellar, verifiable history in multiple cutting edge endeavors (his web page states that over 400 companies have used his designs of over 800 vehicles, products and packages). There's even the jpg color image of one of his somewhat delta-shaped, single pilot fighter aircraft with four serious looking missiles poking out the front, ready for combat, with the word MARINES emblazoned on its airframe, and with his signature and copyright symbol below the rendering, dated 30 Oct 2000. His email address was included in his online information, so I emailed him at 12:46pm EDT that day (10/11/2013):

> Hello Brad Sorensen:
>
> I'm about to publish a non-fiction UFO book, the culmination of years of research. I've read Dr. Steven M. Greer's remarkable book, *DISCLOSURE* (2001), a number of times. One of its most fascinating chapters to me is the one featuring Mark McCandlish's interview, in which he mentions an air show at Norton Air Force Base you purportedly attended back in 1988, in which you were apparently shown some amazing technologically advanced aircraft (if an ARV can even be called that).
>
> If you are in fact the same man mentioned in this narrative, would there be any way I could phone you at your convenience to ask you a few questions? I would greatly appreciate a reply at your soonest opportunity, regardless of whether or not a brief phone interview would be acceptable and/or practical to you.
>
> You can find extensive information about me via my web site. And final note: the list of cars, aircraft, ad infinitum that you've designed is a remarkable thing, in itself. What a fascinating career you've had.
>
> Thank you for your time and consideration in this matter. I hope to hear back from you soon.

Sincerely,

Wade Vernon

At 1:22pm EDT (thirty-six minutes later), I received a single-sentence email message reply from Brad Sorensen, without a salutation or closing, precisely as follows:

Mark McCandlish published without my consent and my non disclosure agreements prohibit me from communicating with you.

The abrupt bluntness of Mr. Sorensen's terse and rapid response took me by surprise. Its impact to me seemed enhanced by the frame of the email web page that only contained his one sentence, as simply worded and starkly declarative as a diamond shaped road-hazard sign on the highway. It reminded me of the slamming door in the face, "Don't ever ask me about this again!" tenor of my friend the Admiral's reply all those years ago in Miami when cornered about hidden UFO information. The similar semantics here don't prove anything, but Sorensen's reference to "non-disclosure agreements" prohibiting further communication with me certainly suits this storyline, just as the Admiral undoubtedly was adhering to his own sworn oath of silence as well. At the very least, this message came from the same Brad Sorensen whom McCandlish spoke of in his *DISCLOSURE* interview. His extraordinary list of high end and high tech design accomplishments listed on his web page further solidifies that conviction. Beyond that, it's up to the individual to draw conclusions. Mr. Sorensen's last name is part of his full email address; his extensive resume can be found online at:

http://www.coroflot.com/Sorensen/resume

The McCandlish chapter testimony, if true (and I believe it is), is nothing short of technologically iconoclastic, especially when taken in the context of theoretical physicist Michio Kaku's pronouncement that it could take "centuries or even millennia for our civilization to develop this technology." Perhaps not, Dr. Kaku. Maybe umpteen billions of Black Budget dollars have done the trick already! That these events reportedly took place in the late 1960's on up to the late 1980's is even more jarring, for one can only imagine what sort of deluxe Sport Model version of the Flux Liner our Billionaire Boys Club fraternity might have developed by now. Pluto might be a half hour away. Travel to other star systems? Can do, folks, can do.

The black divider curtain hanging in that Big Hangar at the Skunk Works on that November day in 1988 represents to me the great social, political, economic and intelligence divide our Jekyll and Hyde power structure has wrought, courtesy of the National Security State (as Richard M. Dolan has said, *the creation of a state within a state*). This is a chasm of such titanic proportions, its wonder makes the Grand Canyon seem a sludge filled drainage ditch by comparison. It eclipses the Grand Canyon with its luminescent majesty. Majestic. Yes. Majestic is an exquisitely chosen word, befitting the unacknowledged, yet ever-present monarchs who rule over its oligarchy of wickedly godlike hidden technology with supreme authority, unchallenged masters of the world, perhaps even the whole solar system by now, answerable to no one but themselves. They have harnessed the field and all its wonders with a technology that handily turns every cubic nanometer of space in the entire cosmos into a power source--one enabling them to tap into the invisible, infinite web of zero point energy permeating all of the ether. The Universe as a glorified array of invisible wall sockets. They can plug in whenever and wherever they like. The force fields are theirs. Their power is limitless. And they're guarding their secret technology and hoarding the resultant riches with a ferocity beyond measure.

Did they even know, did they even guess, before they solved the riddle and realized the Universe is an electrically pulsating, vibrant and unified web of electromagnetically charged force fields, what tapping into and harnessing such limitless power would do to them? Did they know that they would become the living embodiment of the trope that absolute power corrupts absolutely? But alas, this is no longer relevant. The gods of technology don't fret over mere mortal concerns. Jonathan Weygandt's brutal, black jump-suited interrogator was really speaking for them, his superiors, the unseen masters of the Shadow Government, when he shouted those threats in that interrogation room at the hapless, enthusiastic soldier who had seen too much. They are on their own program. They don't obey. They just do what they want. They will do you and your whole goddamn family.

307

So the question inevitably re-emerges: on which side of the curtain do we, the people belong? Do we own the Flux Liner and a myriad of other technological miracles? Or are we simply blindly funding a fleet of them, and other toys beyond imagining, and then some? To quote an expression my late mom used to like to use: Three guesses. Two don't count.

Based upon the material that has been leaked to me by high security insiders, my years of research, my own UFO experiences, and my gut sense of it all, I'm inclined to believe that the McCandlish material in *DISCLOSURE* consists of true and accurate accounts. I wouldn't be surprised if the whole blasted book is the same way. The description of how the 'zero-point engine' of the Flux Liner reduces the mass of the vehicle the faster it goes definitely gives a straightforward explanation of how such a craft could deftly circumvent Einstein's infinite mass problem with superluminal travel.

Regarding McCandlish's comment about it being only a matter of time before we all reap the benefits of this technology (forgive my skepticism about that, Mr. McCandlish), the *DISCLOSURE* book testimony of Dan Morris is relevant here.

According to Dr. Greer, Dan Morris was a retired Air Force career Master Sergeant who for many years was involved in extraterrestrial related projects; Greer interviewed him in September 2000. After the Air Force, he was recruited into the ultra-secretive National Reconnaissance Organization (NRO) where his focus was allegedly on extraterrestrial-related operations. Among his claims: that our military deliberately caused the 1947 ET craft crashes outside Roswell, NM, that we captured one alien and kept him at Los Alamos for 3 years, that Secretary Of Defense Forrestall was murdered because of his intent to release UFO and ET information (not the first time I've heard that), and that Germany was re-engineering crashed UFOs even before WWII. But most telling is his *DISCLOSURE* testimony, wherein he reveals his view of how we're being kept in the dark about the free energy technology that powers the saucers.

Morris alleges that UFOs are both man made as well as extraterrestrial in origin, and mentions how such luminaries as T. Townsend Brown and Nikola Tesla played pivotal roles in the free energy technology breakthroughs that ultimately dovetailed with the antigravity propulsion systems of the ARVs. As for the "why isn't this free energy technology being shared?" question, he lays the blame squarely on the petroleum industry, and how a cabal of the wealth-

iest men in the world control it. Bottom line: releasing this technology ultimately would result in personal vehicles (flying cars?) that
run on free energy, and don't need oil or gasoline, and the oil czars
and Billionaire Boys Club members have grown far too attached to
their multibillion dollar paychecks to let that happen--aside from
the fact that the switchover could potentially crater the global petro-
leum-based economy, to say nothing of the aircraft and automobile
industries. His conclusion? That the secrecy is no longer to keep
us in the dark about the reality of alien civilizations visiting us, but
rather, to keep us oblivious to the truth that free energy is available
to everybody on the entire planet, right now. So the secrecy about
UFOs and alien visitations is fully intertwined with the free energy
issue. To admit to the presence of the former is to have to come
clean about the latter. For now, it seems to me that disclosure is
not a viable option for the leaders of the Shadow Government, men
whose greed and ruthlessness know no bounds.[45]

Dr. Steven M. Greer's book is jam packed with eye opening material. I believe what I've referenced here represents a mere sam-
pling of its powerful impact. If all the allegations presented are true
(and I believe that they are), then this really paints a vivid picture
of the Jekyll and Hyde elements of our government, our military
and the corporations that own them. And the harsh reality seems
to be that our masters have no intention of sharing this free energy
technology as long as there is a single drop of oil remaining in the
ground. It's the power structure. Welcome to our special brand
of corporate totalitarianism, disguised as the dog-and-pony show
called democracy. I prefer to think of it as the *corporatocracy*, as
viewed by John Perkins in his *The Secret History of the American
Empire* (2007)--though I'd advise you not to read his scathing indictment with a full stomach, lest you find yourself needing to puke
your guts out repeatedly at what a globalized, inhumane nightmare
our corporatized humanity has fashioned for itself. As the late celebrated author, intellectual and social commentator Gore Vidal said,
"Although we, the people of the United States are the sole source of
legitimate authority in this land, we are no longer represented in
Congress assembled; our Congress has been hijacked by corporate
America and its enforcer, the imperial military machine." Alas, Mr.
Vidal, the nefarious machinations go even deeper than that. The
alien saucer technology is quietly humming with immeasurable
strength at the core of it all. And the dumbed-down noisy negativ-

ists are blindly, enthusiastically spinning their master's lies.

Further bolstering this viewpoint (about which I am fully in accord) is material I came across in a truly fascinating, engrossing and controversial book, *The Secret History of Extraterrestrials: Advanced Technology and the Coming New Race* (2010) by Len Kasten. I found this book to be a rich smorgasbord of full-spectrum UFO case material, along the lines of Jacque Vallee's classic *Passport To Magonia* (1969), though Kasten's remarkable book makes assertions about developments behind the scenes which surpass some of Vallee's hypotheses (I fault neither of the two researchers for this; they simply have different outlooks regarding UFOs, alien visitations and hidden technologies).

I considered myself an open minded, albeit skeptical UFO inquirer, both before and after my shocking Tyler material violently surfaced through regression work. But I've definitely shifted more towards the cynical, *nothing can surprise me anymore* investigator; my mind opened up to a lot more disturbing possibilities and probabilities after meeting Dr. Jones in Austin, the NASA aerospace engineer who so powerfully interrupted my recounting of the NASA UFO Video with his deadpan, almost world-weary and jaded, "You saw the flyby." Never have four words in all my decades of UFO investigation and research impacted me more dramatically than those four, delivered by a high security clearance insider who clearly derived some quiet enjoyment from monitoring my response. *You saw the flyby* has become the leitmotif of my hidden life, of my willingness to embrace the possible scenarios lurking behind the scenes, to a greater degree than I ever imagined I'd be willing to do. So Kasten's wildest material doesn't faze me in the slightest. Even the so-called outlandish stuff. Truth is stranger than fiction is a cliché which wears quite well for me now. Note to Disneyland devotees: it's not really such a small world, after all.

The Secret History of Extraterrestrials opens with Len Kasten revealing that his interest in UFOs began with a close up sighting, one that may have included a full blown close encounter with him onboard the craft. The Introduction narrative quickly streams past ideas that the Nazi's were experimenting with time travel as far back as the 1920's, and had ultimately developed flying saucers with death-ray weapons that could have enabled them to win the second world war had we not fortuitously bombed their factories to smithereens before they had the chance to use them en masse. Further, that some of their best scientific minds (those who weren't snagged by the U.S. post-war program Operation Paperclip and shuttled out to White Sands, New Mexico and other research sites in America to share with us the war spoils of their cutting edge knowledge; some sources say there were hundreds, perhaps over 1,000 of these imported-and-forgiven soldiers of the dark Reich) managed to slip away to Antarctica during the war years where they established a secret underground base and home for their super-science and flying saucer armada.[46]

This whole hidden Nazi Antarctic base idea has been around for awhile--certainly *The X-Files* has had fun with it--but those who have a kneejerk reaction of derisive skepticism should note that there is a substantial amount of well documented material backing up this scenario. This became clear to me back in my Coconut Grove days when I bought the bestselling book *Genesis*, by W.A. Harbinson (1980), which fully explores in a fictionalized format a series of "what if's" based upon the abundant evidence that precisely such a base was in fact established and was in full operational mode by war's end. I've read *Genesis* four times, and doubtless will do so again. Its well worn and scotch taped paperback binding is proof positive that a 586 page book can be a page turner despite its exceptional length. The *Author's Note* at the end of the book is a full 14 pages long, and the *Sources* another 4, and between the two, the amount of solid documentation cited speaks volumes about how much Harbinson painstakingly paid attention to known facts in fashioning the chilling scenario he depicted.

Fully exploring this line of query would take another book, and then some. But a good number of valid statements about the Nazi's and their flying saucer technology can be made. Primary would be the fact that scientists and engineers in Nazi Germany were aggressively pursuing research into antigravity technology, from Hitler's rise to power in 1933, and more intensively towards the end of World War Two. The cast of characters includes Rudolf Schriever, Klaus Habermohl, Richard Miethe and Viktor Schauberger, and a flying saucer prototype known as the Schriever/Habermohl/Miethe disc.[47, 48]

Nick Cook's utterly engrossing book *The Hunt for Zero Point* (2001), presents a factual, occasionally speculative treatise about the near certainty that exotic and unconventional antigravity propulsion systems were very much in development in Nazi Germany:

the engineer, inventor and radically unconventional thinker Viktor Schauberger was one of the key players, and evidence indicates that a new and cutting edge technology arose around many of his ideas. This technology represented such a quantum leap beyond anything else in its day, that an overt policy of technological suppression emerged to guard it from unwanted eyes (Google: Schauberger-Implosion Motor; Schauberger saucer; Repulsator). Remarkably, during his research, Pope was shown documents by Schauberger's grandson, Joerg Schauberger (who oversaw his grandfather's archives), that revealed his grandfather had constructed what may have been an antigravity device, the core of which rotated a fluid in a high speed turbine, up to 15,000 to 20,000 rpm's and beyond--seemingly a precursor to the aforementioned Plasma Torus Antigravity Centrifuge Engine. And some of the files he was shown by Joerg indicated that a colleague of Schauberger had put the prototype through an unauthorized test run while Schauberger was away from the plant. The device had generated such a powerful antigravity effect that it bolted upward, colliding with the roof of the hangar.[48]

In *Genesis*, Harbinson's background research represented in his extensive *Author's Note* presents a compelling case (in some ways paralleling Pope's material) that the Nazi's frequently were sending U-boats, fully laden with supplies, workers and construction materials, to their 230,000 square mile Antarctic stronghold they'd christened Neuschwabenland throughout the war years, industriously carving out a home base where their newly developed super technology could take hold and flourish. And further, that they had successfully test flown a flying saucer prototype in Germany in the early months of 1945, the culmination of labor measuring in thousands of man hours (likely much of it in the form of slave labor culled from the millions of unfortunate souls in the Nazi death camps). And that finally, in March of 1945, just prior to the war's end, two of their U-boats, U-530 and U-977, packed with their flying saucer research scientists and engineers, along with their most crucial flying disc components, detailed drawings, notes and engineer's schematics for constructing the saucers, and the designs for constructing colossal underground factories and living quarters, successfully arrived in Neuschwabenland.

It is a fact that upon Germany's defeat in May 1945, the Allied soldiers discovered an almost unbelievable maze of gigantic underground factories and research and development centers in Nord-hausen and Thuringia in the Harz Mountains of Germany, and that the embarrassment of high-tech riches, much of it immediately classified, was divided amongst the U.S., the Canadians, the British and the Russians. So the idea that such an industrious and focused Nazi work force could replicate these feats of engineering at the South Pole is not to be discounted. Furthermore, part of Harbinson's horrific theme hinges on the fact that the Nazi's infamously had no reservations about performing human vivisection and medical experimentation on living subjects, and proposes that they could quite possibly have brain-implanted the entirety of their work force, robotizing an underground army of millions of slave laborers who would work 24/7 nonstop joyfully until they dropped dead from exhaustion, briskly removed and replaced by the next human slave automaton. (Back in the late 1980's, I interviewed a professor who was then the director of the Artificial Intelligence department at the University of Miami--regrettably I've misplaced his name--who assured me the technique of electrode brain implantation in humans to electrically stimulate specific neural areas and create desired behaviors and emotions was virtually perfected *in the 1940's*. I asked him to repeat that final statement to be sure I'd understood him. That was a sobering concept, giving credence to Harbinson's thesis.)

The fact is that our country did dispatch a large military operation to Antarctica in December 1946 under the aegis of then Secretary of Defense, James V. Forrestal. Operation Highjump consisted of 13 ships including an aircraft carrier, six Martin PBM flying boats, two seaplane tenders, six two-engine R4D transports, six helicopters, and over 4,000 men. It was commanded by Admiral Chester Nimitz, Captain Richard H. Cruzen, and the famous explorer Admiral Richard Byrd. Undisputed is the fact that the mission was abruptly aborted, and it returned from its intended six month sojourn more than three months early. What is disputed is why. The "official" version alludes to extreme difficulties with Antarctic weather conditions. The unofficial version, as played out in Harbinson's *Genesis*, is based upon unconfirmed though allegedly dependable off the record reports that the intrepid team was met with an attack force of Nazi flying saucers from bases under the ice which confronted them with aerospace and weapons technologies so vastly superior that not retreating would have meant certain suicide for all.

I include this commentary because I believe it merits consideration for anyone who wishes to keep their mind open to all possibilities in their alternate global view of this strange world of ours. In actuality, we may live in a world of complexities far greater than we can imagine. The reality could be that not only does the U.S. possess saucer technology (and perhaps a few other global powers), but additionally, the undercover Antarctic post-Reich Aryans, as well as far more advanced field propulsion systems wielded by a multitude of off world visitors (some reasonably benign, and others, not so much). If you ponder all of these factions gathered together on the same conflict-prone chess board of Sol-3, is it any wonder there are so many UFOs sighted? The Introduction to Kasten's *The Secret History of Extraterrestrials* also alludes to the 1943 Philadelphia Experiment (utilization of high energy electromagnetic fields onboard a Navy ship for teleportation and to engage an invisibility cloak) as having inadvertently opened up a portal that welcomed in a hostile faction of extraterrestrials. So then we might also add into this turbulent mix the intrusion of unfriendly interdimensional visitors. All of these possible factors make for a wild and wooly planet and cosmos surrounding us. As for me, I wouldn't have it any other way. I've always loved the wild west. Life is an ongoing adventure, so relish and enjoy the exhilaration while you're here.

Almost certainly running competitively and in parallel with Germany's aggressive 1940's saucer project, physicist Thomas Townsend Brown (often called T. Townsend Brown, born in 1905), was at the center of the equivalent American program. Brown's name crops up often, in books and on the web, when one is looking into the back-story of antigravity, UFOs and hidden technologies in the U.S. (He was allegedly involved in the disastrous Philadelphia Experiment, according to Kasten's book.)

The Biefield–Brown effect, first discovered by professor and student team Paul Alfred Biefield and T. Townsend Brown, is an electrical effect in which an ion wind transfers its momentum to surrounding particles which are neutral. This effect is more commonly known as electrohydrodynamics (EHD) and is considered a counterpart to the well-known magnetohydrodynamics. It is clear that in the United States in the 1950s and 1960s and decades beyond, an enormous amount of time and money was spent on research conducted with the aim of utilizing this electrical propulsive effect for propulsion systems, during what might be considered the golden era of antigravity research. This timing also corresponds to that period when antigravity research suddenly fell off the radar (1957-58), clearly suggestive that this is when the insiders decided that their potential breakthroughs, heralding quantum leaps in propulsion technology, needed to be buried beneath the hermetic seal of the classified world. Entire volumes have been written about this era, but for the purposes of this book, it is my contention that we now have a good general sense about how America's flying saucer program took off (pun intended), and was quickly locked away.

Kasten's *The Secret History of Extraterrestrials* chapter, *The Politics of Antigravity*, expresses a degree of anger towards NASA for how they seemingly have played dumb with the public for decades about this hidden technology, the thrust of it being, how dumb do they think we citizens are that we believe chemical rockets are the best we can do? This is definitely one of my favorite chapters, which includes an eye opening section regarding four key patents that have been granted in the U.S. which, when taken as a group, appear to be the four key ingredients you would need if you were to name your project, *How To Build An Antigravity Spaceship*. One (of several), granted to T. Townsend Brown on July 3, 1957 (interestingly, right about the time the lid was clamped down shut on this line of research), was concerning the antigravity properties of highly charged transducers. The second, filed by James E. Cox/Los Angeles on July 26th, 1982, was called the Dipolar Force Field Propulsion System; essentially, an electromagnetic energy propulsion system yielding an incredible 1 million pounds of thrust, vs. 29,000 lbs. for a jet engine of the same weight (is this sounding more like *Star Trek* or *Star Wars*?). The third, filed by James R. Taylor of Fultondale, Alabama on March 6, 1992, was for a propulsion system that utilizes electromagnetic energy produced by supercooled, high density electric power.

Kasten's argument is that these three propulsion systems combined together would result in a craft ready for space travel. But if the craft were to use this advanced antigravity technology, it would require a protective bubble to safeguard the pilots and astronauts from the high energy multimillion volt charges surrounding the vehicle. Enter, Ernest J. Shearing of Porterville, California, who filed "Protective Enclosure Apparatus for Magnetic Propulsion Space Vehicle" on September 20, 1991. Suffice to say, this chapter alone is worth the admission price of a copy of Kasten's book, in my opinion. But there's plenty of other exciting and cutting edge material packed into *The Secret History of Extraterrestrials: Advanced Technology and the Coming New Race,* as well.[49]

So if this is all true, and in fact NASA and a plethora of other hidden high-tech aerospace corporations have been lying to us about how far along we are in these exotic field propulsion technologies, why hasn't anybody come forward to leak this information, even just a tidbit? Obviously, I believe I've already met a man who has, in the form of Dr. Jones, the NASA aerospace engineer. And somebody at NASA clearly was offering this glimpse of the other side of the curtain to Felipé of the *Space Science Center* in Santa Fe, and to those of us fortunate enough to have seen his classified video before the *Center* closed down for good.

But there is another man who has been very much in the public eye with regards to NASA's hidden UFO, alien presence and suppressed technologies material. His name is Gary McKinnon.

* * * * * *

UNITED STATES OF AMERICA
v.
GARY McKINNON

This is an encroaching, privatized censorship regime.

--**Julian Assange**
We Steal Secrets

My first exposure to this remarkable story was when I viewed a video interview with Gary McKinnon on YouTube, conducted in London in June 2006, uploaded on November 20, 2007, called *Hacking the Pentagon: an Interview with Gary McKinnon*, posted under the guise of *Project Camelot interviews Gary McKinnon*. The interview was conducted by one of *Project Camelot's* founders, Kerry Cassidy; you hear her voice, but the camera remains on Gary. Both this video interview, and the full text of it, can be viewed at the following web addresses, respectively:

http://projectcamelot.org/gary_mckinnon.html

projectcamelot.org/lang/en/gary_mckinnon_interview_transcript_en.html

(Due to Gary's occasionally strong accent, sometimes his words in the video are unclear. The full transcript at the above web page is worth a visit. The full details in this document are remarkable.)

One of the first things that struck me about this interview was how boyish and innocent Gary McKinnon appeared, at the time a very youthful forty. Along with that, I had the strong impression that he was speaking the truth, his sincerity really came through in his voice and demeanor. He was quite even handed and direct with his words, his gaze was always steady with eyes that didn't shift about, and there were no embellishments or dramatic exaggerations in his narrative; if anything, he seemed low key, even about the topics some people would have trouble handling. The matter of fact delivery of his story unfolded smoothly through the interplay of Ms. Cassidy's thoughtful questions.

As the story goes, starting around February 1st, 2001, from his apartment in north London, Gary McKinnon began to hack into the computer networks of the Pentagon, various U.S. Navy systems, and NASA; his intent: to look for UFO-related material, which gradually came to encompass anything that might pertain to the concept of free energy and/or hidden technology. McKinnon had seen Dr. Steven M. Greer's *The Disclosure Project Press Conference* online in 2001, so he became familiar with its focus on those topics, and the array of credible insiders who had come forward to speak out. He was captivated by Donna Hare, a NASA photographic scientist with a Top Secret clearance, divulging how she'd worked at Building 8 of the Johnson Space Center in the Clear Lake area of South Houston. He was intrigued by her disclosure regarding how her co-workers had blithely introduced her to their technique of airbrushing out UFOs from many of the NASA space and atmospheric photos. As McKinnon told Kerry Cassidy, it was his belief that we should have the hidden technology behind these craft, that it should be used to benefit everybody's quality of life. He cited the perverse tragedy of wars over oil while old British pensioners were dying because they couldn't afford heating their homes because their fuel bills were so high. His passion for uplifting humanity via revealing these technologies drove him to become virtually a 24/7 hacker for years. From the get go, Gary McKinnon was a hero to me.[50]

Using his quite modest 56k dial-up connection, he marveled at how poorly so many of these high-level U.S. government agency computer networks were protected, as he utilized what he referred to as his "graphical remote control" capabilities. Many of the systems he hacked into had no passwords whatsoever--rather shocking, in light of how this was post-9/11. Additionally, he said that one of the passwords he'd successfully used to breach a network (when a password was actually required) was "PASSWORD," which makes it fairly understandable that he'd typed the comment "your security is crap" onto one of the system's screens on one of his outings.

McKinnon stressed that "about 99% of the time" he found nothing. He really hit pay dirt, though, with some jarring, eye-opening material when he hacked into some of NASA's network computers, possibly the very ones in Building 8 that Ms. Hare had alluded to in her testimony. He discovered a cache of photos with enormous file sizes, 200 to 300 megabytes each. Since he knew his meager 56k connection would take untold hours to fully download even a single image of that size (nearly 17 hours for the "smaller" 200 mb photo, assuming the data stream never faltered), he applied some ingenuity to the task. By accessing the remote control program that gave him access to the machine's graphics, McKinnon was able to dial down the resolution of the images to what he recalled was roughly 4-bit color, drastically reducing the file size and the download speed. Then he remotely accessed the desktop of the NASA machine, opened a folder and clicked on the first image he found. Gary admitted he was hoping to see a photo of a flying saucer, along the lines of what Donna Hare had been shown. What began to resolve slowly was what he believed to be Earth from the perspective of space, in greytones. About 2/3 of the screen was taken up by its hemisphere, but as the full image materialized, it gradually revealed an additional element, what he referred to as "a classic sort of cigar-shaped object" (presumably orbiting the Earth or off to the side), but with a number of appendages that looked like geodesic domes on the upper and lower surfaces; it tapered down to somewhat flattened cigar-shaped ends. The object had no seams, nor antennae, nor rivets. McKinnon believed that this was a craft, and that its unusual appearance meant it wasn't man-made. In his opinion it lacked any telltale signs of human manufacturing.

When the image was finishing its slow materialization, and had filled about 3/4 of his screen, it seems that Gary had been found out by a NASA rep at the center. He obviously saw that McKinnon had moved the mouse pointer remotely, and swiftly right-clicked the LAN connection, then punched Disconnect; that was Gary's last view of the high tech, outer space reality hidden from us all. The pivotal moment seemed strange to him, and yet in a way triumphant. He was able to do what he'd set out to do, to authenticate Donna Hare's allegations. But the abrupt moment of elation would quickly devolve into a nightmare consuming more than a decade of his life, with him treading water in a morass of legal misery brought on by the U.S. government.

Even more material with an "outer space" theme via the NASA network surfaced when McKinnon pulled up a now-infamous Excel spreadsheet list whose title was captioned with the highly suggestive: **Non-Terrestrial Officers**, a list which included names and ranks, but clearly not related to any of the traditional Armed Forces branches. When queried by Ms. Cassidy, McKinnon allowed that he couldn't remember details about the number of males vs. females on the list, but that he figured there to be 20 to 30 of these officers tallied. He also came across **Ship-to-Ship Transfers** as well as **Fleet-to-Fleet Transfers**, in some cases seemingly alluding to outer space off-world cargo shipments, and even military operations with regards to other planets; he was definitely convinced that these listings did not refer to traditional ocean going Naval vessels, but rather, to an **Off-World Space Fleet**. Whether this information alluded to our solar system or another star system altogether was unclear. When Ms. Cassidy asked him what he thought when he saw this "Non-Terrestrial Officers" list, McKinnon replied, "Wow, I'm really onto something here!"

On March 19th of 2002, the U.K.'s National Hi-Tech Crime unit interviewed Gary McKinnon at the behest of the U.S. government, and arrested him the same day. The legal wranglings which ensued would take pages to fully detail, but suffice to say, the U.S. authorities were out for blood, and in short, they wanted Gary McKinnon extradited to the U.S. to stand trial, accused him of damaging multiple computer systems to the tune of $5,000.00 for each of the machines he was on ($5,000.00 being the magic number in dollars of what would be required to elevate his charges to a cybercrime), and piled charges on to that so that his potential punishment could tally up to 60 to 70 years in prison. (McKinnon alleges that he never damaged any of the machines the slightest bit.) U.S. attorney Paul McNulty of Virginia stated that "McKinnon is charged with the biggest military computer hack of all time." McKinnon lived in a virtual limbo from that fateful day in 2002, plunging into progressively deeper depression as it seemed more and more like a hopeless case he was fighting. The back and forth legal volleys regarding his possible extradition seemed interminable.

As the ugliness escalated, and it grew more likely he'd be extradited, some British police officials traveled to the U.S. and met with the Office Of Naval Intelligence (ONI), and possibly the Air Force Office Of Special Intelligence (AFOSI) as well. They came back with grim news as to what the U.S. authorities wanted from this case, stressing that it was a very serious matter. Finally an investigative journalist named Peter Warren flew to America to meet with the top officials involved, and queried why exactly it was that they were pursuing McKinnon with such a vengeance--and why his was to be the first extradition hacking case ever. The officials supposedly told Warren that it wasn't just about what Gary had done and what he'd been saying, it was where he might have been and what other things he might have seen that he wasn't talking about which deeply concerned them. This comment during the interview triggered Ms. Cassidy's query as to whether Gary might, in fact, have an Ace up his sleeve; he grinned boyishly and said he'd never say so if he did.

Over the years, as McKinnon's mental health steadily declined, and was further compounded by his Asperger's Syndrome, his lucky break came in May of 2010 when the latest British home secretary Theresa May agreed that his case for denying extradition to the U.S. warranted further consideration due to health considerations. Finally, on October 16th of 2012, home secretary Theresa May withdrew McKinnon's extradition on "human rights grounds." She based her conclusion and decision on medical reports indicating that Gary's continually declining health and chronic depression made him a risk for suicide if he were sent to the U.S. to stand trial. McKinnon's mother Janis Sharp praised Ms. May as being "incredibly brave" to "stand up" to the Americans.[51]

U.S. State Department spokesperson Victoria Nuland was quoted as saying, "The United States is disappointed by the decision to deny Gary McKinnon's extradition to face long overdue justice in the United States. We are examining the details of the decision." (Translation: Our original plan was to toss a coin to decide Gary McKinnon's fate: Gitmo or Extraordinary Rendition for this bothersome Brit who dared to reveal the unadorned truth about our monstrously advanced and carefully hidden technologies we don't intend to share with anybody--certainly not with the everyday tax payers whose tax dollars fund them and keep them locked away?)

Because of how my own *you saw the flyby* NASA related experience has impacted my psyche, and opened my mind to the high probability of the full ramifications of our almost mind-blowing Jekyll and Hyde grand canyon socio-techno-economic chasm operating today, the follow up questions Kerry Cassidy posed to Gary didn't really faze me. She asked him what he thought about the story that a group of 12 Earth astronauts, specially trained, went to a planet named Serpo in the Zeta-Reticuli star system for 13 years, supposedly from 1965 to 1978, as part of an exchange program with the aliens known as EBENS (or Ebens--Extraterrestrial Biological

Entities). (This storyline is expounded upon in Len Kasten's *The Secret History of Extraterrestrials.* It's also alleged that Steven Spielberg used this material at the end of his film *Close Encounters Of The Third Kind,* information possibly gleaned from insider intelligence contacts.) Though I originally thought the Serpo story sounded like disinformation or just an interesting tale, part of me is inclined to side with Gary McKinnon's answer to Ms. Cassidy: "I think all things are possible in these kinds of cases." The thing is, once you've grasped the reality of this super advanced technology being around for decades, and the extraordinary funding the Black Budget provides, I see no reason not to connect the dots and accept the likelihood this is really going on, that these bastards grabbed the ball long ago and ran with it in a really, really huge way.[52,53]

Famous UFO author and researcher Timothy Good (*Above Top Secret,* et al) gave a lecture called *The Secret Space Program Amsterdam 2011.* You can find it on YouTube, as well as a variety of related speaker's videos, such as author Richard Dolan, also expounding upon our ostensibly undercover and super high-tech antigravity secret space program operating right under our noses. A quote which Good relates, and is also in Kasten's *The Secret History of Extraterrestrials,* is one attributed to Ben R. Rich, the former president of Lockheed-Martin's Advanced Development Programs (ADP) group, also known as the "Skunk Works," which he apparently delivered in an alumni speech in 1993 at the University of California, Los Angeles (a revelation he shared just before his death). Mr. Rich told the crowd, "We already have the means to travel among the stars, but these technologies are locked up in black projects, and it would take an Act of God to ever get them out to benefit humanity. . . . Anything you can imagine, we already know how to do." Interestingly, this very same sort of comment came out of the mouth of William Cooper in Santa Fe in May of 1992, during the Friday night introduction to his weekend lecture and seminar I attended. I recall him loudly proclaiming to the hushed audience, "This elitist group I am speaking of, some call them the Illuminati, some the Trilateralists, some the Bilderbergers. Call them what you will. But people, I can assure you of one thing: the technology they possess, and the things they are able to do with it, are beyond your wildest imagination. Their technology would absolutely blow your mind." Food for thought.[54]

Given the nature of our culture's denial about these things, and how the head-in-the-sand naysayers still gleefully bash the pro-UFO, alien and hidden technology believers, I find it apropos to wind this down with the online comments posted by two people who have watched (presumably all or most of) the Gary McKinnon interview conducted by Kerry Cassidy, this one on the YouTube site, (uploaded on November 20, 2007). I've changed their names, but their dialog, copied verbatim (his posted in late August 2013, hers around October 21, 2013), requires no set up:

Joey: He stole a lab top [sic] with access, that's not hacking! and I don't care about that! I said he's full of crap about aliens give me a break. what are you 12?

Delia: If he didn't see those things, the Yanx wouldn't care. But obviously he saw what he saw, and the result is the Yanx trying to lock him up for 60 years. And here you are still clinging to stupid. Go back to your rape porn, you don't belong in the company of decent people.

This semi-comical exchange between Joey and Delia is a prime example of the cognitive disconnect that exists even now, well into the 21st Century, where one side argues that the whole UFO concept is a sham--let alone that aliens exist and are visiting--and the other side, having moved way beyond that as a reality, accepts it as a perfectly logical conclusion that Gary McKinnon's U.S. tormentors want him locked away and silenced precisely because of the paradigm shattering realities he became privy to in his sleuthings.

It reminds me of a statement made by Dr. J. Allen Hynek, who had in his decades of UFO involvement done a full 180 degree transformation in his outlook, from the shameful Michigan swamp gas caper of 1966 to being a firm believer that some UFOs were extraterrestrial in origin. He pondered aloud (I'm paraphrasing), "While some people have finally gotten around to accepting that some UFOs are extraterrestrial spacecraft, a lot of them can't seem to grasp that there are occupants onboard the vehicles piloting them."

* * * * * * *

UFOs & THE REALM OF SPIRIT

Let us look very coolly, as biologists or engineers, at the lurid
atmosphere of our great towns at evening. There and everywhere
else as well, the earth is continually dissipating its most marvelous
power. This is pure loss. Earth is burning away, wasted on the
empty air. How much energy do you think the spirit of the earth
loses in a single night?

--Pierre Teilhard De Chardin
Human Energy [55]

Once on a quiet Sunday in Santa Fe, I was paging through the
classified section of our alternative newspaper, *The Sun.* There was
an ad placed by a local psychic whose name I can't recall, but I do remember an italicized quote at the header of his small advertisement
layout. Its wisdom has remained with me to this day. The quote
read something like this:

We are not physical beings experiencing a spiritual side to our lives.
We are eternal beings of spirit temporarily experiencing a physical life.

For me, the beauty of these words comes from their simplicity
and their truth. They resonated with me. They also triggered memories of a paper I'd written back in 1991 for a philosophy course at
FIU called Phenomenology. My professor, Dr. Kovacs, was hands
down one of the most magnetic speakers I've ever witnessed, the
sort whose words flowed out with deeply passionate, forceful poetry
and wisdom along with the truths he wished to convey, and I rue to
this day that he didn't allow students to tape record his sessions--
they were that powerful. The paper I'd submitted for a mid-term
grade was titled, *Spirits In The Material World.* The thrust of it
was based upon some of the lyrics of the eponymous song written by
Sting and sung by his now former musical group, The Police.

Although the theme of my paper then paralleled the song lyrics and was geared towards the concept of man's futile search for
a deeper meaning in a life so utterly geared towards materialism
and the illusory nature of the physical world, all the while struggling
against duplicitous leaders who subjugate the meek using words of

entrapment, it applies just as well to this book. It speaks directly to our state of being kept in the dark by a hidden cabal of leaders, and our failure as a people and a species to realize the profound significance of UFOs and the alien presence here and their impact on our lives and on our world--which for me is one of the most important issues we face, especially given that the hidden technologies represented by them and their reverse engineered wonder-craft could abolish poverty on a global scale and transform the planet dramatically, practically into a virtual Eden. One of the primary expressions Dr. Kovacs hammered home to me, ostensibly the very foundation of phenomenology itself, was: Legain ta phenomena: *Let the phenomena speak for themselves.* And haven't UFOs been doing just that, for decades? In fact, the phenomenon has spoken to us for centuries, if you include the UFOs rendered in art works dating back to the Renaissance, perhaps even in passages from Shakespeare ("Dazzle mine eyes, or do I see three suns?"--*King Henry VI*, Part 3, Act II, Scene I) and scores of multi-cultural unknown aerial object reports ("flying earthenware") in ancient and medieval times as detailed in Vallee's *Passport To Magonia*, or Ezekiel's high performance Biblical flying machine, and even further back as rendered in some of the famous cave paintings in excess of ten thousand years old. (Evangelicals: please disregard that last part, lest your heads explode from the potent, honest-to-God truth you've been carefully, religiously schooled to reject outright and airbrush out of existence, like the scary Texas State Board Of Education's religious zealots love to do. God bless all their pointed little heads.)

It fascinates me that sometimes a discussion about UFO experiences and abductions, especially when recounted by an experiencer, can naturally flow into the realm of spirit and the spiritual side of life. Such was the case when I spoke with Travis Walton.

Back in March of 2013, I blundered across an online posting touting a UFO event called the *Out Of This World Conference & Festival* in Edinburg, Texas. This was on a Monday afternoon, March 11th, and their pre-event started on Thursday. When I noticed that both Stanton Friedman and Travis Walton would be speakers, I got on the phone and got to work. Although a native Texan, I'd never been down to what is pretty close to the toe of the boot of Texas, so this would be a good getaway. It had to be done.

Speaking with the friendly and helpful Edinburg Chamber of Commerce reps on the phone from my motel room down there, I discovered that the Thursday night event was a Meet-and-Greet, and since I was writing a UFO book, they green lighted me to attend. Their building was originally a railroad depot for the Southern Pacific line in 1927, and became a Texas Historical Landmark as of 1996; the Mission Style architecture of the multipurpose structure retains its innate charm, and houses the Edinburg Convention & Visitors Bureau in addition to the Chamber of Commerce.

I arrived around 6:30pm, and immediately recognized the distinguished man in the suit at the buffet table as Stanton Friedman; rather than ambush him, since he'd barely arrived from his flight, I waited until he was comfortably seated on the back patio, where a talented guitarist played beautiful instrumentals to the scattered groups of attendees. When the conversations between Friedman and others at his table had subsided, I had a pretty lady who was one of the event organizers introduce me. I told Stanton that I considered him to be the Obi Wan Kenobi of the UFO field, one who was solid and trustworthy and worthy of respect, and that I'd loved his books, particularly *TOP SECRET/MAJIC*. My comments were warmly received; in person, I found him to be every bit as impressive and charismatic and full of gravitas as he'd been in his decades of television appearances. What you see is what you get. He agreed to an interview after the event, back at his motel room, which turned out to be one floor above mine.

Having secured my first objective, I returned to the large entry hall of the Center, and immediately recognized the deeply tanned man in the suit at the buffet table as being the world famous UFO abductee from Snowflake, Arizona. Since he hadn't been bombarded yet by the throngs of people who would soon enough surround him, I walked up and introduced myself, offering my handshake and saying, "You are definitely Travis Walton." He smiled warmly, and I reminded him that I'd emailed him and spoken to him on the phone a few days prior, having been given tips on how to reach him by Mr. Friedman. After only a few minutes of conversation, in which he agreed to be interviewed at his book signing table on Friday, the crowds swarmed around him, photos were taken, and hands were shaken. When there was a temporary lull, I sidled back up to him and half-way joked, "Wow, that was something! You must be the equivalent of a rock star in the UFO world."

He grinned and replied without hesitating (I'm paraphrasing), "But you know, I don't see it that way at all; this is really a serious thing, and a big opportunity for me to connect with people from all over who may have had similar experiences." Travis acknowledged that he thought of it as his responsibility to be open to communicating to people who shared an interest in UFOs and the alien presence on our planet, and in the process it allowed him a way to turn his experience into something positive.

As the gathering wound down, on the way back to my truck I reflected on how lucky I was to have met these men, and felt grateful that they'd been so cordial and gracious about blocking out some of their time for me. If I had to describe my first impression of Travis Walton in as few words as possible, the first word would be *humble*. This man is definitely not on a star trip, even though his life altering encounter was made into a big screen movie with big name movie stars. I think humility is one of the most underrated of all human virtues--certainly there's almost a tragic deficit of it in our culture today--and the fact that he so fully embodied that quality meant a lot to me. Other words would be: intelligent, sincere, forthright, and real. I couldn't wait to sit down and talk to Travis at length. In the back of my mind I wondered if or how I would tell him about my Tyler encounter.

* * * * * * *

THE TRAVIS WALTON INTERVIEW

The next day, in the lobby of the massive Edinburg Renaissance Conference Center, I found Travis at his table, surrounded by people buying his autobiographical book, *Fire In The Sky: The Walton Experience* (2010 edition) asking for autographs, and getting photographed with him. When things quieted down, I asked if we could start our talk, and he warmly assented. The following is a direct transcript of our taped interview:

Wade Vernon: One thing I've come across with UFO experiencers, particularly those who believe they've been abducted, is the concept that, if given the choice, they probably wouldn't have chosen to be put through all of this--the fear, the shock and the trauma--but because now it's a part of their history, they wouldn't trade the insights they have now, having had the experience.

Travis Walton: Yeah, that's kind of what I'm left with. At the time, I would have been very reluctant to have had this experience. But by the time I was returned, it was worldwide news. People complimented me for having the courage, quote unquote, to come forward, but I don't deserve that. I tried to avoid the media at first. But I found that, if I didn't do the interview, they'd rely on negative sources, so in order to set the record straight, I found out I had to get out there and answer these requests. And I think now, over the years, I've gradually realized that with the burden comes a major responsibility to try to make some good come of it.

WV: I've noticed there are some people on the internet who seem to conform to Stanton Friedman's *noisy negativists* typecasting with regards to their claiming in very harsh language that you're a fraud, etc., ad infinitum. I'm just curious: did you even know any of those people who were slamming you--one of them seemed to be pretty recent, like a couple of years ago? Or have you even bothered with following it anymore?

TW: People come to me and report this stuff to me. There's one guy who's claiming he knew my kids in school and all of this--

WV: Yes, that's the one I'm talking about.

TW: And the people that were responding to him were pretty well educated themselves, and they said, wait a minute, your facts are wrong. Because, you know, he had all kinds of information that was just not that way. He had Allen Dalis living out east of Snowflake-- he didn't. We all lived in Snowflake. It's impossible for him to have lived out there. So that was his big theory, that the law enforcement people knew that's where I was the whole time. Which is just totally absurd. The whole thing was just totally made up. He claimed to be the county sheriff's nephew. And he certainly wasn't. The county sheriff was Gillespie. He claims to be related to the town marshal, who was not the sheriff. But I asked the town marshal, and he has no idea who this could be. He also said he went to school with my son and nephew, and they could not figure out who this could possibly be, looking through the old yearbook. They have no idea who this guy could be. He also claimed that the crew met with the sheriff at the *Red Robin* diner--and the *Red Robin* diner wasn't even invented for another 20 years after that. So it was one thing after another of these totally absurd claims, that the crew was chased through the woods by an Air Force helicopter. And don't you think those blades would have been hitting the trees? And, you know, this UFO was less than 100 feet away, below tree-top level, and there's no mistaking what we saw for a helicopter. There would have been a huge down rush of air that would have been blowing pine needles everywhere. And everybody knows the sound of a helicopter.

WV: Right.

TW: And he claims it was common knowledge about the Air Force maneuvers. And there are no reports. There were just all kinds of totally inaccurate things. It's just somebody trying to get attention. He's not the only one. Every so often, somebody will come out who claims to have gone to school with me. And it's not always people who are negative. I mean, there's this one guy who is one of my supporters who claims that the first thing I did when I came back was call him. And he wasn't home, so then I called my family. (laughter)

And that's simply not true. The operator listened in on my phone call, and reported it to the sheriff. There was only one phone call, from me to my family.

And there's another one. They claimed they attended that homecoming party for me after I was returned, as depicted in the movie. However, that never happened. There was no homecoming party. So, you know, even people who are positive about this experience, and they're claiming to have some connection, sometimes they're just making it up.

WV: Did you feel the movie was accurate as far as depicting the vehicle that zapped you and then took you?

TW: We described it as looking like hot metal from a blast furnace. Not quite white hot, kind of a golden glow. In the movie it looked more like a cool, molten swirling lava.

WV: How about the waffle iron pattern underneath?

TW: No. No, it wasn't like that. I don't think that their changes made any kind of improvement in the story. But hopefully we'll get that corrected in a future remake of the movie.

WV: Part of the reason I was disappointed is that I thought the actual story of what happened to you was a lot more interesting.

TW: Yeah, and it's not in the category of exaggeration. In the case of the beam that hit me, they watered it down tremendously in the movie compared to what really happened. You know, with Hollywood so obsessed with pyrotechnics and light shows, and that sort of thing, that was an opportunity to show something that was really powerful. The men described that energy as so powerful it looked like I'd stepped on a landmine--like an explosion of energy.

WV: Meaning the guys in the truck who saw it?

TW: Yeah. And Dwayne Smith, one of the crew men, said that he thought I had disintegrated. That it had blown me to pieces. Steve (Pierce) told him, "No, no, I kept watching him until he hit the ground, and he was in one piece." But then while that energy was around me, in this ball of light, it was so powerful that it was certainly what led them to think it had killed me. Flew me back in the air twenty feet. And that's far more dramatic than what was depicted in the movie. The crew said it was so bright that it lit up the woods brighter than daylight. And they could see in all directions from the way that it lit up the forest.

WV: It's been over 35 years since your encounter that night in Arizona. I'm assuming it was very challenging for you, at least initially, to cope psychologically with the fresh memories in those days, weeks and months after this happened. How have you coped with the extreme strangeness of your encounter--is it an ongoing thing that you have to deal with?

TW: I think it's improved over time. It's a lot better. I used to just be drenched in sweat after giving a talk. Even to this day, whenever I talk about it, it's kind of like re-living it. It really kind of hits me with stress, but not nearly as bad as it used to be. But when I first started doing interviews for the movie when it was being released, where there were so many interviews back to back, I thought, man I'm not going to be able to bear talking about this anymore. But it actually sort of desensitized me to a certain extent.

WV: Have you ever done any more hypnotic probing into your onboard time--do you think there might be more material?

TW: I'd be really leery of undergoing any further hypnosis. The theory being, if I was hit by a beam that caused cardiac arrest or actually killed me, and the aliens then had to take me onboard in order to revive me, one theory is that if I were regressed hypnotically back to a time in which I was clinically dead, it might kill me. So that's enough reason to give caution. Little bits of memory have come back in dreams--if they are memories rather than just dreams--but nothing makes any sense, nothing that would follow a narrative of any kind. So, unless it comes back spontaneously on its own, which is unlikely after 38 years, I don't know.

WV: In the decades since your experience, you've had a lot of time to reflect on things. Do you ever find yourself coming back to the simple question, why me?

TW: Yeah, I've thought about that a lot. And that, of course, was a major question in my mind from the beginning. I blame it on myself; I think by being so impulsive and getting close so unexpectedly, I caused things to happen that weren't intended. However, you know, some of the crew members think that I was chosen--that they were waiting specifically to get me, and that they somehow were able to manipulate my behavior and cause me to get that close. But I guess probably I have a personal bias, to prefer to believe that it was just an accident, something that spontaneously happened, because that would imply that it's over, and it only happened because of my actions.

WV: That actually led right into my next question: have you ever wondered if they might come back for you again?

TW: Umm, yeah, you know, people are constantly asking me if they have, because they want to--I don't know, I guess they've made an assumption because they get lots of reports where people are saying it happens regularly to them. But, like I said, I did not voluntarily come forward. And it was world news by the time I was returned. But, if anything more was to happen, I'm not going to pop out there and say, "Oh, guess what?!" (laughs). I've got enough to deal with.

WV: At the risk of sounding like a pun, do you feel that your experience will always make you feel somewhat alienated?

TW: Absolutely, yeah, it turned me into an outsider in a certain way, you know. To know for certain things that a lot of people believe are very likely, but they don't know for an absolute certainty the way that I and my crew mates do.

WV: As a UFO investigator and experiencer myself, I've found that when doing research and working with UFO people in the format of interviews, or utilizing hypnosis to help people who believe they've been abducted, that part

of me very stubbornly plays the Devil's advocate and/or the Doubting Thomas, both in the interest of adhering as much as possible to a more scientific format, as well as for the simple fact that I know some people make things up. So I guess you could say I'm a bit guarded about that. Have you found this to be true for you, as far as meeting other people who profess to having had these types of close encounters?

TW: Yeah, yeah, and I think there's a thing called healthy skepticism. Knowing the truth of what happened to me does not make me completely "open-minded credulous" to the point where I accept all stories. There are some cases that impress me as quite valid, and some that don't. But I keep my opinions to myself, because I was so offended by people judging me in the absence of having done any investigation. So, until I investigate a case, I'm not going to point fingers about which ones I think are good or not. I do encourage people to get the facts and do the research before they decide about anything. Because, you know, healthy skepticism is a good thing-- but blind skepticism is just as dumb as blind credulity.

WV: Did you have any particular religious and/or spiritual beliefs before, and have they changed, or are you pretty much the same place you were before?

TW: I wouldn't consider myself religious, but I do consider myself spiritual--and I think religiosity interferes with spirituality. Religion is an organized, well ordered organization and spirituality is an understanding of everything. I think they're actually different categories.

I've had people from just about every religion come to me and say, "I can't believe this, this contradicts my religion." But then people from a similar religion say "this falls right in line"--and I'm talking about Muslims, Catholics, Mormons. I come down to it that there's about 7 billion different religions on Earth--one for each person. They take whatever denomination they're a member of, and then put in their own personal views in there--

WV: A little bit of this, a little bit of that.

TW: Whatever fits their own set of beliefs and biases, and that affects people's thinking to too great of an extent.

WV: To me, what's so beautiful about this whole thing is that, when you grasp the reality of there being multiple alien cultures, probably countless alien cultures, not just in our galaxy but throughout the cosmos, it's all so enormous, and it just makes the whole big picture that much more magnificent, that there's so much creation in God's Universe.

TW: Yeah. The way I put it is they're selling God short.

WV: Exactly!

TW: You know, God was only good for making life on this little speck--

WV: Exactly! Believe me, I'm with you, I'm an ex-Catholic (laughter)--I agree with what you're saying.

TW: Well, Catholicism is a good example. My son was dating a Catholic girl, and her pastor brought in this guy that gave this talk, and called her in and told her she should break up with him because UFOs are something that comes from the Devil. But since then the Vatican has come forward and declared that there's nothing wrong with believing that there are other civilizations out there.

WV: Wasn't that good of them to do that--to give us permission?

TW: Yes, well if you look at Galileo and that whole historical episode in the past, the Vatican had their own inquisitors. So now, they're hip to the situation, but they haven't always been.

Our interview drew to a close. I would walk away from this thoroughly impressed with Travis Walton's plainspoken sincerity and his articulateness, coupled with a lot of eye contact. He expressed important things, using words that were succinct and unadorned by exaggeration or hyperbole. In short, he came across as absolutely real.

Before I stood up to head into a lecture by Stanton Friedman, I told Travis that I'd been on a table onboard a craft too, but that mine was in Tyler when I was three, it was extraordinarily traumatic, and that it was extremely rare for me to own up to it to anybody, let alone to people I've barely met. I handed him the last 2 pages or so of my manuscript (then in full re-write mode), just to give him an idea of where I had taken the narrative. He read through it and offered me some good positive feedback. I bought a copy of the new edition of his book, and he inscribed it simply with, "Wade, keep up the important work! Travis Walton."

Travis made a number of significant points in our interview, and I'll be coming back to them. Both Travis Walton and Stanton Friedman proved to be powerful public speakers, and their talks were well presented in the Renaissance Center's state of the art auditorium.

Mr. Friedman's lecture that weekend particularly impressed me with his impassioned indictment against the insanity of our country building a gigantic arsenal of nuclear weapons, using nuclear fusion for unbelievably destructive H-bombs instead of focusing on creating fusion reactors to give us almost unlimited cheap energy.

Thus far it seems the biggest stumbling blocks are being able not only to achieve the 100 million degrees Kelvin temperature required for the fusion of hydrogen nuclei to occur in a contained environment (reactors don't have the advantage of the incredible pressures at the sun's core to help induce fusion, ergo much higher temperatures are required), and also to create a container that can withstand such astronomical temperatures for more than a few seconds--unless of course the Billionaire Boys Club is lying to us about fusion reactors and already has scores of them operating in their underground domiciles, which wouldn't surprise me in the least.

Perhaps one day soon the super-genius Taylor Wilson will circumvent the greed of The Boys and grant us all affordable home fusion reactors. Now here's a story for you:

http://en.wikipedia.org/wiki/Taylor_Wilson

* * * * * * *

THE STANTON FRIEDMAN INTERVIEW

My interview with Stanton after the Meet and Greet went well. His generosity regarding his time really impressed me, he'd had a long day, and we spoke past 11pm, and he had a full day ahead the next morning. Here is the essence of our talk:

Wade Vernon: In the book, *UFOs and the Alien Presence: Six Viewpoints*, **edited by Michael Lindemann, you were interviewed by Ralph Steiner back in 1990. You made some excellent points about "why a government cover-up" after the Roswell crash, how it probably started because it was deemed ill advised to admit to the public that aliens from another star system had crashed their vehicle, we weren't sure about their intentions, but we knew that we no longer ruled our skies, so possible widespread panic was averted. And further, it was a simple fact that the first nation to harness the power of their off-world technology would have an enormous military advantage, so all the more reason to cloak it all in a veil of secrecy. It seems to me that the ongoing cover-up now, in the second decade of the 21st century, is likely a lot more about hiding the super technology behind it all as opposed to wishing to avoid panic. The public has had over 60 years to become conditioned to an alien presence, certainly TV shows like** *Alien Nation* **and the** *Star Trek* **television shows and films,** *The Day The Earth Stood Still, Close Encounters, ET, Independence Day*, **ad infinitum, to say nothing of readily available YouTube video clips of UFOs in Earth orbit and zipping through our skies. So my gut sense is that this recovered and/or back engineered alien technology, which has clearly mastered antigravity and field propulsion principles, would unquestionably make an internal combustion engine obsolete overnight--and it should be put to the public good. But the oil czars are way too happy with their multi-billion dollar annual gross receipts to let that happen. Better to keep us in the dark and handcuffed to the stone age technology that petroleum driven internal**

combustion engines represent. Where do you stand on this?

Stanton Friedman: Well there's an enormous difference between knowing how something works and being able to produce it at a reasonable cost. So I think there's more to the reason for the cover-up, the standard response to "why the cover up," is, well first we need to know how the damn things work. They would make wonderful weapons delivery and defense systems. So you'd have to set up a secret project--you can't tell your friends without telling your enemies. Secondly, you don't want the other guy to figure out how the things work before you do. Third scenario: two highly trusted individuals around the world--the pope or the queen, for example--make an announcement that indeed some UFOs are alien spacecraft. One of the biggest things that would happen would be a push by the younger generation (which unlike mine, was never without a national space program, and we've always viewed ourselves as Americans, Canadians, Greeks, Peruvians, etc.--the national ethnic regions are attached) to have an earthling mentality vs. nationalism. Sounds great, but the problem is, no government wants that. Nationalism is the only game in town. The kicker here is that people in power don't want to give up power. I don't care who they are, or where they are. Okay, now. The fourth problem is due to the mindset that, strongly stated, all the intelligent life in the universe is right here on Earth, and how the world was created in 4004 B.C. And all of this UFO stuff is the work of the Devil. So what are you going to do with that? And for quite awhile these religious fundamentalists had quite a strong following in the United States.

WV: Trust me, they still do here in Texas. (laughter)

SF: The fifth problem is different. The fact that they're coming here and we're not going there means they're more advanced technologically than we are. We'd very quickly have a whole range of energy companies that would be going out of business; also ground and air transport, car and plane companies would go out of business, then computers and communications businesses would be going out of business. There would be economic chaos. My only answer to that is to say that look, we used to have horse and buggy manufacturers. We don't anymore. We have to make room for new technologies, that's part of the reason for progress--competition, to the winner go the spoils. The sixth reason is . . . different. I have on seven different occasions had people quietly tell about cases in which pilots chasing UFOs have never returned. And if I've heard seven, there's a lot more than that.

WV: Sure, that makes sense. Could you get more specific about your sources?

SF: Retired mechanics, people who are at the military bases when the jets headed out in pursuit and never came back. In Frank Feschino's book *Shoot Them Down!* (*Shoot Them Down! The Flying Saucer Air Wars of 1952* (2007)), he points out that there were over 200 cases of military pilot's deaths and accidents between 1951 and 1956.

WV: I think I've seen that book on the web.

SF: And the kicker is that *The New York Times*--these were all in *The New York Times*, that doesn't mean that there weren't other cases reported elsewhere--these were aircraft that were disintegrating, disappearing. And five of the 200 pilots had over 100 air combat missions in Korea, against MiGs. These were very good pilots on these missions. And they come back to the United States, no MiGs, and then they have fatal accidents. Rather peculiar. And these were cases in which the truth did not get out. So, the military doesn't want to admit to that kind of thing, as you can imagine.

WV: It seems like I've seen at least two cases mentioned in which there was something reported along the lines of a disintegration ray being used against the pilots and their jets. So can we assume that some of these cases represented hostile aliens?

SF: All these cases seem to be in self defense. There's a Cuban MiG case. Our guys in Florida, Boca Chica Naval Air Station--they monitor the Cubans of course. I talked with a guy who was in the NSA, and they're listening--that's what they do--and the Cuban radar station notes that there's a flying saucer approaching.

WV: Do you know approximately when this was?

SF: Around 1967--I've got the specific information at home.

WV: Okay.

SF: It approaches, and the word gets back to the Cuban headquarters, which was run by Russians at the time. And the reaction was, What should we do? They scramble two MiGs, they get up there, they approach it, they radio back, "It's a spherical object with no appendages." The pilots are instructed to "tell it to get the hell out of our airspace." They do deliver the message, and nothing happens. "What should we do now?" The guys on the ground say, "Shoot it down!" Okay, so the MiG pilot gets a radar lock on, and he's getting ready to shoot it down, when suddenly the other MiG shouts over the air waves, "The first aircraft just disintegrated!" Now the UFO, which had been coming in at like a thousand kilometers an hour (rough numbers), and heading southwest, goes straight up, it was at 30,000 feet and goes up to 90,000 feet, heading southeast at Mach 7. The guys at the NSA were listening to this--that's their job at the NSA--and they send in a transcript to NSA headquarters. The information comes from the original tapes--which is very unusual. So I asked the guy, "Do you think they were trying to figure out more details about what happened during that period of time?" He said, "I don't know." So, when I get a case like that, I can't "fill it out," until I get confirmation from another guy--and I did from another guy at the NSA a couple of years later, he knew about that story too. So, if I've heard of 7 such cases, there's a hell of a lot more of them.

WV: No question.

SF: What people sometimes forget, is we like to think in terms of a democracy, one vote per person, that kind of thing. But who speaks for planet Earth? I'm not opposed to giving out information if we could get the Russians and Chinese giving out their information, too. If we could all agree on that, we'd be way ahead of the game. And so, that would be a reason--you get people saying, "Of course they'd tell their families"--No, they wouldn't tell their families. It doesn't work that way. There's a book, *By Any Means Necessary*, by William Burrows. He talks about the deaths of 166 military crew members in airplanes whose job it was to trip over radars in North Korea and China right after the war. Fly straight in and see how quickly they come after them, see how good their radars are. And unfortunately they shot down a bunch of our planes. Just as we've shot down a bunch of Russian planes. And the families were told, the plane was lost at sea, sorry about that. And finally they called a conference in 2001, gathered the families, gave them the medals their sons had earned, and told them what had happened. The point being, if they were willing to go that long, 50 years or so, without telling the families, then they're willing and able to keep UFO secrets too.

WV: That brings to mind a question. Didn't you meet Philip Corso before he died? (author of *The Day After Roswell* (1997), the controversial book whose late author not only confirmed the Roswell UFO crash, but asserted that he'd been responsible for 'farming out' to numerous U.S. corporate R&D firms the high-tech artifacts from the crash, supposedly spurring on the development of and/or creation of such things as lasers, fiber optics, Kevlar vests and the integrated circuit chip--wv)

SF: Of course.

WV: Well I've read his book a couple of times--which is just like yours, I've read all of your books two or three times. Corso asserts in his book that we were fighting a "skirmish war" with the aliens. Any thoughts on that?

SF: Corso had a big mouth, he made big claims, but without substantiation. You'll find that information on my web site (**stanton-friedman.com**): *Fraud In Ufology*. You'll find information about Corso and Bob Lazar. Corso, just to give you an example: he claimed in a sworn statement to Peter Gersten--an attorney in Arizona--that he had been a member of the National Security Council--which is a very high level job. So I checked with the Eisenhower Library, and not only was he never on the National Security Council, he was never at a meeting of the National Security Council. I sent information about that to Peter in a letter, and Peter showed it to Corso, and said, "Don't you think we ought to take that out of there?" "No." In a second case he said that he saw an alien body brought in on July 6th at Ft. Riley, Kansas. In the first place, that's a grotesque violation of security. He didn't really need to know about that for any reason. This would be like guys with atom bomb parts going into a McDonald's and leaving their truck parked outside. It just doesn't work that way. And when I asked how he knew it was July 6th, he said, "Well I know when I was transferred there." Well, that was actually in March or April. And it makes no sense, I mean why would you put a dead alien on a truck? Roswell had plenty of airplanes. It would have been put on a B-29.

Corso also took credit for things he wasn't involved in. He started working at Trudeau's office in 1960. The trouble is, the work he supposedly passed along into industry just didn't add up. Just did not compute.

WV: I really enjoyed the 1994 Larry King UFO special, "Live From Area 51." You had some insightful things to say about that base, sometimes referred to as Dreamland, and Bob Lazar's allegations of having worked there and at a supposed nearby super secret, partially underground facility to the south called S-4. (Lazar, he of the "Sport Model" alien saucer allegations [I first saw him on TV around late 1989, presented by the Las Vegas KLAS-TV award winning anchor George Knapp] described watching a test flight of an alien disc at S-4 similar to Billy Meier's Pleaidian beamship, and how he'd snooped around onboard it, seeing seats that were only large enough for children; and various claims about the propulsion system, that it used "Element 115" as a means to trigger gravity waves to propel the ship via field propulsion, with a whole host of other claims made to both the media and in Michael Lindemann's *UFOs And The Alien Presence: Six Viewpoints* book, e.g., we were reverse engineering them at S-4. Element 115 has been referred to by Lazar's detractors as "unobtainium."--wv) During King's broadcast, you expressed serious doubts about him being for real, partly because you'd never been able to establish his credentials. It seems that now you've concluded without a doubt that his story was a hoax, is that correct?

SF: Well I tried, I contacted MIT (Lazar claimed attendance there) and talked to 5 people; none of them knew anything about him; CalTech, nobody knew about him; I talked to his high school, they called me back about his graduation, that he'd been held back a year. All he'd taken was Chemistry. He was ranked 261 out of 365. Bottom third. You don't get into MIT by being in the bottom third. (The list goes on and on, posted on his web site. Friedman is convinced Lazar and his Area 51 and S-4 allegations are a total hoax.)

WV: Another guest on that program was Dr. Steven M. Greer. To me he came across as being just as level headed and articulate as you. Since then, I've read his books *DISCLOSURE* and *Hidden Truths, Forbidden Knowledge*, and have viewed numerous times his online video segments of the Disclosure Project forum in Washington in 2001. Typical of the web, I've also seen disparaging remarks about him, claiming he's a fraud. He's never struck me that way. Have you had much interaction with Dr. Greer?

SF: Oh yes, I knew Steven.

WV: You knew him really well?

SF: Well, we drove up there together in the same car, and had plenty of time to talk. Some people have made claims he was a fraud. He is a medical doctor. He has made a lot of claims that weren't substantiated. He said he briefed somebody--they were at the same dinner party, that's not the same thing. He was telling people what stocks to buy to invest in free energy, that sort of thing. He's a charismatic speaker, but that doesn't mean he's telling the truth. He also thinks all alien visitors are benign. Nobody else I know thinks that's true.

WV: I'm glad to hear you say that, because I don't either. When I was circling around at the Meet and Greet tonight there were some people saying they were convinced that all of our off-world visitors are friendly, and I said, "Well I'm glad you think so."
    Last question: I've always thought of you as being one of the most learned, clear minded and scientifically ori-

ented pundits in the UFO community, sort of our 21st century J. Allen Hynek, if you will. In all of your tireless research in government archives, and the people you've met, and the sites you've traveled to, and the inside information you've sometimes been privy to, can you make any comments about your take on the UFO phenomenon right now? I guess what I'm asking is, if you had to think of this enormous amount of information you've been exposed to as a large equation on a vast blackboard, and you drew an equals sign after all of it, what would this add up to, in your mind?

SF: I don't think there's any doubt that alien spacecraft are visiting the Earth. That we have recovered the wreckage of alien spacecraft, multiple times. Not just 2 in Roswell in 1947. But also others, one in Aztec, New Mexico in 1948 [Friedman mentions *The Aztec Incident: Recovery At Hart Canyon*, an exhaustively researched book by Scott & Suzanne Ramsey, with Dr. Frank Thayer & Frank Warren. I bought a copy from Stan (he wrote the Foreword), and it definitely sold me, based upon the evidence this married couple of avid truth seekers dug up in decades of extensive research, document location and interviews--including some of the people who were at the crash site--that this was a real event.--wv] No question that the government's covering up. The only people who have objections to the flying saucer reality are the people who haven't studied the evidence. So, given all that, I still don't believe we've been able to build a fleet of our own flying saucers, so to speak. I'm not convinced that we've made as much progress as people wish we had. Because I believe we would have used it in warfare. (Interesting side note: Friedman's reasons not to disclose in 2013, all of them well measured, are virtually identical to those of 1990; a true commentary on how we are stuck in the muck--wv.)

Before I left, I showed Stanton some of the documents I'd be presenting in my book, including the article from *The Santa Fe New Mexican* establishing the existence of the *Space Science Center*, and the interview with its former owner Felipé, mentioning the NASA classified UFO video. One of the great things about being a contributor to this UFO body of knowledge, is that it's okay to agree with some things colleagues say, but to disagree with others. Obviously, Mr. Friedman and I are not on the same page with regards to the development of superluminal vehicles derived from back engineered alien technology. And that's okay, I respect his right to have his professional point of view. But just out of curiosity, having brought my copy of Dr. Steven M. Greer's *DISCLOSURE*, I managed to find a few moments that Friday in March (the morning after this interview) when Stan wasn't mobbed by fans at his book signing table in the lobby of the Renaissance Center, and I opened the book to the page showing the schematic diagram of the ARV, and said to him, "I was just curious to know if you'd seen this, and what your thoughts are about it." He looked over the image and the accompanying text, and judging by the expression on his face, I'm guessing it sort of caught him off guard. It made me wish that we'd had more time to talk over this ARV issue; barely seconds later, a new crowd formed around him, and I took my book back and thanked him for his time.

In the interest of simply stating my own findings (and not insinuating that Friedman is necessarily wrong about Bob Lazar, or the argument we've not used back engineered technology in war yet): First, in Len Kasten's *The Secret History Of Extraterrestrials*, pages 147-148 refer to our sleek black B-2 bomber as "the Antigravity Bomber," referencing electrostatic elements to both its exhaust stream as well as its "wing-like body's leading edges," cited by both a 1992 edition of *Aviation Week* magazine (information supposedly leaked to them by rogue West Coast engineers and scientists risking their lives by disclosing) as well as the physicist Dr. Paul A. LaViolette.[56]

Second: in Dolan's *UFOs And The National Security State, Volume Two*, in over a dozen pages he cites numerous allegations from military and intelligence sources that not only corroborate Lazar's tale of alien craft at S-4, but go way, way beyond it. Things that almost make Lazar's material seem like a cold, stale bowl of porridge.

In his lecture in 2011, *The Case for the Secret Space Program*, Dolan mentions Bob Lazar and makes a good point: that intel operatives may have used him for spreading disinformation, specifically: Lazar really was at S-4, and really did see the Sport Model saucer in a stationary hover mode outside its hangar. But the need for "Element 115" ("unobtainium") fuel may have been a bogus bit of disinformation used to discourage competing global powers from pursuing reverse engineering the vehicles, since unobtanium is so difficult to obtain (when in fact the Sport Model saucers may well have operated the same way as the ARVs, via tapping into the limitless Zero Point energy). Dolan's lecture is via *Exopolitics Great Britain* (*exopoliticsgb.com*) The YouTube video of this talk is at:

http://www.youtube.com/watch?v=Nn67VeaZNAQ

I firmly believe anyone with even the slightest interest in the truth about UFOs should have both (and soon to be three) of Richard M. Dolan's *National Security State* volumes in their home library. *Volume Two* is like a stick of dynamite to me, and I haven't even read all of its nearly 600 pages yet. Put it this way: the comment made by Dr. Jones, the NASA aerospace engineer I met in Austin in the early 2000's, confidently revealing--apparently with justified bravado--the indisputable presence of a Super Science hovering at the periphery of our dumbed down psyches (we, the sheeple), dovetails nicely with the technological thrust of many of Dolan's *Volume Two* revelations: "Right now at NASA we're working on things . . . that are almost beyond your imagination." For me that's an easy thing to grasp, considering there are supposedly over 120 Underground Bases and labs in the U.S. today. Check out *Project Camelot's* information, and pay particular attention to the full ramifications of the Tunnel Boring Machines, particularly the nuclear powered **TBM** [**NTBM**] that melts solid rock and leaves behind glass-like walls:

http://projectcamelot.org/underground_bases.html

Or, Google: DUMB (Deep Underground Military Bases). Be prepared to wake up with a start, as if a bucket of freezing cold ice water has been dumped on your head.

Simple formula, folks: Take decades of unaccountable, National Security cloaked mega-billion Black Budget dollars, combine small armies of ruthlessly enthusiastic mad scientists, and let 'er rip, boys. Underground labs toiling madly, the unseen Ubermen soldiers of our Empire Of Technology. They work while we sleep. Again, like my Alabama buddy said, shaking his head while watching Linda Moulton Howe's *Strange Harvests 1993* video with me, "We're all being played for fools."

ALIENS AND THE HUMAN SOUL

The value of the soul, from an alien perspective, is highlighted in Shirley MacLaine's bestseller, *Out On a Limb* (1983). MacLaine's composite character, David, waits until they are out in the wilderness of Peru, in the area of Machu Picchu, to begin lecturing her about his metaphysical beliefs. Among them is his assertion that in his travels he'd met an alien from the Pleiades, a female humanoid named Mayan. It takes some time and effort on her part, but once MacLaine gets past her skepticism about his encounters with his so called off-world friend, she allows her mind to open up to the possibility that David's experience was real. And as she does, Mayan's lecture material to David begins to make sense to her (the wording of these passages is almost poetic in presentation; reading this rite of passage book is time well spent). David tells her that Mayan explained that the energy, the cohesive glue that holds protons, neutrons and electrons of the atoms together, is the Source energy--a kind of ocean in which everything floats. (In our physics, it would be referred to as gluons--which are packets of energy--the energy mediators, in fact, for the strong nuclear force which holds the clusters of protons together in the nuclei of atoms.) And Mayan stated that this energy (represented by gluons) is the energy our souls are composed of. So one could say according to Mayan's assertions, that gluon energy = the Source energy = our soul's energy. And Mayan asserted that this Source energy is the intelligent energy that organizes life; and this intelligent Source energy is everywhere, and in its totality, comprises God. So, as in many spiritual disciplines, each of us carries within us a small spark of the greater, universal light, the energetic whole known as God.

Mayan told David over and over, that in all the Universe, there was nothing with a higher value than the value of one living soul, and that vice versa, within the value of one living soul was the value of the entire Universe (interestingly, this sounds somewhat like the Holographic Model of the Universe). Furthermore, that all of mankind is on an upwardly moving evolutionary progression, even if it seems to us that we are stagnant and stuck in a limbo state. Each and every rebirth, followed by its eventual transition into the next afterlife of each soul, brings with it the totality of mankind onto a slightly higher spiritual plane. And herein is the intrinsic value of each soul: that each living soul affects the upwardly lifting, mechanical motion of the Universe. So that, in Mayan's philosophy, each soul has significant and equal importance, because each one contributes to the spiritual evolution of the cosmos.[57]

Years after reading *Out On A Limb* (I bought the new hardback in 1983), I was having a talk about UFOs and spirituality with a friend in Miami; he was completely open to most of the concepts about alien visitations, and he shared an interesting concept with me. He'd heard through a friend who was into trance channelers that information had come through saying that one reason some of the aliens visiting are so concerned about humanity and our self destructive habits is that the nuclear fireball created by the detonation of a hydrogen bomb destroys souls--just obliterates and disintegrates souls. And some of the aliens had a vested interest in our wellbeing, and in the preservation of our souls, since they were somehow intimately involved in our spiritual evolution (again, this is channeled information, and therefore conjectural, though thought provoking). So this dovetailed nicely with the philosophies Mayan had espoused, seeing as how the Source energy is both the energy of our souls, as well as the energy that holds nuclei together--the strong nuclear force. Since it's a given that the superheated plasma of the nuclear fireball disintegrates the atomic structure of anything within a certain radius of its ground zero detonation, this makes perfect sense, that this obliteration would include the Source energy, and therefore the souls of those in close proximity.

Mayan also commented to David that one of the primary shortcomings of our western science was its failure to acknowledge the importance--let alone the existence--of the spiritual realm. Western science ignores the existence of the soul, therefore it's a given that it overlooks the spiritual roots of the Source energy. (This harkens back to my interview with Dr. Wheeler, the theoretical physicist at UT, and his comment about how he couldn't help but think we were missing out on tremendous insights in our failure to look at the Universe and its beautiful mysteries in a more poetic fashion.)[58]

This is a large part of the thesis of the late Dr. John E. Mack's book, *Passport To The Cosmos: Human Transformation and Alien Encounters* (1999). I'd become aware of Dr. Mack's association with Budd Hopkins back in the early 1990's, and added his 1994 book, *Abduction: Human Encounters with Aliens*, to my personal library. His status as a professor of psychiatry at the Cambridge Hospital of Harvard Medical School, as well as the founding director of the Center for Psychology and Social Change, and ultimately, the Program for Extraordinary Experience Research (1993) via a Rockefeller grant, gave the field of UFO and abduction investigations some added weight and credibility. (Sadly, he was killed in a traffic mishap in London on September 27, 2004.)

In Mack's *Passport To The Cosmos: Human Transformation and Alien Encounters*, he has woven a marvelous narrative with thoughts on the subject of UFOs and abductions from a global panoply of thinkers, among them American abductees, African shamans, American Indians, and wise people from myriad global cultures.

Malidoma Some´, a shaman of the Dagara people from Burkina Faso in West Africa, believes that what we perceive as a thing in the material world is simply the result of an element in the spiritual realm manifesting itself on the physical plane. Mack also references natives of North America, among them Sequoyah Trueblood, who points out that Native people live in a world of "spirit and meaning," vs. the white man who lives in a world of "science and facts."[59]

Spirit and meaning, vs. science and facts. The polarization of our Western scientism--the outlook which posits that science has primacy over all other interpretations of life--has indeed split us apart, severed us from the holistic, larger reality underlying all existence, and therefore from the broad scope of the UFO and alien phenomenon. Perhaps nowhere does this schism reveal itself more clearly than in the information from Mack's work with a Zulu shaman and self-described abductee Credo Mutwa. Mutwa's philosophy centers on his assertion that Western civilization is based on the lie that mankind is the crown of creation, that we're alone here, that there's no possibility that more evolved or more intelligent life forms exist beyond Earth, with the biblical, sometimes vengeful God ruling over our planet and nowhere else. (The astoundingly spread out, shimmering mega-sprawl of intergalactic space is merely a garishly overdone light show to entertain us down here.) But Mutwa, like Sequoyah Trueblood, believes in a cosmos populated by a vast number of advanced beings. He opines that if the governments of the world were to announce that aliens really are visiting, that the masses would see behind their crooked power structures and challenge "the corruption, the governmental lies" and all of the "rotten industrial systems." Mutwa says that the *mantindane* (aliens, or "sky monkeys" to the Africans) he knows of consist of eight or ten sorts of creatures from the stars, and that they keep a curious and watchful eye on us as though making adjustments to humanity's progress, for their own reasons.

Credo ruefully sees humanity as "shamefully disunited" and seemingly incapable of focusing our energies collectively to deal with the serious issues facing us. He cockily asserts that he would welcome the *mantindane* taking over, conquering and subjugating Earth, that it could be the best thing that could happen to us, showing us what being oppressed is really about, and that fighting off the common enemy would bring us all together.[60] His outlook makes me wonder if he'd ever seen or heard any of the several instances in which Ronald Reagan had publicly voiced similar sentiments, most famously before the U.N. General Assembly on September 21, 1987, "We often forget how much unites all the members of humanity. Perhaps we need some outside, universal threat to make us recognize this common bond. I occasionally think how quickly our differences worldwide would vanish if we were facing an alien threat from outside this world." (There are actually allegations in the UFO literature claiming that Ronald and Nancy Reagan had a close encounter with a craft and occupant, including some sage off-world advice imparted to the then-Governor of California, presumably because the aliens knew in advance via access to a future timeline that he would serve a two term presidency. Considering Mr. Reagan's comments, I see this as a possibility.)

Other abductee cases cited in Mack's *Passport To The Cosmos* reveal opinions ranging from the sublime to the mercenary: that "the grays" are overseers of the Earth, and are here to help us to grow and mature, with the realization that something has gone terribly wrong with us; or they're here to check in on their investment; or to steal from us the body materials their mysterious agenda requires; or to create a hybrid race as an insurance policy that some bodies will remain on the planet if we do succeed in self-annihilation.

These dramatic differences in outlooks clearly define the dilemma those in America and other western civilizations are confronted with when dealing with the UFO cover-up. As Mack relates, "For the native peoples I have interviewed, including many who are close-encounter experiencers themselves, the universe is filled with life, or entities of various sorts, and some of them have the capacity to show up on the material plane. But for abductees with a Western scientific background, it is often the mind-shattering terror of these encounters that forces them to acknowledge the reality of the beings."[61]

Quite true for us. The men in charge behind the curtains, MJ-12 or whatever identifier you assign to them, are all about hiding the nuts and bolts of the phenomenon and its hidden technological wonders, while eschewing any involvement with the deleterious psychological and spiritual effects experiencers are confronted with, partially enabled by a brainwashed public fully engaged in 2/3 of the Triple-D mindset, derision and denial.

Among those convinced off-world visitors are present, a broad spectrum of beliefs regarding their motivations exists, from benevolent beings here as saviors of the human race to the darker scenario of sinister aliens bent on colonizing our planet with a new species of alien-human hybrids right before our (blinders securely fastened in place) eyes--a storyline played out in great detail in David Jacobs's book *The Threat: Revealing The Secret Alien Agenda.* Jacobs received his Ph.D. from the University Of Wisconsin, Madison, in 1973; the topic of his doctoral dissertation was UFOs, and a revised edition of it was published by the Indiana University Press in 1975 under the title *The UFO Controversy In America.* This is a deeply learned and intellectual man, one whose decades of relentless research transformed his stance from being a flying saucer agnostic all the way to the other end of the continuum; he now firmly believes that there is an alien race visiting here that is coldly conducting an alien-human hybrid breeding program via abductions. What I find fascinating about him is that he makes no bones about the fact that he knows some people think he's crazy (and he's coached his family members to "play dumb" about his books and research so as to avoid the pain of ridicule and character assassination), but his well measured arguments remain firm, based upon his tireless investigations, much of it involving the hypnotic regression material gleaned from hundreds of UFO abductee cases. Because of the respectability and solidity of Dr. Jacobs's background, I keep my mind open to the possibility that this disturbing scenario is part of our behind the scenes reality.

Linda Moulton Howe, as mentioned earlier, is a long term, relentlessly inquisitive investigator and highly regarded researcher whose published material, via her books, videos and comprehensive web site (*Earthfiles.com*) often reflect her successful cultivation of well connected intelligence community insiders who seem happy to impart to her what at times is clearly classified information; she appears to have her hand firmly placed on the pulse of the UFO and alien visitation zeitgeist, offering us some of the more detailed, albeit sometimes disturbing, answers to the oft asked questions, What are their plans, what do they want with us, what's their agenda? Are they benign, are they malevolent, or somewhere in between? Once you've gotten past the hurdle that they exist and that they're here, the facts do reflect that multiple types of beings are visiting, possibly including both benevolent and malevolent non-physical ones; and it seems inarguable anymore from their behaviors that we're experiencing a full range of intent, from dark to light. Ms. Howe's publishings certainly give one plenty to ponder in that regard.

In her fascinating and extensively illustrated book, *Glimpses Of Other Realities, Volume II: High Strangeness* (1998), Howe presents an expansive array of information, some of it rarely, if ever seen elsewhere. A few scenarios revealed strike me as sounding so far out they make some of the most outlandish episodes of *The Twilight Zone* and *The X-Files* come across like innocent chatter lifted from a first grader's *Weekly Reader* in comparison. High strangeness, indeed. But given the paradigm shattering bizarreness of what happened to me in a wide open, deserted field on the outskirts of Tyler, Texas when I was three, I don't discount the possible presence of the truth in many of the allegations Linda and her intelligence sources bring to the table. But you might want to keep a shot glass and a chilled bottle of your favorite tequila handy if you plan on an extended reading session from this volume. It can get intense.

One of the qualities Ms. Howe has that I've grown to appreciate is her sense of wonder in the face of the world--a phrase I learned at FIU in Dr. Kovacs's Phenomenology class, as we students were taught to cultivate this outlook, deemed the best way to *let the phenomena speak for themselves.* Look with wonder, and let the source of wonder (or the terrifying, or the bizarre, or the nearly gut wrenchingly unacceptable) speak for itself. Linda clearly is completely fascinated with every facet of the UFO phenomenon, from the dark to the light, and presents every conceivable scenario imaginable in her writings, in the interest of being a thorough investigator who dutifully reports her findings; her innate intellectual curiosity and desire to find the truth seem to compel her to expose every angle, oftentimes introducing a concept with "Perhaps this is what is happening . . . ” or “Perhaps this is what was meant by . . . ” and letting that particular vignette, concept or scenario stand on its own and speak for itself. In short, I like her style. It’s intellectually rigorous, lively and full of life; she has established herself and her persona as one who tenaciously and rigorously seeks the truth. She’s won my loyalty. Some of her research is fascinating. Her oeuvre is a testament to a rich intellect at work.

To read Howe’s *High Strangeness* is to indulge oneself in a colorful spectrum of otherworldly vignettes, hinting at the presence of multiple hidden alien scenarios embedded in an invisible fabric of interwoven threads, each comprised of far out potential realities. Days could be devoted to the topics presented here: highly intriguing results from high tech lab analyses of metals allegedly taken from crashed saucers, courtesy of Ms. Howe’s peerless connections; 6-fingered hand imprinted metal control panels from one of the crash retrievals (like metallic versions of the Graumman’s Chinese Theater movie star cement hand castings), the precise workings of which were deemed so beyond our technology that even after decades of research and analysis all that could be deduced was that they were highly advanced neuron-based computers enabling the aliens a direct interface/thought control guidance of their craft via knuckle and fingertip sensors; that some of our alien visitors might be what one intelligence source referred to as neither benevolent nor neutral (ergo potentially malevolent) NHE’s (Non Human Entities), who, although non-physical, are able to perform cattle mutilations and even project false images of UFO craft and the mysterious black helicopters so often witnessed around the dead carcasses (a la Vallee’s *Messengers Of Deception*), and that the ultimate, core truth behind this is something the public “really could not handle.”[62] (The evangelicals have really had a field day with this concept, the internet is bursting at the seams with assertions that all alien visitations, vehicles or no, are demons here to do Satan’s bidding. This all or nothing mindset so many religious fundamentalists and zealots employ has never worked for me; alas, this anti-intellectual, empty headed and often damaging rhetoric is typical fare of the Texas State Board Of Education. I’ll pass on that, thanks anyhow, folks.)

Perhaps because I admire the military sense of discipline and the extremely rigorous training that goes into making even the most basic soldier, I found that Military Voices, the first chapter of *High Strangeness*, intrigued me the most and offered me the sort of material I can sink my teeth into and cogitate its content. As you dig further into the book, there are scores of pages devoted to the ostensible lessons learned by abductees under the tutelage of their abductors. Some might call it unfair of me to feel this way, given the highly strange nature of the childhood encounter I've presented as a real experience in my rite of passage narrative, but I've found that the skeptic in me doesn't have much of a taste for another experiencer's material if it seems as though the multiple-abductee has been comprehensively schooled in the alien's language and the fully expanded version of their Magna Carta over the course of multiple abductions. It simply doesn't sit well with me, and triggers my heightened sense of skepticism, especially when said accounts refer to a non-physical, out of body, astral world visitation and experience. I don't dismiss the material outright, but I don't devote a lot of time to the astral realm aspect of the UFO genre. Perhaps I need an even more open mind, since I'm asking others to open theirs to my very strange but true story.

Aliens have created the human race via genetic engineering over eons, and we are in actuality human androids, worker drones for our off world makers, the greys. Such is the gist of one of the scenarios offered, this one being a speculation from Ms. Howe based upon her research. Howe quotes "Sherman," a composite character of two of her intelligence sources, as saying that one of the Ebens stated, regarding humanity, "We made you, we put you here, but now you have to live it." My immediate gut reaction was, Wow, wouldn't it be too rich for words, if all of us troubled, dumbed down tribal warriors perennially on the rampage down here, convinced we're the best the Universe can manage by way of superior life forms (as we scheme up increasingly creative ways to subjugate, torture and annihilate ourselves) are in fact the misfit toys of toymakers who have grown weary of our childish antics and have abandoned us to our own devices? As if they're telling us, "Guys, we screwed up on your nucleotide sequencing, you're a mess, the human race is a wash, do what you can with the time you have left, try to do your best. You have free will, and a semblance of autonomy. Do the best you can. You're on your own now." (I suppose we shouldn't feel completely abandoned, as witnessed by the sporadically performed experiments during their abduction rituals, and the organ harvesting of our animals. Although the sloppily discarded, ravaged animal carcasses do sort of read like a Great Big

F-U to our poor hapless ranchers.)[63]

Howe's speculation does, in fact, coincide with some of the material leaked to her by Sherman. The information Sherman imparts to Ms. Howe I found compelling, some of my favorite conversations in the whole book. Although I don't blindly accept everything he says, I definitely resonate with a good bit of it.

Sherman relates that the complexity found in our abduction literature is partly due to the fact that some of the many varieties of beings we encounter are biological androids, robots delegated to perform a myriad of planetary tasks, and can be made into any form or design their makers desire--essentially, we were told, "an advanced race uses robots;" that one (or more) of the captured Ebens shared a wealth of information with us, both about our history and theirs; and that word has spread off world that their human DNA manipulations down here haven't been going very well, inadvertently attracting dark factions to our planet, the ones doing potentially harmful experiments on us, and the benevolent aliens haven't always been able to keep them in check. (This is a mere cursory account of salient points; the full details are quite engrossing.)[64]

One of the passages that really grabbed me was the exchange in which Howe queried Sherman if there was "a Devil out there somewhere?" He replied that there is an evil force which "even the Ebens talk about;" that there is a Supreme Being and there is an Evil One; and that the Ebens feared that Evil Entity as we feared the Devil. When Howe queried where the Devil was, Sherman said he didn't know, but that, "If the Ebens are scared of him, then I'm damn scared of him, because the Ebens can control everything," seeing as how "they have been around this universe and done everything."[65]

The passages alluding to how there definitely are malevolent aliens coming here to do dark deeds made me glad for what I'd heard from multiple sources, one of them Dolan's *UFOs & The National Security State, Volume Two*: that our Strategic Defense Initiative (SDI) or Star Wars defense system was in fact installed to defend Earth from intruding hostile aliens, not to protect the U.S. from Soviet missiles. Back in 1993, an L.A. pal in the film business gifted me a VHS tape with footage showing one of NASA's Shuttle missions that had videotaped what definitely looks to be a type of particle beam weapon blasting from the ground out into space, targeted at a large glowing orb that was streaking down into Earth's atmosphere but then suddenly engages a full reverse and zips back into deep space to escape the beam, blazing away at an estimated speed in excess of 200,000 miles per hour; a lesser version of it can be viewed on YouTube at:

http://youtu.be/je7-eT7fFyg

One fellow who posted on this page (among the current 100+ comments) wrote, "Eat lightning you alien scum."

Because there is so much material on the web regarding possible alien and joint alien-human bases, some of them underground, I searched the Index of Dolan's *UFOs & The National Security State Volume Two* encyclopedic tome and found that it has numerous pages of information about underground, hidden bases.[66] Alien underground bases and tunneling had been mentioned by Linda Howe in her *Glimpses of Other Realities, Vol. II: High Strangeness* book too, in which Sherman stated that government officials had been concerned about how much tunneling had been done by the Ebens, that they were worried about possible earthquakes from all the subterranean activity, but were assured by them that, "we know more about this planet than you do. So don't worry about where we dig. . . ." This whole concept dovetails with another comment Ms. Howe quoted from a retired military officer, that "the alien technology is so advanced and the beings are so strange that no one would believe it. Keeping the public and media away from what's really happening isn't difficult. It's a story that no one wants to tell, that no one knows how to tell. The truth is stranger than fiction."[67]

Upon reaching the final stages of my book research, I came to realize that some of the most important (and somewhat disturbing) material about UFOs and the alien-human interaction was coming from Linda Moulton Howe, so I sought an interview with her. My reasoning was simple: you'd be hard pressed to find a more convincing form of irrefutable physical evidence of UFOs and alien activity on Earth than from the bloodless carcasses of thousands of mutilated cattle (and myriad other animals) over a period of decades, all around the world--ultimately realizing and accepting the stark reality that the perpetrators behind these high-tech operations are not human--and certainly not predatory animals, either. That dozens of respected law enforcement officers and salt of the earth ranchers Ms. Howe has interacted with over the years have told her they believe this to be a fact is sobering, to say the least. What really clenched my conviction that I had to interview this remarkable and fearless investigator was the response she gave to Art Bell in his October 3rd, 2013 interview with her on his *Dark Matter* radio program, when he queried if the cattle mutilations have diminished or "trailed off" at all in recent years. She replied:

> No, it's changing latitudes and longitudes. Latitude and longitude are the keys. There are cycles of mutilations at certain latitudes and longitudes, over and over and over again, throughout the decades, probably centuries. And right now, from 2002 to 2013, there literally have been five or six thousand mutilations in Argentina alone.

Linda Howe has a very warm and engaging personality with a wonderful sense of humor, coupled with a mind as sharp and deeply perceptive as any of the top tier college professors I've ever been wowed by--and that is a fact; she kindly granted me an extended phone discussion on December 23, 2013, presented in the next section.

* * * * * * *

THE LINDA MOULTON HOWE INTERVIEW

December 23, 2013

Wade Vernon: I came across the recent radio interview Art Bell conducted with you, posted on YouTube on October 3rd, 2013, *Art Bell's Dark Matter:*

http://www.youtube.com/watch?v=TCtrR19TsXE

In your full body of work, your oeuvre--not only in your books and the YouTube videos, and the *A Strange Harvest* and *Strange Harvests 1993* videos, and your *Earthfiles.com* postings, and then add to that the recent Art Bell interview--my gut sense of all of this material lumped together is, if you step back and look at the big picture of what's going on, especially when you factor in the cattle mutilations often correlated with UFOs in close proximity, it seems to me that this is a real morality play, of Good vs. Evil, and to me goes to the concept of Universal Right and Universal Wrong. In one specific instance (in the Bell interview) you alluded to the case of a cow that seemed to have been thoroughly brutalized, that the whole body was paralyzed while its organs were taken, except for its head, which dug itself into the ground, obviously in agony, while its organs were removed. And there was something so unbelievably malevolent about that to me; I'm a huge animal lover, and that just really gets me going. I agree with your comment, it was like they were flaunting their brutal stealth technology, as if to say, "Look what we can do." So, if you want to roll with that, I'd be interested if this seems like a morality play of Good vs. Evil to you.

Linda Moulton Howe: What I'd like to do is place the context that I myself have. In September 1979 on, I began trying to investigate what was happening to all these animals in Colorado and the surrounding states, because there was another uptick--animal mutilations had been reported in the United States and Canada from the early 1960's on, there were other reports in Australia going back to the beginning of the 20th century.

So this was a similar removal of organs: ear, eye, tongue, jaw flesh, internal organs, sexual organs and rectal/vaginal areas, and the same patterns of tissue from animal to animal around the world, and I was focused on what was happening in Colorado, where literally there were dozens of similar bloodless, trackless animal mutilations in the summer of 1979. In September, I drove up to Sterling, Colorado to begin research with Tex Graves, who had just retired after 20 some years of being Logan County Sheriff. He showed me Polaroids he had taken of mutilated animals in the field. Several of the Polaroids were of a steer that had the classic animal mutilation excisions of an ear, eye, tongue, genitals and rectum cored out. The steer was lying on its right side, with four legs as straight out from the body as you could ever imagine, with the hooves, top and bottom, perfectly put together. The ground underneath this steer was like face powder, it was so dusty. And there was not a trace of a track, there was not a single movement of those four hooves, meaning that the front legs were perfectly lined up hoof to hoof, the back legs were perfectly lined up hoof to hoof, with no signs of anything, no signs of struggle, nothing around that body; and yet, the head of the steer had moved and had dug an eight inch hole while these tissue and hide pieces were removed.

Sheriff Graves and I talked about this, and he said, "The only conclusion that I can bring to this, Linda, is that something had the ability to paralyze a large 2,000 lb animal on the ground, in place, while the incisions were made, leaving the head alone to dig that hole. And then he and I went to the heart of the question I had then, and the heart of the question of what you have now, namely, what would be the purpose? I remember saying to Sheriff Graves, with this particular animal, it seems as if something is flaunting in your face that it can do this. And he said, "Yes, I agree." After examining and investigating over 200 animal mutilations, Sheriff Graves himself had come to the conclusion that something had the ability to do excisions to animals that humans could not do, and he was the first person ever, to say to me that day in Sterling, "I have come to the conclusion that the perpetrators of animal mutilations are creatures from outer space." That was his quote. And over the next 9 months, I began to produce, write, direct, edit and finally present the report in a TV special that first aired on May 25th, 1980--a 90 minute special for the CBS station in Denver, entitled *A Strange Harvest*. I had come to the conclusion (finally, after those 9 months), that all of the animals that I saw by myself first hand with my TV crew made it appear to be a harvest, since the pattern of excisions was the same, animal to animal. The bloodlessness of the excisions automatically indicated a technique that was not being used in the 1960's, the 1970's or the early 1980's. It was indicative of a procedure that could instantly cauterize in ways not known to our technology at the time, as confirmed by veterinarian, pathologist and hematologist Dr. John Altshuler. And finally, the issue of agenda: if this was a genetic harvest, then what was the reason for the tissue and the fluid being taken from these animals?

Now, this is one of the most repulsive facets of the interaction of non-humans on this planet from a human point of view.

WV: Absolutely.

LMH: And eventually I had to come to terms with, what am I covering? What am I reporting? And what is the impact on humanity? This is what I wrote in my first book, *An Alien Harvest* that came out in 1989, nine years after the broadcast of my documentary film *A Strange Harvest*. So by then, I had a decade under my belt of having been exposed during that period of time to dozens of other people in law enforcement who had the same conclusion, that we're dealing with creatures from outer space. There were also hundreds and hundreds of men, women and children in the human abduction syndrome, who were having parallel interactions with non-human intelligences in the sense that they were being taken into craft, they had excisions of tissues taken from their bodies, but they were returned alive, while animals were killed and mutilated.

WV: Right.

LMH: And in all of that, in that decade, what I had to come to terms with is that whatever we're dealing with, it is not indicating in any way the wholesale intent to eliminate humanity. In fact, humanity, Homo sapiens sapiens, appears to be a serious subject for their own interaction, for the non-human interaction, and so do the animals, but they are handled differently. Humans are treated without death, and the animal life is treated with death. And once you begin to put that on paper, once you begin to acknowledge the fact that humans are being handled differently, you begin to re-think whether or not this non-human interaction with the planet is repulsive and negative, or is perhaps related to an issue of positive survival for Them and us humans, which is what I wrote about in my first book *An Alien Harvest* (1989), and is further discussed in depth in my next three books [*Glimpses of Other Realities, Vol. I: Facts & Eyewitnesses* © 1993; *Glimpses of Other Realities, Vol. II: High Strangeness* © 1998; *Mysterious Lights and Crop Circles, 2nd Ed.* © 2002.]

But if you then begin to look at this through the lens of survival--whose survival? And after 33 years of trying to get to the bottom of the animal mutilation phenomenon around this planet, I personally have come to the conclusion that we are dealing with a survival issue. And that the survival is something else, that there is something affecting the non-human intelligence, and threatening its existence, and that they are somehow linked to whatever has been the evolution of life on planet Earth. And that perhaps they are returning to this planet in the 20th and 21st centuries because this was the last time on this planet that any genetic material could be harvested that would be healthy enough to sustain whatever is the survival problem on the part of the non-humans.

Meanwhile, the non-humans and the Homo sapiens sapiens have some kind of a genetic link--that's why I think there is this harvest. The animals appear to be the closest that they can come to a large extraction of genetic material from this planet, without harming humans. So, to come to the philosophical questions: are we dealing with non-humans that are evil? Are we dealing with non-humans that are divine? Are we dealing with the yin and the yang of good and bad? I think, that in December of 2013, we are dealing with at least 3 non-human intelligences that have been terra-forming and interacting with this planet for millions of years, not just the last few centuries. And I think that these three competing intelligences, over all these millions of years, have had agendas to use this planet, mixing and matching genes, and ultimately one of them was responsible for the manipulation of DNA in already evolving primates to create Homo sapiens sapiens. Once you come to that, which has been in several government documents, one of which I was allowed to read with my own eyes at Kirtland Air Force Base, you then step back to a much larger box. Which is that, at least three competing alien intelligences (competition does not mean that the competitors are necessarily evil, or good--but it means competition) are at work here. And if we, the Homo sapiens sapiens have been created by at least one of these competing, non-human intelligences, then we are on the Petri dish, and they are looking at us through the microscope lens. Those that are on a Petri dish under a microscope never have an objective view of the eye and the mind that are looking down on the Petri dish. The best we can do is continue to try to study what the non-humans do on this planet--which includes the mutilation of the animal life, all kinds of animal life, not just cattle and horses. And that we humans are taken in a different way in the human abduction syndrome, subjected to the removal of sperm, subjected to the removal of eggs, subjected to the introduction of hybrids, that are presented to humans in abductions while being told, 'your sperm or your eggs were used to create this child.' That is something that general humanity would say was perhaps a scary hybrid, but could be where the evolution on this planet might be headed, toward a future in which the hybrids become the new surface humanoid Earth life.

And if it is this complex, if it is this big, the animal mutilations are only a facet, and an insight into the genetic harvesting by three competing non-human intelligences that have been interacting with this planet millions of years--and suddenly something has happened in the 20th to the 21st centuries that falls into the category of urgent survival. And that urgent survival includes the non-humans, and struggling, evolving Homo sapiens sapiens. And that is why I cannot characterize with any certainty that the animal mutilations are something with evil intent.

WV: Okay. And what you touched upon while answering dovetails with my second question. Again, from the Art Bell interview, you shared that you'd met with a DIA agent in 1999, and I'm assuming that's where you got some of the material you just now gave me, because one of the quotes from that transcript was, that he said "My job for 23 years was to monitor and analyze the geopolitical territorial conflict of three competing extraterrestrial factions on this planet." So is this correct, that some of the information you just related came from this DIA source?

LMH: I had already been doing my research and my work as an investigative reporter through the lens of what I laid out. And he'd asked through a World Bank person to set up this meeting with me, and it took a month to do it. He brought into focus for me, even further information that had already been coming to me through the human abduction syndrome, and through law enforcement trying to deal with the animal mutilations; and some of it was clearly coming from government Intel operators who have been crossing my path and delivering information to me since 1979, with all the while my being suspicious that some of those path crossings by people working for the government included those who were delivering lies of counter-intelligence.

WV: Right.

LMH: So, in 1999, the meeting with the man who had retired from the Defense Intelligence Agency was a 7 hour focusing more sharply, and everything that I had been exposed to prior to 1999 was everything I just said to you in this first long answer.

WV: Okay. My guess is that your DIA source probably covered a lot of material in 7 hours. Am I correct in assuming that this might come out in another book of yours? Or did he mostly just confirm what you'd already known?

LMH: (pauses) I had never dealt with the information he shared with me publicly, because frankly, it was shocking to me, and is shocking to me to this day. And I'm not using that word in a titillating way, I'm saying that, I think that the size and the scope of the truth (pauses) ultimately is probably going to come to humanity through quantum physics, and concepts that Michio Kaku has been talking about for the last decade, that we are not only in this Universe, in this solar system, but that this Universe is part of a multiverse, of an infinite number of universes, separated by some kind of field. And that eventually we may begin to learn that dark matter and dark energy are related to exactly that fact. It's not a singular Universe. It's a multiverse. And in that multiverse you could have antimatter universes juxtaposed to matter universes. And if there are an infinite number of these paired matter and antimatter paired universes, what is the relationship between those universes? And is that where the origin of the yin and yang, positive and negative, evil and good, Divine Field and Hell--is that where there is actually a physics behind those concepts, and that we humans today, in 2013, are like 2 month old babies when it comes to comprehending the actual cosmos that we are life forms in? And we're not alone, not in this Universe, and not in any of the other universes, and that we are so far removed from the reality of the Universe that we are in, and kept deliberately dumb and blind by not just human political forces, but also by non-human political forces. And it will only become, I think, open knowledge to humans when it is decided to be revealed by some of the forces keeping these policies of denial and lies in place. Disclosing the fact that we are in a cosmos that is teeming with life, with many mixed agendas. And that only when it serves the interest of the power-greed groups on planet Earth to disclose, will humanity be exposed finally to some of the bigger truths we've been lied to about for at least 5,000 years.

WV: That precisely coincides with convictions I've expressed in my book, that one of the primary remaining reasons for non-disclosure is greed. And part of that is because, there's no question to me, especially from this NASA aerospace engineer I've met with, that we definitely have ARVs--

LMH: Right.

WV: There's no question in my mind we have those, and in fact, some of the things he alluded to made it sound like the ARVs were more like Volkswagens compared to some Bentley Turbo's that they have now. The ARVs are just the beginner model, maybe even phased out by now. But what I'm getting to along that line is, that I think that correlates with what you said, that there might be some information that could potentially be so overwhelming that the government people believe its disclosure would result in not just the *economic discombobulation* effect, as Stanton Friedman says, but also psychological effects, and religious ones, and a myriad of power structures that could crumble with disclosure. But it could also be that some of the ultimate truths, especially what you said about the

multiverse, might themselves be so overwhelming that they could interfere with people even wanting to get out of bed in the morning and live their lives.

LMH: But the thing about it is, I am living proof that a person can hit this material innocent, and hit the roughest parts, the mutilation of animals, and go to it as an Emmy award winning TV writer producer editor and director, and not know what I was getting into, survive, do a television program, keep on going around the world investigating, going out and collecting tissue, and fluid, and soil and plants trying to deal with forensic evidence, and doing so for 33 years, and still surviving. And continuing my work through the internet and television and radio and books, and still survive. So that I have come to a peaceful place inside myself that I would like to extend to all people. That we are on a planet in which humanity's power-greed structure has kept the masses from knowing all kinds of truths. But if you extend yourself to do research, to find out, to read, to look at ancient texts and modern books, you realize that the truth is all around us, and the government cannot keep truths hidden forever. I don't think the Universe works that way. And eventually we humans are finally going to learn that we do have allies--I'm convinced we have allies, advanced intelligences.

WV: Absolutely.

LMH: And if there were some wholesale intent to eliminate Homo sapiens sapiens, it would have been easy to do at any point. And the very fact that humans are alive and expanding and populating the Earth in the 21st century while non-humans continue also to interact and harvest from the planet, I think sets up the theme or the metaphor for what is going to happen to the rest of the 21st century, that we will eventually be introduced to the fact that we're not alone in this Universe or other universes, and that we are going to move into a whole new phase. Once all the governments of the world finally get onboard with telling the truth, then piece by piece other aspects of what governments are holding in their treasuries will come forth. It will be a completely different planet, but it has to go through the generations that are now going to pass into other afterlives, and the new generations are going to be much more accepting of all of the new information, not clinging on to the policies of denial from the past. And so one way or the other, the truth is going to join up with this planet; it's the question of how and what happens as that transition unfolds.

WV: The final question I have for you is one I asked Whitley Strieber years ago. The first part of it is, have you ever read *Childhood's End* by Arthur C. Clarke?

LMH: Three times.

WV: (laughing) How funny! I've read it five times.

LMH: (laughing) It was the most impactful book of my childhood. I remember reading it for the first time somewhere in junior high, and then I read it again, and again, and I still feel it is one of the most astonishing books because it seemed to me he was writing from knowledge.

WV: Yes.

LMH: And that whether the knowledge was because he was a human abductee, or the government wanted to use Clarke for getting out these difficult ideas and concepts in science fiction which would be safe, it would be talking about the replacement of the world population by something that would fall into the category of hybrids--I think that has been on the table for all these decades.

WV: Absolutely. Well, what really got me was, also, that it included the concept that the Overlord ships would be hovering over world capitals for fifty years before they showed themselves, meaning their physical bodies, which is in a way a metaphor for how we're being gradually conditioned all these decades, to accept not only that they're here, but to gradually ease us into their appearance, because some of them are sort of scary looking.

LMH: There are also major technology components involved in non-human technologies: 1) invisibility. 2) The ability to neutralize and manipulate gravity. 3) The ability to project 3-dimensional holograms, which the human mind and eye cannot distinguish from anything that naturally exists on planet Earth. Those three components of the past, present and future non-human presence on this planet are key. Non-humans could have a 10-mile-long ship in high Earth orbit, but our human eyes would not be able to see it.

WV: Sure, absolutely. But the other thing that struck me about *Childhood's End*--and this is partly what I discussed with Whitley--was the concept that the Overlords visiting here were essentially cosmic midwives. They were not able to mutate into a higher life form the way we were. We were able to mutate and join the Overmind, and they were not, so one of their primary roles and duties was to assist us in the merging. And this theme struck me because my gut sense from my research and my experiences is that some of the visitors don't have souls.

LMH: Right.

WV: Whether that's because they were genetically engineered that way, or because it simply happens in the Universe that some beings come to life without a soul. To me that makes us profoundly interesting, and of interest, and valuable to the off-world intelligences, that somehow we are able to mutate, and that we have a spirit that can transcend. And some of them cannot.

LMH: That is one of the key subjects I talked about with Art Bell on that radio broadcast, probably with the most depth for the first time in my career.

WV: That was my impression, and I recognized some of it from your *Glimpses Of Other Realities, Vol. II: High Strangeness* book. And it prompted my next question-- since you mentioned Michio Kaku's theory that some alien overseers may be tasked with monitoring the progress and status of entire solar systems. My sense was that this concept dovetails with *Childhood's End*. (Aside from the fact that you said there are three races that are in competition here.) So my question is, do you concur, that some

of the concepts from *Childhood's End* are what may well be going on right now?

LMH: When I started working on *A Strange Harvest*, if you consider all the things that have happened to me in relationship to government insiders, and you think back on *Childhood's End* as a brilliant work of science fiction, as it was presented to the world, then you might say, Who is feeding Arthur C. Clarke all the information for that book--that would be published safely, as science fiction? That's how counter-intelligence operates. That which is very, very difficult, and you're supposed to guard, but you want to condition the Earth population for the eventual day that the different truths will be presented, that's the way to do it. So, in terms of ever being able to prove that Arthur C. Clarke ever received some sort of intelligence officially from the government, we can't do that.

WV: Right.

LMH: But in terms of that book being a *spooky* anticipation of what was going to evolve over the next decades, Clarke's book is incredible. And there are some of us who know for a fact there are government and military operatives, who have leaked material through television, Hollywood films, books, radio, conferences and other public outlets. Such assigned operatives have created various vehicles in which truths about the non-human phenomenon have been delivered to general populations without the risk of trauma because the leaks are not "official government proclamations." Such leaks to condition people with the concept of non-human entities interacting with Earth have been occurring since at least World War II.

Our talk wound down from there. Regarding Linda's October 3rd, 2013 radio talk with Art Bell, I found that one of the most striking comments she made to him was:

> Today in 2013, my own hypothesis about all of this is: that we are the latest model of a genetically tweaked, evolving, standing up primate for reasons not clear to anyone, and probably not our government either. This DIA man told me in 1999, he said, 'You might as well think of humanity as the 37th model of the Hoover vacuum cleaner.'

It's funny, this DIA man's comment harkened back to a part of Whitley Strieber's *COMMUNION* (1987) in which he shouted at his female Visitor that they "had no right" to conduct the invasive medical procedures on him that they put him through, and the reply was "We do have a right." Rather dramatically humbling, to think that we really might be the property of off-world genetic engineers (as mentioned by the DIA man, and previously quoted by Linda Howe's source Sherman, and myriad other sources as well), perhaps still struggling to produce "the next generation of Hoover uprights."

The other hard impacting part of Bell's interview with Howe was when Linda alluded to an alleged briefing delivered to President Reagan in his early days in office in 1981, in which the President was informed that multiple types of aliens were visiting. Her words:

> Assuming that there is maybe total truth or some truth in this transcript, Ronald Reagan is given first a kind of superficial briefing of, 'We're not alone, there are different types, they've been coming here a long time, we know where some are from, we're in control'--that's how the briefing went.

Supposedly the Intel men told Reagan that one species, referred to as Trantaloids, was a hostile insect-type of being, and Howe relates that the Intel men stated, "We do have some concern about this insect type." Howe elaborated that the alleged briefing document used the word HOSTILE to describe the Trantaloids.

My curiosity being what it is, I Googled the word *Trantaloid*, and one of the links that popped up was for a blog called *The Alien Visitor Evidence*, with a featured document on the page titled "The 5 Species Described in Serpo Releases 23 and 27A ." (There is a lot of "Serpo Releases" material available online; what I've seen and read so far has been fascinating, particularly the Intel briefing with President Reagan, who comes across with the same unflappable, spontaneous sense of humor he was known for--which lends a definite air of authenticity to the transcript material and the possibility that this really happened.) **The Serpo Releases 23 and 27A** may be found online at:

alienvisitorevidence.blogspot.com/2010/10/5-species-described-in-serpo-releases

If you scroll down to the **Serpo Release 27A**, the Trantaloid species are described as VERY HOSTILE. This **27A** material is rather disturbing, so I'll leave it up to the individual to decide if it should be read in full. Let's just say that it confirmed my hunch about the hidden agenda of the Star Wars/SDI system, and made me very glad that I followed my gut instincts back then and voted for Mr. Reagan, both terms.

There are myriad YouTube videos of Linda Howe giving thought provoking UFO lectures and presentations; as mentioned, she's a natural born public speaker, and her razor sharp intellect shines throughout each of them. High up on my "must see" list is one she presented regarding updated material about the genuinely bizarre UFO encounters which occurred in Rendlesham Forest, U.K., in December 1980, at the joint U.S.-U.K military complex (then a nuclear weapons storage facility, also) known as RAF Bentwaters. This lecture was presented via MUFON (Mutual UFO Network) at their *41st Annual International UFO Symposium*, held in Denver, Colorado in 2010. Ms. Howe's presentation, running about 98 minutes and well worth viewing in full, was on Saturday evening, July 24th, 2010. It can be found on YouTube at:

www.youtube.com/watch?v=Rulsob9htY8

The Rendlesham/Bentwaters case is a complex one, certainly presenting many experiences and instances of what Ms. Howe terms "high strangeness," made all the more compelling by the fact that the experiencers are military men and officers, some of them with high ranking. This case was also well covered in Leslie Kean's *UFOs: Generals, Pilots And Government Officials Go On The Record*, with accounts from both the seasoned UFO insider Nick Pope--relating that Lieutenant Colonel Charles Halt used a hand held tape recorder and recorded his anxiety-riddled impressions of the UFO in real time, during the encounter--as well as Sergeant James Penniston, whose journal book of sketches he'd drawn during the sighting are fascinating, to say the least.

An analysis of the full impact of this series of encounters at Bentwaters would require more time and space than is available here, so I focus on the last fifteen or so minutes of Ms. Howe's presentation, in which she proffers the hypothesis, via all the data presented, that the triangular black "glassy metallic" craft witnessed in Rendlesham Forest by multiple military men was, in keeping with their accounts, in actuality a time travel vehicle from Earth's future--40,000 years from now, to be exact; and that one of the primary reasons for their somewhat desperate time jumps back here is to "borrow" genetic material from us, because on their timeline in the future, humanity has devolved into sickly and dying creatures, desperately in need of genetic rejuvenation, for lack of a better descriptor (my terminology--wv).

Linda Howe's fascinating video finale immediately triggered for me the memory of my verbal exchange with Budd Hopkins during our interview in 1994, when I mentioned that Jacques Vallee had derided the concept that aliens might be harvesting genetic material from us, Vallee's reasoning being that an alien civilization capable of utilizing a stardrive propulsion system would be so advanced with their genetic engineering capabilities that they would only require one abduction for their needs--and simply clone umpteen barrels of the material for whatever their strange projects may be.

But then, when correlating this whole highly strange scenario with my own encounter in a Tyler field at age three, particularly when regarding the issue of a child's blood products being filtered out of him, a burst of insight came to me: suppose the issue is not only one of replicating the material? Suppose the true core issue is the *vitality of the original, living material?* Suppose the very real possibility is that cloning, while offering a brilliant way of reworking and tweaking the living-energy matrix, results, at best, in an ersatz sample of living material? One utterly bereft of the luminescent vitality of the original living material it has replicated? In terms of rejuvenation, perhaps cloned material just won't cut it. It simply lacks the *vitality* of the real, original material.

Another line delivered by one of my favorite science fiction characters, Roy, the Nexus Six replicant of *Blade Runner* fame, seems germane here. Perhaps one of the real UFO storylines truly is a desperate survival issue, and Roy--in the guise of one of our future selves--is coming back from a sickly, dying off future Earth to tersely growl at his makers, "I . . . want . . . more . . . *life*, fucker!"

As mentioned, I'm a big Whitley Strieber fan. I've already cited and paraphrased from a few of his UFO books, both fiction and non-fiction. I'm fascinated by how he can take what seems to be well documented research material and weave it into an interpreted and extrapolated "what if" fiction narrative, as he did so beautifully in his almost poetically realized *Majestic* (1990). In my view, he's done it again with his sci-fi/horror novel, *Hybrids* (2011).

Drawing upon rich veins of material about aliens, the Dulce labs, and possibly genetically hybridized humans and animals to be found in the UFO non-fiction literature, Streiber's *Hybrids* has a ring of what might be part of the truth for me, both regarding my Houston UFO close encounter *white-out* (with its seeming information transmission), as well as the possible implications of my bizarre blood transfusion encounter in Tyler. The relevant passage, regarding alien genetic engineering, cloning and tampering with humans via utilizing some of their ultra-advanced high-tech machines:

> Was the duplicate person that resulted the same, or somebody different? The aliens didn't care. They used this device to achieve immortality, also to alter their memories, even to implant purchased memories of things they had never done and places they had never been. Among the many activities they carried out on earth, aside from attempting to hybridize themselves with human beings, was to use sophisticated implants to gather desirable human memories for later resale. They did a lot of things here. They mined Earth for some sort of material that repelled gravity. It was found embedded in lead, iron and uranium seams, from which it had to be extracted atom by atom. They took human eggs and semen to other planets and presumably created their own human beings there . . . all of these were reasons why they demanded that their presence be kept secret. There had been threats, the first one delivered to President Eisenhower in 1952: Do it our way, or you risk the death of Earth.[68]

This brief excerpt, offering a mere taste of what an ancient alien civilization may have in the way of miraculous technology, leads me back to pondering what may lie in store for us as we approach the event the brilliant inventor and author Ray Kurzweil refers to as The Singularity: the moment in our history at which machine intelligence exceeds that of its makers. Sentient machines. Machines which become the Alpha to our Omega. Machines which will quickly be put to use in the design, manufacture and implementation of ever advancing robotic machine intelligence. The Singularity may well be merely the first initial swells that gather strength and merge into a tsunami of cyber intelligence: super androids, ultra-machines.

Perhaps *Star Trek: The Next Generation* neglected a possible scenario: humanity itself is the Borg collective, when rapidly advancing android ultra-machines force us to become their cyborg subs.

Resistance is futile.

UFOs:
IN THE DARK AND IN THE LIGHT

Our whole life is startlingly moral. There is
never an instant's truce between virtue and vice.
--Thoreau

With all of the research I've done for this book, and over all the years of my searching, and my gradual, ongoing process of coming to grips with my personal UFO and alien encounter material, I've come to the conclusion that at the deepest level, this is about the human spirit interacting with non-human intelligences. It also comes down to a soul's coping mechanisms, particularly with a severely traumatic childhood abduction.

As the late Budd Hopkins said to me when I interviewed him in Santa Fe back in 1994, regarding his long term investment in the UFO and alien abduction phenomenon, "I've been at this for eighteen years . . . I've seen a hell of a lot of sadness in eighteen years. How can you accept the idea of frightening a child? I've seen children just absolutely terrified. I don't think they intend to scare a three year old, but that's the actual effect . . . suicides and breakdowns have occurred." When Budd said those words to me, he could have been voicing a narrative of my post-encounter path. I did have something like a breakdown in my thirteenth summer, bundled together with the raging hormones of adolescence kicking in, and it is my certainty that a huge part of that psychic trauma and steady descent into a deep depression was from the pressure of all that Tyler material deeply locked away for a decade, stuff that wanted to surface, to purge. In my original book about this experience, completed late in 1993, I had written it as a work of fiction, because I found it too unsettling to own my material. But I am owning it now. This passage, Phase I of my screaming kamikaze dive into adolescent madness, comes from our family reunion with my father's kin in Virginia, a week long vacation spent in beautifully scenic river front cabins, in July 1970:

By night, the marina of the Rappahannock River Yacht Club felt safe, as much a harbor for pleasure boats as it was for me. I had a kinship with the marina the same as I did with airports. Airplanes and boats both promised escape and speed, a means to get somewhere else, and get there fast, anywhere but here. And back then anywhere but here was the place I wanted to be.

It was the fourth night of our vacation. By then I'd devolved into a deaf mute, and thankfully most of the relatives left me alone. I knew my parents were concerned, but what could I do? I couldn't possibly have explained to them the sort of inner psychedelics my mind was treating me to. The only upside to my plunge into depression was the concomitant passion play unreeling within. I could stare at a green leaf for an hour and find the Universe within its boundaries. Seventy millimeter Panavision should look so good. It was my first acid trip, and I didn't even have to buy a hit.

After dinner at one of the cabins, night fell, and my feet hurried me from the noise and the lights to the darkness and the water. As I approached I was welcomed by the familiar gurgling sounds of water lazily slapping against the dock pilings and dripping down from the underside of the decking. Nudged by warm sea breezes, boats of every size and description tugged at their moorings. Somewhere in the distance a buoy clanged its bell. The air felt swollen, sensual, impregnated with salt and the odor of fish. It made me smile. At the far end of the dock was Uncle Danny's cabin cruiser. He'd suspended marine lights over the side, and they formed an underwater sphere of visibility amidst the dark waters. A small group of relatives and locals stood staring down at the underwater tableau.

The illuminated water churned as though boiling. Seemingly endless chains of crabs swam in circles, round and round and round, overlapping and crossing each other, but never colliding. They appeared trapped on an underwater Ferris wheel. That they had no idea they were being observed lent a clinical atmosphere to our gathering. Drawn into the light, their movement was a dance of madness, made even more poignant by the fact that they paddled madly to their doom. Uncle Danny, Uncle James and my cousins Brad and Bart took turns with the net, effortlessly plunging it into the fray and coming up with one or more of the hapless creatures. With a smooth and easy arc they were tossed into a cold igloo full of ice, destined for the boiling pot and tomorrow's dinner table. For every crab that was taken, another took its place in the procession. It was a scene at once beautiful and cruel.

Under other circumstances, it wouldn't have seemed so intense. But in my frame of mind, I read into the scene some grand tragedy, a lesson from nature that transcended the cold facts of the food chain, of how the bigger and smarter creatures devour the smaller and the less intelligent. There was also something about don't get too close to the light, or you'll get taken.

It was then that I began to realize nature can be both very beautiful and very cold. Asking only for respect and acknowledgement, she requires no thanks for what she gives, and offers no apologies for what she takes away.

I stared at that scene of the hunter and the hunted for a long time. In their death march the crabs possessed a rhythm which kept perfect cadence. In my cracked-open head I could hear the music they moved to, made with a flute and bass and drums, whose beat and frequency was ever increasing, like the sound of the heart of a runner running ever faster. It was the sound of nature in progress, of life going through the motions that define life itself, the sounds of a seedling budding and bursting through the soil to reach for the sky. In my madness not only could I hear the music the crabs moved to, I could see scenes in fast-forward of the blooming and blossoming of nature in all its Technicolor pageantry, of flowers and blossoms unfolding, of the hidden becoming manifest. All these things I saw superimposed over the real event before my eyes. A year later, I heard the song *Cross-Eyed Mary* by the group Jethro Tull from their incredible album *Aqualung* for the first time; the instrumental opening with Ian Anderson's mad, rhythmic flute playing was precisely what I'd been hearing in my head. Funny how that works.

All of this was in my world, while a few feet away my uncles and cousins saw nothing more than the dumb luck of scooping up and trapping tomorrow night's dinner.

Later, dazed, I stumbled back to the cabin to bed. As soon as my head hit the pillow and my eyes closed, my loony tunes band kicked in. I had no control over the volume or the programming, so I tried to relax for the show. While *Cross-Eyed Mary* jammed in my head, I plunged into an underwater dreamland where roses bloomed and burst open next to thrashing sea-anemones. Dazzling fragments of light splintered over coral reefs. Waves crashed in my ears. It was all so amazing and crazy that I longed to share it with someone. But the nearest sanitarium was many miles away, I suspect.

I must have opened the doors of perception too wide; madness happens when too much information floods through the sensory register. I was doing my best to ride the crest of my own cerebral tidal wave. Where it would take take me, God only knew.

Where it took me, barely three months later on a cold October night in New Orleans, was the second phase of the Budd Hopkins scenario: in my rebellious madness I had been running with a pretty fast crowd, so it wasn't that difficult to get my hands on a bunch of pills, some downers. At the very tail end of a downward spiral of depression that took me to the dark bottom of a well, putting both myself and my family through hell, I swallowed pills by the handful, and just for good measure, a bottle of aspirin and the remaining half of my Caution: Drowsy Inducing warning-labeled allergy tablets. They kicked in pretty steadily, and I tucked myself into bed, with no expectation of seeing the morning sun. The physical sensations were, remarkably, as though I was being slowly frozen, and somehow a tranquilizing coolness gently washed over me, from my feet upwards through my body, as though my body and mind were slowly, steadily submerging into dry ice vapors. From the 1993 former version of my story, before I was willing to own it:

> Shut down time, the kid's leaving for good, be sure to shut off all the lights. Cold vapors, how strange, yet how soothing. Years before the advent of cryonics, I was freezing myself to death. Gently. I became dimly aware of tears trickling down my face. Cry. Cryogenics. Cool vapors. Time to die.

> My torso and neck went next, and then all that was left was my head. All that remained of me, who I'd been, who I was, who I might have been, my hopes and fears, my dreams, my joys, my tears, gathered up into a tight locus of brilliant energy up inside my head. I'd become something like a collective of sparks, and me, the boy I'd always known, was encapsulated in those sparks. Then, in an instant, like the last great switch had been thrown, there went my head. But wait . . . what? Instead of total darkness, I had the immediate awareness of lights, and strange surroundings.

> I was floating in space. There was blackness, but stars were everywhere, and I'd left my body somewhere far away. I, the spark entity, floated around amidst the glory of countless stars, their light pure and unwavering in the vacuum. If this was purgatory or hell then that was hard to believe; this was way too beautiful and serene. Pain was gone. Mental anguish was gone. There was nothing but heavenly splendor, peace and silence.

> Then all at once I knew that I was not alone. I felt it before I saw it. Up above and to the right of me there was a presence. And when I saw it, its appearance matched how it felt. A glowing, radiant orb.

> It was an oblong white light of unimaginable beauty. It shimmered and shone with splendid effulgence. Its beauty transcended the capacity of words to adequately describe it. And its light radiated feelings of love, profound love, compassion, empathy and boundless understanding. I could sense that it was alive. It had sentience. It was a radiant being of light. Presently, it moved towards me.

With no warning, no prelude, in the next instant I was within it, and it was within me. We had merged. The overwhelming radiance of its light streamed through me. Though I was sure I was out of my body, I could hear myself gasping for breath and sighing at the sudden shock of release, the shock of unconditional love loving me. White light was everywhere. I was completely unprepared for its intensity and its goodness and warmth. Knowing that it was alive, that it was sentient, I began to talk to it. I explained how I got there, why I was there, that I knew with my upbringing I could never be who I was supposed to be. My Catholic catechism teachings had nothing but condemnation in store for boys like me. I repeated my reasoning like a mantra of sorrow. I'm not growing up right. I can never be who I should be. I can't be. I can't. . . .

The pulsating light shifted around me as though its elements were rearranging themselves. It was concentrating and crystallizing itself. In a few moments I was suddenly within a diamond of white light, a white diamond whose myriad facets glistened and sparkled like ten thousand suns. And just as quickly as the diamond had formed, it transformed yet again into something else. Facets of the radiant gem separated symmetrically around me. They became glittering petals as the diamond opened up. I was in the midst of a phosphorescent white diamond rose in bloom. As it burst open, the diamond petals bathed me in their warm and loving light. I heard a deep, resonant voice begin to speak. As the diamond rose blossomed open around me in all its radiant splendor in dazzling slow motion, the voice said with great strength and utmost warmth and gentleness, "Don't be afraid . . . to bloom differently."

I heard and felt the simple, wise, healing words reverberating inside of me. They cleansed and charged every part of my soul with their power. I sighed with a sense of absolute peace, relief and understanding. The diamond rose's light stayed with me for a long time. And gradually, my awareness faded out, and I had silent solace. And it was good.

Incredibly, I awoke the next morning, got dressed in a daze, and stumbled in a groggy stupor on my walk to school--no escape. I have no way of explaining my survival, other than a miracle; I weighed barely 110 pounds wet in eighth grade, and I'm certain I'd downed enough handfuls of chemicals to end a full grown man. I guess it wasn't my time. The title of my 1993, fictionalized, unpublished version of my UFO rite of passage: *To Bloom Differently*. I keep the tattered manuscript in a box at home.

It wouldn't be until five years later that I would buy the newly published *Life After Life* (1975), the groundbreaking work by Raymond Moody, M.D., and be exposed to the terms Near Death Experience (NDE), Out of Body Experience (OBE) and Being Of Light. It amazed me to recognize so easily this pure being of love and light that I'd witnessed firsthand, appearing in the pages of Moody's book. I think of mine as having been a Guardian Angel. When I merged with it, perhaps a purification and purging of the chemicals in my body ensued. I have no other way to explain being here now.

I debated for a good long while as to whether or not I should include this troubled act of my life's play; after all, this is a book about UFOs, aliens, abductions and hidden technologies. But I realized that including it is important, and relevant, for a few good reasons. One, because it illustrates how repressed psychic trauma from an early childhood close encounter can play havoc with a young soul. (I've sometimes thought of its impact as being like having a tall sky-scraper lifted off of its footprint and dropped on you. Absolutely crushing the bejesus out of you in an instant.) Two, because of the word soul, itself. My NDE and OBE gave me firsthand, experiential real life proof that there is life beyond the physical, and that the existence of the soul is a reality. So I have no problem with Mayan's philosophies, or those of all other belief systems that recognize its existence. Three, because during my OBE when I popped into what appeared to be outer space beyond Earth, I was introduced to an exalted being, or perhaps I should say that the exalted being introduced itself to me; it reminded me of the ideas presented in Jim Marrs's fantastic book *Alien Agenda*, that there exists a hierarchy of beings overseeing our planet and its evolution, and the most elevated ones are, essentially, angels. I definitely met mine.

And finally, my OBE is important because the Being Of Light I met was bar none the most beautiful and pure embodiment of goodness, love and light conceivable. It taught me that the radiant purity of love that exists out there beyond the physical plane is almost beyond human comprehension; and it left me with the sense that polar opposites, beings of darkness, lurk somewhere out there as well. As for the Beings Of Light, it's clear to me why many of Moody's NDE and OBE experiencers said they didn't want to return to resume their physical lives after experiencing the bliss of being in the Light.

Since I'm also quite clear about how much evil exists in the world today (for me to make a list right now would be to write another book to accommodate it), this has opened my mind to the possibility that at the deepest level, an ever-present conflict in life engages us, the age old battle between good and evil--also expressed as the archetypes of Good and Evil. The comments in Linda Howe's book *Glimpses of Other Realities, Vol. II: High Strangeness,* about the Ebens being afraid of the Evil One, really hit home with me--the words almost jumped off the page. A powerful testimonial, to hear that aliens from another star system share this fear and revulsion of a dark and evil being. Perhaps some of them have souls, too. Perhaps some of our visitors don't. That's definitely the direction in which I'm leaning.

I'm an ex-Catholic, partly because the Church's hypocrisy towards members of my alternative tribe got to be way too much for me--aside from all the altar boy rapes (See: documentary film: *Deliver Us From Evil*/Amy Berg, director). One of the few things I do retain from my former Catholicism, aside from a firm belief in God, angels, saints and the Devil (or an evil being or collective of dark beings some call by the same name), is one of the teachings and practices that was taught to us in Catechism in which we renounce Satan. I say the words aloud every now and then, to clarify it for myself, and send my convictions out there: *I do renounce Satan and all of his works, and all of his false promises.* It feels like the air clears once the words are spoken, especially if out loud. With all the tens of thousands of altar boy rapes the Catholic Church is guilty of (they are legion), and therefore the souls they have stolen and damaged irreparably, I've definitely asked myself more than a few times, does the Catholic Church truly renounce Satan, too? Just thinking out loud here, padres and Cardinals. Talk amongst yourselves. Discuss.

I have no desire whatsoever to hang out with the Dark One or his legions when my soul finally does leave this planet. I hope to join with more Beings Of Light, and with my departed family and friends on the other side, and do good deeds out there. I will never understand willfully choosing the dark side. And I'll never need or want anybody to try to explain it to me. That's one of the few things in this amazing Universe I'm not the least bit curious about.

* * * * * * *

High above town in my own lab
I don't care what anybody thinks.
I adore what I make--these images
These freaks.
Whatever my feelings say
My monsters kiss and dance . . .
As the villagers approach my tower
With their clubs and their torches

from "The Presentation of Self
in Everyday Life" -- **Bill Zavatsky**

In the process of writing my B.A. thesis for my Research Methods course taught by Dr. Wendy K. Silverman at FIU in 1990, *The Presence Of Specific Personality Factors and their Effect On a Person's UFO Belief Systems*, I came across a surprising amount of sober minded UFO-related published material in some respectable psych journals, even back in 1990 when I was doing the research. One of them which stood out for me in bold relief was a paper by one of Budd Hopkins's colleagues, the psychologist Aphrodite Clamar, titled *Is It Time To Take UFOs Seriously?* After administering a battery of psychological tests to her subjects who were claiming UFO and/or abduction encounters (including projective tests, intelligence tests and the MMPI) her psychological report stated, in part:

> To summarize, while this is a heterogeneous group in terms of overt personality style, it can be said that most of its members share being rather unusual and very interesting. They also share brighter than average intelligence and a certain richness of inner life that can operate either favorably in terms of creativity or disadvantageously to the extent that it can be overwhelming. Shared underlying emotional factors include a degree of identity disturbance, some deficits in the interpersonal sphere, and generally mild paranoid phenomena (hypersensitivity, wariness, etc.)

I certainly see myself in Ms. Clamar's conclusion. The road of one who has been taken can be a very solitary path. Starkly so for me, other than my bouts of promiscuity. Perhaps at those times I've merely emulated what I was taught. In my own way, over the years

I too have been a visitor, and an outsider; I have abducted scores of men (although no force was used), many a fine specimen at that, put them on a table, implanted them, and then set them free. If the orange and brown skinned ones who snatched me up from that field in Tyler really did implant me and have always known of my whereabouts, I hope they've enjoyed the show. I've aimed to please.

I also fit the profile of what Kenneth Ring referred to in his book *The Omega Project* (1992) as one who manifests an "encounter-prone personality." *The Kirkus Reviews* (1992) really tore into Ring's tome, partly for suggesting that "UFO abductions are not alien encounters per se but some kind of symbolic event, a conclusion that puts him in the 'soft-core' camp spearheaded by Strieber and Jacques Vallee. . . ." While I did enjoy Ring's book, I thought *The Kirkus Reviews* had a point. I can accept that possibly some of these encounters are representative of alien intelligences or "mind at large" interfacing with us; but I'm also convinced that a lot of the "legitimate" UFOs, a/k/a flying saucers, are solid, honest to God nuts-and-bolts alien spacecraft, as echoed by Friedman and Howe.

Though I differ with some of Ring's conclusions, my life experiences certainly fill the bill: UFO encounters, NDE's, psychic phenomena, some ESP skills. I even lived for six months in a haunted house in Houston, refusing to move out because it was rent with option to buy, until it became clear that I was compromising my health by remaining in the midst of unhappy spirits that didn't want me there--another book in itself. My take on it is that my highly intense encounter in Tyler at three, aside from its obvious psychic traumas, opened up my psyche--some would say opened up my Third Eye and my Crown Chakra--to an expanded view of reality. I believe that all of us who have been taken have been opened up this way. Sometimes intense encounters force open the doors of perception, priming the experiencer for enhanced psychic acuity.

Further along this line of reasoning: some Florida relatives of mine once shared with me in 1991 the details of a UFO sighting of unusually long duration they and all of their neighbors had back in the 1960's, they said the date was in 1965-66, in Gainesville, Florida. My uncle, Rick McAllister, a prominent doctor and gifted sculptor, and his wife, Annie, my father's sister, a nurse and a very talented artist herself, related how in mid afternoon on a clear, sunny weekend day, they were outside in the yard, and suddenly noticed that about 100 or so yards away to the east, a classic double-disc

shaped flying saucer was silently hovering. Its shiny metallic surface reflected the afternoon sunlight. It was stationary at a couple
of hundred feet in altitude, and quietly stayed put for quite awhile-
-Annie said for over an hour, possibly closer to two hours. When
Rick retrieved his field glasses, they both noticed through the viewer
that the double-disc had what looked like portholes or small windows in a circle around the upper half of the craft. Both Rick and
Annie commented on how interesting and strange it was that no jets
ever showed up to check it out, considering the extended duration
of the sighting. From their account, it sounded like over a dozen
neighbors up and down their block had come out to view the oddly
stationary intruder. It was so clearly visible for so long, everybody
could agree on its size of 30-40 feet in diameter, its unconventional
shape, its silence and its strange motionlessness. When it finally
departed, it slowly cruised away, then abruptly shot straight up and
disappeared in a flash--a fairly typical flying saucer departure.

When I mentioned this to their son, Nathan, on the phone when I
contacted him for more background information about his family's
sighting (much later, in 2006), he said that he'd had a close up UFO
sighting as a youth of eleven or twelve (he's not sure which) in the
early 1960's, also in their Gainesville, Florida neighborhood; he was
roughly 7 years older than me, so that would be about right. Hearing about this came as a surprise, since I'd seen Nathan at a couple
of family reunions before our phone call, and he knew how obsessed
I was with nuclear physics and astronomy and UFOs, and had never
mentioned his sighting to me.

One early afternoon, possibly spring, he was outside, and suddenly saw a large object fly overhead, moving so fast that it was there
and gone. It swooped over their garage, and tore off into the direction of the deep woods beyond their property. It streaked by so fast
that he thought it was going to crash. Without hesitation, he sprint-
ed after it as fast as he could, keeping a close eye on the ground so he
wouldn't trip on a rock or an uneven patch of terrain as he dashed
through the woods on his football player legs.

At this point in the story, at first Nathan said that he'd been so
intent on watching his feet on the ground to make sure he didn't
trip or fall, that it was almost as if he ran headlong into the object,
and stopped abruptly just in time to avoid colliding with it; it was silently hovering low above a grassy field surrounded by trees. Then,
on reflecting, he said that maybe actually what happened was that

he reached a large open clearing, and then he looked to the far end, and that's where he saw the large disc. It was dull metallic in color, like pewter, perhaps thirty feet in diameter, hovering silently above the grassy ground. Barely a second later, or so it seemed, the flying saucer took off skyward, streaking upwards so fast it practically dematerialized before his eyes. He ran to the spot where it had been, and saw a large, circular swirled pattern of flattened grass, the classic "UFO nest" formation, where the disc had been quietly lurking a few moments before. He squatted down to touch the flattened blades and study them a moment.

On impulse, Nathan ran back to his house, and told his father the full story while catching his breath. Sadly, his dad laughed it off, asking him where he'd gotten such a vivid imagination. He wouldn't even accompany Nathan back to the field to see the swirled, matted down circle of grass. When he recounted the experience to me on the phone, he added that he'd felt a deep sense of betrayal, that his father wouldn't trust his perceptual skills enough to at least accompany him to take a look. I told him I could relate. Familial ridicule for reporting a valid UFO sighting can be both hurtful and infuriating. Especially when you know you're right, and their ignorance and fear, masked by derision, are blinding them to the reality of the encounter you've related, and the elusive phenomenon behind it.

When he abruptly changed the details about how he came upon the hovering disc, it brought to mind Whitley Strieber's words to me regarding my Houston solo sighting, "If you were that close to it, then *you're in.*" I don't think I ever repeated Whitley's unusual comment to Nathan; it seemed like something he wouldn't have wanted to hear.

As mentioned before, when my cousin Nathan McAllister was attending UNO in 1970, he came over for dinner to our house in New Orleans, and introduced me that night to the Rider Waite tarot cards. He was a gifted reader, and told me years later that he'd quit using them because he'd thrown the cards a few times in a row for a self reading, and the exact same ones fell in the same card-spread positions repeatedly. The improbability of it sort of freaked him out. Our similar tarot card skills are an intriguing feature from our pasts. It doesn't prove anything, but enhanced psychic abilities are things UFO experiencers sometimes share.

I think it was in the Budd Hopkins material where I first heard about the theory that sometimes family blood lines are studied in the process of what could be called tandem abductions. It seems pretty far out to me to imagine these beings could track us like that, but then, if their civilization really is as ancient as some intelligence sources say, who knows what their capabilities might be? If there's really a million years separating our species (or even more astonishing in scope, a billion years), their dazzling interstellar propulsion systems might be one of the lesser skills to be found in their box of technological magic tricks. It seems clear to me that our Shadow Government boys jumped a few centuries ahead with field propulsion technology with the help of aliens and some of their crashed vehicles, and some of the (presumably early prototype) ARVs were apparently flying barely 20 years after Roswell. The technological gap between 20 years and a million years would be staggering. Almost incomprehensible, really. Indistinguishable from magic. And how about a thousand times beyond that, a billion years?

Nathan's verbal editing of his encounter made me wonder about how close up he actually might have gotten to the disc. The fact that the object reportedly departed almost the instant he got close to it is inconclusive, too. The visitors do seem to be able to wipe out memories and re-splice the inner video in our heads, so who knows how long they lingered. In that kind of experience, without hypnotic probing, who knows?

When I told Nathan back in 2006 that I'd like to include his encounter in my book, he enthusiastically consented, even going so far as to say I was welcome to use his real name, his profession and location. But based on the unconventionalities in my story here-- my alternative lifestyle, my bizarre UFO encounters, and the speculating I've been doing here about Nathan's encounter--I decided it would be best if I kept him disguised. He has a solid, upper level white collar career in the South. There's no justification for potentially compromising his job or his personal life. I love him and his incredibly talented sister and brother and family, and always have, so they'll remain incognito.

Remain incognito. What better terminology to use for the extended cover up, since one of the primary m.o.'s of the Billionaire Boys Club is to remain hidden--incognito. In 2013, Dr. Steven M. Greer and Richard M. Dolan did their very best to try, once again, to force the government into disclosure about UFOs, alien visitations and back engineered technologies. Along with their esteemed colleagues, the *Citizen Hearing on Disclosure* came together in another attempt to try to pry open the doors of secrecy. Here is the web site for it:

http://citizenhearing.org/index.html

The *Citizen Hearing on Disclosure* was an unprecedented event in the matter of size, scope and the presence and participation of former U. S. Congress members. Running from April 29 through May 3, 2013, an esteemed panel consisting of researchers, UFO witnesses (military and others), activists and political witnesses appeared in Washington, D.C. before six former members of the United States Congress to testify to their certainty about the presence of UFOs and an alien intelligence engaging us here on Earth. More than 30 hours of testimony from 40 witnesses were presented over five days. The event was heralded as the most concentrated collection of evidence regarding the subject of extraterrestrials ever presented before the press and the general public simultaneously.

In Dr. Greer's allotted time before the panel, he presented a veritable laundry list of top secret military installations and their presumably not widely known activities, as written up in a folio containing research and intelligence he and his associates and contacts had gathered. (These are all allegations, some of which already have a fairly developed basis in factual evidence; with my background of experiences, I don't see it as much of a stretch that all of this information is the truth.) My favorite locales and details of their covert activities: Los Alamos National Labs, connected underground via high-speed tube shuttle to the Dulce Labs (Archuleta Mesa of Northern New Mexico) where "the biological work is being done" (the web is jam packed with horror stories that this subterranean locale is the nexus of a vast, multi-level underground joint alien-human operated complex of experimental labs, supposedly chock full of horrific genetically mutated human/animal monstrosities if you should blunder down to the lower levels housing Nightmare Hall, etc. Google: Dulce Base. Just be prepared to read and see stuff that seems lifted straight from a Clive Barker horror movie); In southeast Arizona, there is the Army Intelligence Center at Fort Huachuca; Greer states it is the site of a UGB--an underground base--where one of his witnesses will testify he'd worked on 9 separate extraterrestrial vehicles that had been brought down with electromagnetic pulse (EMP) weapons, and that there are several different species of extraterrestrial biologicals stored at this facility. See:

http://huachuca-www.army.mil/index.html

Also included: the Redstone Arsenal, outside of Huntsville, Alabama, where one of Greer's contacts allegedly was working for IT&T on a "trans-dimensional system."   And one of my favorites: in the Cheyenne Mountain complex (presumably part of NORAD): witnesses on Greer's team will testify that they've tracked an alien craft in our solar system measuring 26 miles in diameter. Hope they're friendly.

By far the most impressive speaker to me was Richard M. Dolan, which is fitting, because he closed the proceedings with his speech on May 3rd. One of the most moving points he made was that, "the National Security apparatus has created a global culture that has suffocated the truth." Beautifully put. Dolan ruminates that in the event of full disclosure happening, society would ask, well, if the President isn't in charge of this, then who is? I like how he refers to the elitist cabal who are actually running things behind the scenes (in my view, the Billionaire Boys Club or the Shadow Government) as the "Breakaway Society," its head honchos known as the leaders of the Black Budget. Dolan notes that, "if the answer is anything along the lines of my own research so far, it will show that the cover up has long ago gravitated away from formal presidential authority, and into international and private hands." He astutely notes that one of the problematic issues might be that, if the newly "enlightened society," after first querying, "how have you managed to keep this secret all these years?" then asked "who are the aliens, where are they from and what are their agendas?," the Breakaway Boys might not be able to answer the alien agenda questions with certainty. Can anyone trust creatures almost incomprehensibly advanced?

If you compound these issues with the myriad of cataclysmic changes potentially brought about by initiating a full disclosure--industrial, economic, technological, sociological, religious and spiritual, just for starters--then in some ways Dolan's impassioned rhetoric brings to mind a line used a lot by battling courtroom attorneys in the classic TV series *L.A. Law*, namely *asked and answered*. Meaning, while giving all the reasons full disclosure is necessary (and I agree with them wholeheartedly, obviously), he also lists how such a powerful event in the wink of an eye could result in the total collapse of our society as we know it. And who knows how long it might take to re-structure it? And perhaps the most God-awful prospect of all: if people like the thugs and goons at the NSA, CIA NRO and DIA are in fact pretty close to the core of it all, who's to say that they'd allow for even a 10%--let alone 100%--dissemination of what they know and have held onto for decades about UFOs and aliens and back engineered alien technology? Who's going to play David to their Goliath? The President?

I like to think that my attitude towards life is often positive and optimistic, though the harsh realities out there (some of them seen, some unseen) make me wonder if the human race has a snowball's chance in Hell of digging itself out of this Jekyll and Hyde framework in which the National Security state has us trapped. In my mind, the facts about the Shadow Government's adamant insistence on non-disclosure could go even further than has been discussed or speculated. How about if the ugly truth is that the Boys sold us down the river in the 1940's? That the aliens made the fellows who pounced upon the wreckages and alien survivors--calling them their own--an offer of technological wonders beyond compare if they'd keep quiet about the abductions, cattle mutilations and establishment of permanent alien bases, both underground and at the ocean's bottom? (I used to think this an absurd conceit, that their technology meant they didn't have to ask for that which they could take by force. But Corso's *The Day After Roswell* [despite Stanton Friedman's unfavorable assessment of him] made a great argument: that we threatened the aliens that we would turn the Earth into a radioactive wasteland with our atom bombs [and soon to be exponentially more powerful fusion H-bombs] if they mistakenly thought they could conquer or own us. If we couldn't have Earth, neither could they. Clearly, this exquisite gem of a planet is a cosmic prize in this neck of the galactic boondocks where we reside. They wanted their shimmering jewel intact, not burnt to a cinder. So a compromise was floated, and a dark alliance was forged.)

And so how about if, after decades of this wholesale betrayal of the human race, all those underground bases scattered across the U.S. and the rest of the globe (as mentioned and detailed in Dolan's and Howe's books) are now populated with a bizarre mish-mash of humans, aliens, cloned humans, human-alien hybrids, human-animal hybrids, and androids that make Data seem like a simpleton--and surrounded by a technology that matches, or surpasses Jean-Luc Picard's galaxy-class Starship *Enterprise*? Could it be that the very basis of their never wanting to divulge this is because the Boys fear that the resultant outrage felt by the common man (numbering in the hundreds of millions) at this titanic betrayal might never be fully placated? Like, never ever? (This scenario overlooks what could be the penultimate betrayal of all--that the Boys might know that some of the visitors may even have dibs on our souls for some unknown alien agenda, and their million--or even billion--year old technology enables them to remove the Source energy from our bodies with the ease of a peach pit plucked from a ripe peach.)

In order to take this scenario to its ultimate conclusion, let me introduce a man who is one of the most singular fellows I've ever found on the web: John Greenewald, Jr. I first blundered across his web site, *The Black Vault* (**theblackvault.com**) back in 1999. It appeared to be an interesting compendium of classified information culled by its industrious, ingenious and courageous founder. If you watch this YouTube video you'll be able to listen to this remarkable character smoothly deliver a talk about how he came to learn how easy it is to get bushels of classified documents from the U.S. government, provided you have an incredible amount of patience--which clearly this ballsy man has in spades. The video is him lecturing in Washington, D.C. at a meeting called *The X-Conference / Live Speaker Presentations* (the date isn't given or posted, but verbal referencing in the video indicates it was in 2005):

http://www.youtube.com/watch?v=oOWZxIAkkE4

Or you can Google: The Black Vault & UFO Secrecy: John Greenewald, Jr.

Greenewald is a natural public speaker, and paints a comical Keystone Cops portrayal about how stupid and stupidly run our governmental agencies can be. A lot of the material he compiled and eventually uploaded to his web site (it still looks to be free content, in keeping with his goal to provide free information of all things hidden) is documents he was able to snag via relentless mailings of FOIA requests to all the acronym agencies, beginning when he was in his early teens. He hilariously recounts here that the CIA, NSA and FBI definitely knew who he was from an early age when they received 30 FOIA request letters at a time from him. I urge all UFO inquirers to watch this roughly 55 minute video, if for no other reason than because Greenewald establishes beyond any doubt whatsoever that not only is the government covering up tremendous amounts of UFO material, but that this practice is, in his reckoning, a bona fide conspiracy, and they're up to their eyeballs in it. Not a hard sell for me, Mr. G.

This is all well and good. Obviously, I'm onboard with every word this remarkable guy says about UFOs. But while perusing his now sleek and sophisticated web site (please donate so he can maintain it), I found some very disturbing information that fits like the final, perfectly matched puzzle piece for this multicolored puzzle I've laid out. It has to do with the ultimate final frontier, superseding even Captain Kirk's *space*. It's the frontier known as *the human mind*.

Years ago, browsing *Half Price Books* in Austin, I snagged a used copy of a non-fiction expose about the CIA's program called MK-ULTRA. I plowed through the horror stories fairly quickly (mind control experiments, LSD dosed to unwitting subjects, like that--and those were the lighter fare), and some of it was so creepy that I speed-read it just to get past it. By the last page, I mentally filed it in my Maybe So/Maybe Not box. Mostly, I refused to believe that human beings could do such horrific things to each other, in the name of "advancing science." I threw out the book, a rare thing for me. It was that awful.

Scanning recently through Mr. Greenewald's *Black Vault* articles on MK-ULTRA, I was amazed that I could actually page through the transcripts of the Congressional hearings exposing some of the astoundingly evil things the CIA did to people. This wretched stuff gave great credence to the passage in *Psychic Warrior*, the David Morehouse autobiographical Remote Viewing expose, in which he said an insider commented to him about the Agency, "it is an abomination." I second that emotion.

Closing out that section of the site, I came upon one other mad scientist header, listed as follows: **DARPA MIND CONTROL PROJECT.** You can find it here:

www.theblackvault.com/m/articles/view/DARPA-Mind-Control-Project#.UoHuu-KaLvY

If it's possible to feel affection for an acronym, my heart goes out to DARPA: Defense Advanced Research Projects Agency. Always have loved the sound of that. While I can't fault this collective for allowing the DARPA-NET to blossom into the Internet (who can fault anybody for birthing our global mind?--unless it becomes our undoing), this current and ongoing Mind Control Project makes me wax nostalgic for the bygone, more innocent days of MK-ULTRA. If you drop down below the main heading of **DARPA MIND CONTROL**, the next line elaborates, "*Toward Narrative Disruptors and Inductors*: Mapping the Narrative Comprehension Network and its Persuasive Effects." The idea here is that these mad scientists conducting this research at ASU--Arizona State University--are doing beta-testing on a system that, once in full operation, will essentially allow them, via directed magnetic fields, to wipe clean specific "narrative networks" in a subject's mind and replace them with others, meaning they can take chunks of our belief systems, erase them, and replace them with whatever they see fit. This article begins with how rumors of this program surfaced in July 2013, but quickly became established as a reality via a whistleblower. The program overview poses this question:

> What if the government could change people's moral beliefs or stop political dissent through remote control of people's brains? The aim of the program is to remotely disrupt political dissent and extremism by employing **Transcranial Magnetic Stimulation** (TMS) in tandem with sophisticated propaganda based on this technology. **TMS** stimulates the temporal lobe of the brain with electromagnetic fields.

My reply to their "what if" question would be, What if we don't want the government changing our moral beliefs? What if we don't want the government tinkering inside of our heads at all?

The link on the *Black Vault* page leads us to the original source web page, called *Activist Post*, found here:

activistpost.com/2013/07/secret-darpa-mind-control-project.html

Activist Post elaborates on **TMS**:

> Once it is determined that disruption of certain portions of the brain can enhance persuasive messaging, individuals can be persuaded to do things they normally would not do and believe things they normally would not believe. This could include something as simple as telling a closely guarded secret, to believing in government propaganda, or even committing a violent act. The

group writes that, "once we have produced a narrative comprehension model [i.e., how individuals comprehend stories and persuasive messages], end users [aka the government] will understand how to activate known neural networks (e.g., working memory or attention) and positive behavioral outcome (e.g., nonviolent actions) nodes with strategic communication messages as a means to reduce incidences of political violence in contested populations." The group will investigate "possibilities for literally disrupting the activity of the NCN [narrative comprehension network] through **Transcranial Magnetic Stimulation**."

The amorality of this system, seemingly already on its way to being mainstreamed into use, brings to mind a pivotal scene from W.A. Harbinson's *Genesis*. Aldridge, the evil genius leader of the New Aryans of the Antarctic, tinkers with a recently abducted key government official from the Cheyenne Mountain complex (NORAD), partially drugged and reclining in a lab chair in his undersea, bottom of the ocean high-tech domain. Over his captive's head hangs a surgical canopy with a halo of bright lights, a stereotaxic skullcap, and a cluster of dangling electrodes ready to be implanted. Aldridge coldly and clinically explains that the desired brain targets have been fixed, and he's now painlessly drilling into the helpless victim's head; as the hair-thin electrodes are being placed, he's told that he'll experience a brief panic, soon to be replaced by feeling "nothing." Aldridge implants the final electrode, icily informing his newly created human automaton, "I am taking your mind now."

The one primary difference between the mind control of *Genesis* and this cutting edge **Transcranial Magnetic Stimulation** is that our high tech Boys in the real world don't have to drug you and place you in a lab chair and fiddle with electrodes. They can zap their victim's brain with an electromagnetic field from a distance, the ultimate in stealth brain surgery and behavior modification, and the target goes on about his day, unaware that "they are taking his mind now." He won't even consider it odd when he abruptly switches political parties. Killing opposition leaders or conscientious objectors might suddenly seem acceptable, inspired even. As Aldridge explains to his newly created human robot, about to set him free and turn him loose at NORAD:

Man is thus just a machine, to be utilized, controlled, operable by simple laws of give and take, without will of his own . . . The mysteries of the human mind, its creativity, its moral imperatives, have been reduced to a set of components which we endlessly play with. Man is not a magical creature--he is a container of various impulses. These impulses can be rearranged to a pattern that will change his behavior.

By stimulating the appropriate areas of the hypothalamus with submicro-electronic electrodes, I can regulate your blood pressure, your heart rate and respiration; your sleep, your appetite, even the diameter of your pupils; I can place you in suspended animation or make you work till you drop. Do you understand that?[69]

I wish I could say I see a way out of this. It seems to me that this **TMS** system was custom made for the Boys who don't want to disclose. A world full of happy, blank idiots driving around the surface world in their petroleum combustion powered cars, oblivious to the astounding technology surrounding them and the free energy that could make their lives immeasurably easier--that would be the ultimate victory for the Shadow Government. A few magnetic pulses here, a few pulses there, and voilá, a robotized society is theirs, with none of the surface dwellers any the wiser. I'm inclined to think the Boys may have already used it on Congress, in the same way Aldridge and his high tech saucer men would kidnap key officials and steal their minds as a means to influence key decisions.

There's a particular senator I'll call "Senator Manchurian" who recently, whenever given the chance to help reign in the NSA's ever-expanding ultra surveillance reach, retreated to chambers to hammer out changes in a bill, ostensibly to benefit the populace, only to come back with a new version even more oppressive than the previous one, happily touting it. It makes me wonder if Senator Manchurian has been zapped with a few good **TMS** pulses. Every day is a new day in the Empire Of Technology.

It's real life horror stories like this one that make me wonder if the late William Cooper--he of the paranoid manifesto, *Behold A Pale Horse*, which I read and found horrifyingly plausible--might not have been right on target about the whole big picture. He always stressed that the final stage of the global chess game the Illuminati are playing, ushering in their New World Order of One Planet, One Government, would be a totalitarian regime in which the entire populace is brainwashed and robotized. If Mr. Cooper hadn't been killed in a shoot-out with government agents in November 2001 (some say he brought it on himself, but I don't know, I wasn't there when it happened) I would look him up and arrange to have a sit down interview. Though some wrote him off as a paranoid mental case and a troublemaker, so many of his warnings have come to pass that I'm reminded of the old expression, "Just because you're paranoid doesn't mean they're not out to get you." When I shared some of Cooper's assertions about the evil intentions of the Shadow Government with my brother Aaron, he replied, "They can just put me into a pod and make me dream." (See: **Zeitgeistmovie.com**)

This wicked DARPA program also reminds me of a Q & A I had with my first UFO therapist, Dr. Edmond M. Bazerghi in Austin, Texas. When I knew it was our last session, in the late summer of 1979, I told him that I thought he was one of the wisest and most insightful men I'd ever met, and I valued his opinion. The full measure of his sage, humanitarian act of kindness, stubbornly refusing then to unlock my padlocked subconscious mind, is a gift I carry inside of me always. I asked him if he could make a comment about humanity, about how he perceived the state of affairs with regards to mankind on Earth. Though his words were spoken in August of 1979, I believe that they apply even more so today.

Dr. Bazerghi briefly paused to gather his thoughts, then looked me in the eyes and solemnly stated, "I'm concerned that we might destroy ourselves because we don't understand what we are."

His words still ring true for me. I think our potential for self destruction isn't just because of our not understanding what we are, though. I think it's also about the things we might be willing to do with our scientism--our ultra-advanced scientific technology utterly divorced from spirit, from morality, from accountability. A few years back, sometime between 2003 and 2006, there was a cover story about genetic engineering and experimentation in one of the weekly newsmagazines. An ever so innocent passage written in a small box off to the side of the main article casually alluded to a genetic engineer who had spliced human DNA with cow DNA in his lab, and his resultant hybrid "took." The newfound proto-life form began to grow, and its healthy cells divided steadily. The final comment was that "it was terminated before reaching full maturity." No doubt.

My gut sense tells me that the deeply buried horror show of the Dulce Labs is real, Nightmare Hall is too, and marooned somewhere way down in those lower levels, there lives a profoundly tortured, godforsaken, insanely howling cow-boy who doesn't understand what he is. But he knows that his makers are monsters.

FROM THE FORCE FIELDS
TO THE CRASH FIELDS

In the seven some odd years I lived in Santa Fe, I'd never had enough time off from work to make the trip down to the UFO crash fields outside of Roswell. Just a few short years ago, I'd have said "alleged" crash fields. Not anymore. See: Stanton Friedman, *Crash At Corona* (with Don Berliner), ad infinitum. . . .

I spent Christmas 2012 in Santa Fe, at the mountaintop home of a long time friend and client. It was wonderful to see friends I hadn't seen in many years, and to discover that I still have my ski legs. I carved some serious S's in fresh powder down the mountain side of the Santa Fe Ski Basin. Fresh powder is good for speed demons.

I've always had a deep feeling of affection for wilderness, certainly due in large part to my formative years on the edge of my Tyler fields. I love the sense of rugged isolation and desolation that can be felt while traveling through such spaces. So two days after Christmas, I bade my friends goodbye and drove my truck south on I-285, down into the heart of New Mexico's wide open spaces. First stop: Corona, New Mexico.

Driving through the village of Corona, pretty much a "blink twice and you'll miss it" kind of place, I found the Chamber Of Commerce on the main road. I went in and introduced myself to the nice lady behind the desk, and told her I was writing a book about UFOs, and hoped it wouldn't seem like a strange question if I asked her if she knew anything about where the purported 1947 flying saucer crash had been, as I'd heard it wasn't far from Corona. She stood up, went to a wall map, and said, "No, that's not a strange question at all. The crash was right about here," alluding to an area of private ranch land southeast of Corona. The funny thing is, if you Google Corona, NM, you'll find a statement about it right there next to the state map, including that the town, "is the closest habitation to a purported UFO crash in 1947 about 30 miles to the southeast." It fascinated me that a local official took it all so matter of factly. I also found it interesting that the Historic Marker sign on the road nearby stated that Corona was established in 1947. That certainly makes sense, given that a sudden influx of Army Air Force and intelligence men descended upon the locale in the early days of July that year.

Driving southeast out of Corona, down towards Roswell, I was treated to achingly beautiful stretches of wide open ranch land, topped off by cerulean and turquoise skies, sometimes open and flat in every direction to the horizon, save for mountain peaks to the north and northwest. The desolation was reminiscent of *Christina's World*, as though Wyeth might have drawn the inspiration for his painting from the endless dark straw and flax colored fields surrounding me. Motoring on a single lane road for awhile, I kept my speed down; I was in no hurry, and the vistas were beautiful. What I didn't count on was the powerful feeling that came over me, and grew steadily, the farther out from Corona I got. It was a feeling I can only describe as the presence of Other. I knew that, given my extremely unusual background, and my heightened psychic sensibilities, I might be able to pick up on something "different" about the locale; what I hadn't anticipated was that by the time I got to the approximate location of the crash--looking virtually identical to how the land looked in the Showtime film *Roswell*--the residual, indelible psychic energy of off-worlders, and the remnants of their demolished craft, were overpowering for me. It was a haunting feeling. And it was as real as anything I've ever seen or held in my hands. More than 60 years later, and I could feel their presence, as though their fateful final trip had been yesterday. I pulled over near a ranch entry gate. I got out and stretched my legs, then grabbed my camera and shot a lot of digital photos in every direction.

It was only mid-afternoon, so I had time to spare before motoring into Roswell by dark. I walked around awhile, and gazed out at the crash fields, and reflected on what it must have looked like to the locals more than 65 years before. What a different view of the world they must have had back then; there were 40 million radios in the U.S. in 1947, but only 40,000 television sets, with probably about 2/3 of those in the New York area (info via Wikipedia--wv). I wondered what a man in 1947 would have thought of my iPhone in the back pocket of my Levis; what would he have thought, confronted with a device barely credit card sized that can stream sounds and full color images and video from the ether, and place me in voice and visual contact with anybody on the planet, wirelessly?

Staring wistfully out at the crash fields, picturing the scores of men it must have taken to gather up the pieces, and the larger remains, and the bodies, and secret them away, it occurred to me: in my own way, I am shattered, I am fractured, I am scattered like the pieces of alien craft wreckage that were strewn across these wide open and empty fields between Corona and Roswell. My whole life has been an attempt to put myself back together again, to reassemble the pieces from an experience that shattered me at age three. I think I've done a pretty good job of holding things together, but the thing about being taken is, once they have you, you're theirs and there's nothing you can do about it. Even if you never meet again, the experience leaves you marked for life. You are forever different than your peers. And with what I know about what was done to my blood, even more so. Like Kimberly said way back in 1991, "in some ways, you are your parent's child, but in some ways, you're not." The scenario always reminds me of what I'd heard about baby birds, that if one falls out of its nest and a human picks it up and places it back, it never fully integrates with its own kind again. It is forever marked from its experience of Other. It is forever an outsider.

This sense of fragmentation also brought to mind the discourse between Dr. David Bohm and Renée Weber in *The Holographic Paradigm And Other Paradoxes*. Bohm had commented about how mankind is fragmented and splintered into endless pieces, divided by nations and religions, the fragmentation and division carried down all the way to families, and even within the individual. And how this fragmentation gives way to chaos, and the chaos makes it all the more difficult to grasp that we are all one, as determined and defined by the very reality of the unity of all matter and energy, that "even matter is one in the vacuum," harkening back to the concept of the zero point energy. And how all this fragmentation and chaos and conflict contributes to a deeply corrupted energy in the non-manifest reality of our existence, what he termed, "the sorrow of mankind."

As I gaze out and picture all the pieces, I also reflect on what an indescribably powerful and critical moment that early day in July 1947 was in our history. We were at a crossroads, and our government and military men took the low road: to cover up, to lie, to deny, and ultimately over decades it monstrously metastasized into the Triple-D, and beyond. It brings to mind what Budd Hopkins had said, that the government is kind of stuck, because they can't very well get on the TV and tell the world that we aren't in control of our skies, and we don't know for sure whether or not some of the visitors are malevolent. And perhaps Credo Mutwa, John Mack's insightful Zulu shaman from *Passport To The Cosmos*, had a good point about what might be preventing disclosure, that as it stands, people already have a jaundiced attitude towards the government, and with disclosure, people would revolt against the power structure for all its lies and its "rotten industrial systems." Why would the Boys ever disclose, when it's so much easier to cling greedily to the astounding power they've harnessed, and let the masses continue on the path of fragmented madness?

Walking the crash fields also made me reflect on how the precursor to Project Blue Book, known as Project Sign, had in 1948 prepared an "Estimate Of the Situation;" although no copy has ever been found, Capt. Edward Ruppelt, for a few years the head honcho at Blue Book, claimed to have seen one of the originals. Regarding the Top Secret Estimate, Ruppelt alleged that it was presented bluntly, "The situation was the UFOs; the estimate was that they were interplanetary!" Apparently the Estimate made its way up the chain of command, all the way to Air Force Chief Of Staff, General Hoyt S. Vandenberg, who outright rejected it due to its shocking nature. The report had concluded that aliens were conducting a full scale observation of the Earth, but that an attack didn't seem imminent. Allegedly Ruppelt had told Donald Keyhoe privately, "The general said it would cause a stampede. How could we convince the public the aliens weren't hostile when we didn't know it ourselves?"[70] More than 60 years later, it seems clear to me that even the Shadow Government, with all its resources, still can't answer that one with certainty. Even the Boys who have used them as seemingly benevolent technology tutors might not feel absolutely sure about the real intentions of the off-worlders. Are the "nice ones" in actuality our Overlords, here with a hidden agenda? Are we, in fact, their genetically engineered experiment gone awry? And are we their pawns on a vastly complex, interstellar game board?

As for the concept of aliens surveying the planet, how can one think otherwise, knowing all the facts? The Lubbock Lights, an ongoing group of as many as 18 or 20 craft flying in formation, making up to a dozen appearances between August and November 1951. Exeter, New Hampshire: that entire area of the country utterly blanketed with sightings in August and September of 1965, some of them for extended periods, some of the glowing objects up to 90 feet in diameter, many of them witnessed by police, military men, scientists and professionals, the culmination of sightings all over the country that year. And some pundits believe the Others may have caused the great Northeast Blackout of November, 1965, a real life *The Day The Earth Stood Still* without a Washington landing. Michigan, 1966. Pascagoula, 1973, the Hickson/Parker abduction with robot androids conducting the examinations. Nuclear missile silos, Malmstrom Air Force Base, 1975, some objects, glowing orange and silent, as large as football fields, deactivating multiple ICBM warheads in the wink of an eye. Fast forward: The Phoenix Lights, March 1997. And on and on and on.

Meanwhile, the Boys are still clinging to their technological booty, and I don't see them sharing it with us anytime soon. It has all come down to one of the most destructive of all human frailties, one of the Seven Deadliest: Greed. Pure and simple.

So here we are: decades into the Twenty-first Century, Sol 3. It's possible that our days are numbered, if you combine the NSA's advancing, egregiously intrusive surveillance game on all 330 million+ Americans, with the itching-to-be-implemented, thinly disguised transnational corporate power grab, **TPP**--*Trans-Pacific-Partnership* (a/k/a the Corporatocracy steadily closing its strangulation grip around our throats), with the Boys wielding their fancy new **TMS** (Transcranial Magnetic Stimulation) ray gun that's primed and ready to wipe us all clean and rewrite us into their happy idiot puppet monkeys. The grim truth is, this type of psychotronic research has been going on for decades; go to this web site and try not to get creeped out to the max:

http://nanobrainimplant.com/

Try not to get even more freaked out by the mind-control links on that page. Alas, the human race as a group of free thinking, sentient beings. Those were the days. . . . We really might not have too much time left for laughter, folks--at least not the genuine kind that comes from deep within spontaneously, vs. programmed from a distance via targeted electromagnetic energy pulses aimed at our craniums. Some evidence suggests that this has already been an active practice via NSA satellites bathing us with focused beams; living in the shocking wake of recent intelligence revelations courtesy of Snowden, Manning and Assange, is that such a stretch of the imagination? Yesterday's paranoid rant is today's reality. So perhaps it's time for a little alien joviality?

Regarding *The Day The Earth Stood Still*, still a classic more than 60 years after its debut in 1951 (who can forget *klaatu barada nikto*?, or the giant chrome alien robot with the high-energy disintegration ray eye?): could it be that aliens watched that film on the really, really big High Def screen on their Mother Ships, and got the idea for a Washington, D.C. saucer invasion the following summer? One year later, that's when they made their move (July 1952). To paraphrase comedienne Judy Tenuta, It coulda happened!

Almost equally funny, but also bordering on tragic, is the human stupidity that still comes into play today, the Triple-D that still poisons the masses via media outlets, such as newspaper columnists, those who still cling to their ignorance and deride UFO experiencers by writing ill informed pieces on subjects they know next to nothing about. By way of a pseudo-intellectual rant, Caitlin Dewey's May 14th, 2013 column for *The Washington Post*, timed to coincide with the end of the Citizen Hearing On Disclosure, is a prime example of cluelessness in full bloom (*The fear that drives our alien belief*). Read much, do you, Caitlin? Seen any of Linda Howe's thoroughly researched documentary videos or books? Exposed yourself to Stanton Friedman's eye-opening *TOP SECRET/MAJIC*? Thumbed through Richard Dolan's epic *National Security State* volumes? How about Leslie Kean's exhaustively researched *UFOs: Generals, Pilots and Government Officials Go On The Record*--ever hear about that one? Those are the only questions I pose to any of these rubes anymore; if they can honestly tell me they've seen and pored over Howe's and Dolan's works, studied Friedman's leaked government documents, or read Kean's masterpiece of solid research and journalistic integrity from cover to cover, and they're still stuck in stupid mode about UFOs and the alien presence, then truly we have nothing to discuss. Go back to your wooden blocks, your crayons and your coloring books. The Boys and the National Security State must surely view Dewey and her ilk as their 21st century Golden Children, able bodied enablers of our country's and our planet's gargantuan metastasized lie.

Meanwhile the Boys, a/k/a Richard Dolan's *Breakaway Civilization,* patrol the Earth onboard their magic black triangles, rule the solar system with the mastery of their super-technology, and perhaps even travel amongst the stars in their miraculous antigravity vessels manufactured right here on Sol-3. How does one get on the wait-list to become an **Off-World Flight Officer**, anyhow?

POWER, WISDOM & LOVE

If the Universe collapses, it might trigger another Big Bang. Maybe that's already happened, and we're just one in a long line of universes. Personally, I believe in continual genesis, that is, there's a never ending process whereby universes collide, split apart and give birth to new universes--perhaps with different laws of physics within each universe.

--Michio Kaku[71]

Before I departed from Miami to head to Santa Fe, I took a day off in late February of 1992 to meet my artist pal Gustavo for a day at the beach. He had a nice upper floor apartment on South Beach, home to his live-in studio, and we walked from there to Ocean Drive and 12th street, a beautiful stretch of south Miami Beach. It was an idyllic day, breezy and warm with clear blue skies, the perfect weather to relax and swim and get some rays.

To me, the vibrant colors Gustavo applies to his widely collected canvases are a reflection of his eyes-wide-open view of the world, and his perceptions of the multidimensional realms of reality, what some call the paranormal. We were very much on the same page in our beliefs about UFOs and aliens visiting, so it seemed natural that we should get along so well. I had always known Gustavo to be very metaphysically tuned-in, and he was quite accepting of and receptive to my emergent abduction material when I shared it with him. In fact it didn't faze him in the slightest. Perhaps knowing that I was departing South Florida soon was the impetus for him to impart to me some fascinating information. His spontaneity was part of his charm, and he sprung some pretty wonderful information on me that day, out of the blue.

After taking a swim and unwinding in the sun for awhile, we got into a deep discussion about aliens and what they might be doing here; Gustavo shared with me some channeled material he'd come across in his metaphysical quests; I think he said he was actually there when the medium spoke and the material came through. In his own words, he told me that the channel had stated the following:

When the Universe exploded in the Big Bang, three factions of sentient life emerged from the primordial fireball: Power, Wisdom and Love. Of the three, Love was the most rare, and ultimately, the most sought after across the Universe (reminiscent of aforementioned concepts, with some UFO insiders professing that our souls [viewed as fonts of Love] are highly valued, perhaps even a rare commodity in the cosmos, which sounds congruent here--wv).

During this epoch, or Time Of Man if you prefer, Earth and its solar system are on the threshold of moving through a part of the Milky Way galaxy where the magnetic and gravitational fields are different--in a way that will be potentially tumultuous for life on our planet. Earth is an experiment in Love, and therefore, the Power and Wisdom factions are here in large numbers at this time, because Earth's experiment is rapidly approaching a crucial point, when these unusual force fields will make everything go haywire, perhaps even make life for us humans an extreme challenge. The Power factions are here because they want a piece of the action when the Love experiment unfolds. The Wisdom faction is here to help with the transition.

Gustavo's recounting of this material resonated with me then, and more than two decades later, it still does. (As far as the astronomical mechanics and cosmological view of it, it's believed that the Milky Way itself is in motion, and rotates on its axis once every 225 to 250 million years, and our solar system is simultaneously making its way in a revolution about the galactic core, at a respectable clip of about 420,000 miles an hour, and our galaxy and our nearest big neighbor, the Andromeda Galaxy, are rushing towards each other at about 87 miles per second, and we and our galactic neighbors of the Local Group of galaxies are collectively moving towards the constellation of Hydra at about 400 miles per second; and of course don't forget that Sol-3, our 8,000 mile wide ball of confusion, is rotating at 1,000 miles per hour, day in, day out. So next time you feel like you're stuck in the muck, and down in the dumps because you don't seem to be getting anywhere, just remember: you're actually hauling ass all the time!)

Gustavo's channeled material sounds to me like part *Childhood's End* with a dash of the alien-hybrid concept on the side. Pondering it always reminds me of the view of Earth as the pale blue dot, most recently seen and enjoyed at the end of the superb Julian Assange documentary film *We Steal Secrets*. The way the camera slowly and steadily tracked away from our fragile home, gradually revealing it to be the beautiful pale blue speck imbedded in the river of stars in the Orion Arm of our galaxy: it seemed clear to me that even from a vast distance, the pale blue orb was a tranquil beacon, and in fact that is how I imagine the Earth is viewed by aliens. We're a homing beacon that draws in the off-worlders from near and far, and perhaps from very, very, very far away. We live on a planet of almost unparalleled beauty, heartbreakingly, achingly beautiful; so why don't we see that truth with crystal clarity, and treat her as such, when so many visiting off-worlders can see that, some of them seeing with eyes that don't look at all like ours?

Our Universe is so vast, and so full of wonders, and clearly so full of life--magnified immeasurably by the potential reality of the multiverse. A recent viewing of *How The Universe Works* (Discovery Channel/2010) beamed into my home living room via the miracle of Netflix streaming (certified junkie here) treated me to some new far out factual bits: it took 200 million years after the Big Bang for the first stars in the cosmos to ignite; at the 1 billion year mark, the first galaxy formed; then over the next 8 billion years, countless more took shape; then, about 5 billion years ago, in a remote corner of one of those galaxies, gravity began to pull together dust and gas, and gradually as the cloud coalesced, it gave birth to our star, Sol. Nine billion years after the Big Bang, our solar system sprang to life, and with it, planet Earth. And speaking of scale: the Andromeda Galaxy, our nearest galactic neighbor, is about twice the size of the Milky Way--about 200,000 light years across; and the largest galaxy we've discovered, *IC1011*, is *6 million light years across*--sixty times larger than the Milky Way. Perhaps the most amazing quote of all: we've learned that space is quite big, at least 150 billion light years across. It might be infinite. It might literally go on forever. Despite all this, the mysterious force of Dark Energy may continue to accelerate the expansion of the cosmos, and the force of the receding galaxies, until the Milky Way becomes a lonely outpost in a vast and empty blackness. All the while, the stars near the Milky Way's center dance and jitterbug at a million miles an hour around the enormous black hole at the core--like most galaxies we know of.

When it comes to the cosmology of our reality, a big part of me still believes what he believed as a young fellow his father sometimes called *the budding scientist*: perhaps every star in all the Universe, and across the Multiverse--trillions, or quadrillions or quintillions of them, or all the way to a vigintillion (one followed by 63 zeroes; 120 zeroes in Great Britain), or beyond--are in fact the neurons of God's brain. And Sol-3 is merely a meager synaptic junction. So we exist inside a phenomenal cascade of electricity. The Universe is God's brain. And we're always in His thoughts.

So, in the midst of all of this almost incomprehensible enormity, it made sense that in some of the Serpo material I read through, I found a passage saying that our astronomers puzzled over the mystery of our alien visitors: *how did they find us?* Yet to me it's a given: Sol-3 is the glittering jewel of Orion's arm, beautifully radiant even from a great distance. And though we are out in the boondocks, 25,000 light years (some say 28,000) from the galactic core, Earth has the drawing power of an up and coming Hollywood ingénue. We are an experiment in Love; yes, that definitely works and rings true for me. And yes, I'll admit we're not doing such a bang-up job of it, but we do have souls, and we do have the power of love, and we do have spirits that can transcend, and we do yearn to do so. God, are we ever good at yearning. We've got the yearning thing down cold.

I go a lot on gut instincts, I guess you could say this is part of my psychic Third Eye. And my psi skills tell me that Mother Earth's power of Love is also the force that draws Dark forces to us--just as surely as our pale blue beacon shimmers and sings her exquisite notes in the silence of space. Even if the alien species known as the Trantaloids are merely a dreamed up myth of disinformation woven into the *Serpo Releases*, I know with certainty in my gut that there are dark and hostile ones that have come here--perhaps because, being the polar opposites of Love, they wish to extinguish and destroy our light. What better prize for an alien species that is made of pure malevolence than to descend upon us and vent unlimited hostility, seeking to snuff out and destroy this experiment in Love? What could taste more sweet to them than to savage that which they despise? Again, I'm glad I voted for Mr. Ronnie SDI Reagan--twice.

When I reflect upon life on Sol-3 and what it's really all about today in the early decades of the 21st Century, I resonate with this concept of the Power, Wisdom and Love factions being here and interacting--and in the process, seemingly drawing a kaleidoscopic panoply of life forms to our pale blue dot, as though we're the last drive-in movie picture show for the next two thousand light years of dust filled space and star filled void. To what end, who can say for sure?

Could it be that some of them are simply jostling for good seats, vantage points from which to watch dispassionately as our awaited Apocalypse unfolds?

EPILOGUE

THE SUMMING UP

There is nothing new to be discovered in physics now.
All that remains is more and more precise measurement.
--Lord Kelvin, 1900

Special Theory Of Relativity ($E=MC^2$) is published.
--Albert Einstein, 1905

When I interviewed Travis Walton in March of 2013, one of the most on-target statements he made to me was, "Healthy skepticism is a good thing--but blind skepticism is just as dumb as blind credulity." I couldn't have said it any better.

In the matter of my encounters, I've stated my case plainly, I've told the truth about everything, and I leave it for those who read my story to make up their own minds. If my abduction material triggers an instant gut level response of skepticism, I support that; but, in keeping with the thrust of Mr. Walton's comment, I'd like to return for a moment to focus on a specific aspect of my 1991 hypnosis sessions with my Miami psychotherapist, Ellen Lyon, MSW.

Because I am a certified hypnotherapist with a B.A. in Psychology, and I've conducted a few hundred hypnosis and hypnotherapy sessions with clients over the years, and I've been on the receiving end of direct suggestion hypnosis and hypnotherapy sessions numerous times myself, I'm quite familiar with hypnotic trance states and the different levels of trance depth, so I'm able to speak with some authority about them.

In the extant psychological literature regarding UFOs and alien abductions, some writers, sometimes in the mental health field, have professed that it's an easy thing to do to create false memories about these experiences via direct suggestion hypnosis. Since this is not a textbook investigating a possible dark side of hypnosis and hypnotherapy, I'll leave that for others to research and report. I do acknowledge the existence of the term False Memories, and the work of the well known psychologist Elizabeth Loftus (some of her fascinating publishings and theories were required reading during my studies at FIU). But rather than venture into a debate about these issues, I'm going to focus on one thing, because it happened to me, during my session with Ms. Lyon: hypnotic abreaction.

A hypnotic abreaction is an intense emotional reaction or response, often brought about during time regression hypnosis, in which material from a previously suppressed traumatic memory or event surfaces. Some have compared the process to the lancing of a wound, to release poisons, or in the case of the psyche, to release repressed memories the subconscious deemed too intense to deal with at the time. In my case, it was in our second session near the end, when I was suddenly aware of a previous kinesthetic effect, the heavy pressure on my arms, and pressure on my body, coupled with a perception of an electrical current tingling in my arms, with all of these sensations occurring simultaneously in real time--and I hadn't felt this specific combination of sensations since my recurring nightmares in Tyler, Texas more than 30 years previously. This singular event was, and remains today, a powerful marker for me.

I can appreciate that to an outsider, this might not seem that significant, or even a powerful enough occurrence to have merited my stating with agitation, and then shouting out, "I'm just realizing this is all tied together . . . THIS IS ALL TIED TOGETHER!!" This was a spontaneous outburst, and although the words emerged via my conscious mind, the realization which triggered the verbal outburst came from a very deep level of my subconscious, where the deeply buried traumatic material had been tightly locked away long ago, and was surfacing. When the door finally opened, all of it burst forth, and the kinesthetic effect that came bundled with it instantly engaged, just as surely as plugging in a digital clock instantly makes numerals glow on its face. Electrical currents are very powerful things, certainly an underlying theme in my story. The Universe is charged with electricity. So are we. We are electricity.

In my hypnosis training at the Omni Hypnosis Training Center in Ft. Lauderdale, FL late in 1991, part of our schooling included information regarding a man named Walter Sichort; if you Google him you'll find he developed a technique of hypnosis trance state he called Ultra Depth. As part of our curriculum at Omni, we were given tapes as well as written hypnosis scripts detailing precisely how to induce and achieve Ultra Depth--which is just like it sounds, it's considered a very, very deep hypnotic trance state. I latched onto this quickly, and using my microcassette recorder at home, created an Ultra Depth self hypnosis tape, utilizing a self-designed physical trigger I could use solo to initiate and facilitate the plunge to the depths, if you will. (An example of a physical trigger would be a hypnotherapist in session lifting a client's hand at the wrist and telling him that when his hand is released and drops to the armrest, he can allow that hand dropping to be a trigger for the subconscious mind to relax even more, and go deeper . . . and deeper . . . into hypnosis. One of the tools of our trade involves these Deepening Techniques. Another tool is Reinforcements.) One of the big advantages of achieving this depth is that you eventually go so deep that you access what I call the river of the subconscious, and while you are floating or swimming in it, you have to a great degree cut loose from the conscious mind, and the subconscious mind is in charge, in a very gentle, non-invasive and relaxing way. This is obviously the ideal trance state to be in for accessing repressed memories, since they reside in the deepest channels of the mind's subconscious. And while Ellen Lyon was using her deepening techniques on me in our sessions, I was activating my Ultra Depth self trigger, as well. The point here is: I was in a profoundly deep hypnotic state when I was in Ms. Lyon's therapy chair. The deepest part of my subconscious mind was fully engaged. I was floating and swimming in a subconscious river. Whatever material appeared down here wasn't dreamed up by the conscious mind; it existed on its own, and finally revealed itself. The phenomena were speaking for themselves.

This moment in my second hypnosis session is what I view as the very heart of my alien abduction certification, if you will; it was what hit me like a sledgehammer, and resonated in my soul, to the very core of my being, and even down to the marrow of my bones, even as the pulsating electrical current in my arms was tingling me to my bones, and telling me without question: *this really happened.* Think of it this way: a simultaneity of three scenarios, all three connected together, or enmeshed. The three connected scenarios:

Scenario 1: I am comfortably reclined in a comfortable recliner chair, with my feet up on an ottoman, in a silent, dark room--the treatment room my therapist uses for hypnotherapy--with my eyes closed. In my deeply relaxed state, I am aware that I've just moved away from a vividly realistic scene of myself as a three year old boy, myself as that young boy who'd just walked into the entrance of and climbed onboard a strange vehicle on the ground, and was feeling fearful and possibly feeling the need to cry; and as I'm standing there in that dark corridor, I'm aware that my therapist is using hypnotic patter to gently nudge me back to the present day and present time, to pull me away for now from the potentially shocking material (Ms. Lyon possessed a mastery of therapeutic techniques, in case that hasn't become evident yet). At this moment, as though a mechanical switch was thrown, I began feeling the physical sensations as described, heavy pressure on my arms, and heavy pressure all over my body, coupled with a pulsating electrical current tingling in both of my arms, bringing to the fore an instant shock of sensory recognition, recognizing that I had experienced these combined sensations before, many, many years ago. Electrical current.

Scenario 2: I am a young boy lying flat on my back in my bed in my bedroom in the dark in Tyler, Texas; it's late at night or the wee hours of the morning. I am once again feeling a sense of dread, because every time I have this nightmare that I'm in, I recognize that it's happened many times before, over and over, with the same scene played out each time, and as it begins I silently say to myself, *not again:* I find myself laid out on my back and paralyzed on a table, with very bright lights around me, and shrouded faces above me, scary figures in motion that are definitely alive but don't look like anyone or anything I've ever seen; and in this moment in my dream, I'm aware of the heavy pressure on both of my arms, and on my body, and of the peculiar electrical current pulsing through my arms, tingling. Presently, the scene becomes too real to bear, the appearance of the beings becomes more clear and horrifying, and I begin shouting and crying out for my mother, and I force open my eyes and scream out loud, over and over, as loudly as I am able to scream.

Scenario 3: I am a child of three, laid out on my back on a table on an alien spacecraft. This is not a dream. I am unable to move, and there is a bright light overhead, and I am surrounded by figures that are alive but they are not from here; they are hideous looking and have skin with a mottled orange and brown coloration. I feel a heavy pressure on my arms and on my body, and a powerful, steady tingling in my arms, as if from an electrical current. I don't want to believe I'm here, I want it to be a nightmare, but it's not. I scream out as loud as I can for my mother and father, but they can't hear me. No one can, other than my captors. They ignore me and go on about their business.

Now take those 3 scenes, Scenario 1, Scenario 2, and Scenario 3. Connect them together, with an understanding that all three were functioning in real time, with simultaneity as I reclined in that chair in my therapist's treatment room. The Here-And-Now reality of the dark treatment room, connected in Real Time to the identical Tyler bedroom-nightmare, connected in Real Time to the original scene, onboard an alien spacecraft, each one activating in sequence, 1, 2, 3. It was like an invisible tether connected all three, the full grown man in the therapist's office connected to the Tyler bedroom, connected to the original event, onboard a vehicle from somewhere else. The tether? It was the heavy pressure on the arms, and the heavy pressure on my body, and the electrical pulsations tingling in both arms. The kinesthetic effect was the tether--coupled with the muffled sounds of bubbling liquids.

When I met with Kimberly several days after my final hypnosis session, and she pulled out the sealed document containing the page on which she typed what she'd seen, I wasn't prepared for how bizarrely it would match with the pressure and electrical tingling on both arms. I got chills when I read what she'd written. I read it several times to be sure I was understanding it. It had the air of truth. Absolute truth. And with the other peculiarities about my blood, and the things two different doctors had said about it, nearly 18 years and 1,200 miles apart from each other, it occurred to me that I had everything but an iPhone snapshot and an mpeg film of my time onboard. It became that real for me. It still is.

What I've laid out here is an analysis of how the mind functions with repressed memories--or how mine specifically did--and the nature of memory itself, and how repressed memories can be accessed via deep state hypnotic trance work with a gifted hypnotherapist. One of the key words here is *memory*.

I've always possessed an uncanny memory, just like my father. I have on several occasions amazed my mother with my very early recollections. For example, years after the fact, I remembered vivid scenes from the first house we ever lived in, the specific floor plan of the ranch house; scenes of the pine wooden bucket with the red apple painted on it that Mom kept her clothes pins in for the laundry, and helping her hang the wet clothes out to dry on the clothes line, sometimes sandwiching myself between rows of starched white bed sheets flapping in the sun and wind, pretending they were the sails on my sailboat. I was barely a year and a half old then. Even at age one, I was very aware of my awareness, that is, I knew that I possessed a consciousness. That awareness of being aware fascinated me, the curious little boy in me. Life was full of wonder. Life was meant to be explored. Life was meant to be celebrated, and savored.

One of the puzzle pieces that fell neatly into place after Kimberly's psychic visions led me to Ellen's hypnosis sessions is one of my earliest Tyler memories, a recurring one that had always been shrouded in mystery, and whenever recalled, made me wonder why it was there at all, why it kept coming back. For years, as I grew from boy to man, a vivid childhood scene would surface from time to time, always unexpectedly; it appeared sporadically, at different times and in unpredictable situations, and would simply bubble up from the depths spontaneously, without warning. It always puzzled me, because there was an element of intrigue about it, and I never quite understood what it was that my subconscious was trying to tell me by replaying it so many times. It came and went of its own volition, and I never understood why. Now I do.

It's what I'd been calling my Mystery Memory, and it goes like this: I am a little boy, just past age three. I am suddenly aware that I am crouched down low in a field on my hands and knees, in a clearing of grass and wildflowers, a stand of towering pine trees in the distance surrounding me and my huge circle of open ground like an arena; I am clutching the Earth, huddled protectively, like a wild animal hunkered down in self-protection mode. The fingers of my hands are dug down into the dirt, as though I derive some reassurance by making sure it's really the ground I'm clawing into. I am far away from home, and sunset approaches. I know that I am far away because I am beyond and out of sight of the cluster of unfinished houses, the wooden house frames that always defined a distance I should never venture past, a distance that was too far from home for young little boys striking out on their own.

It takes me a moment to gather my senses and to appreciate the strangeness of my situation. I seem to be emerging from a paralysis, something beyond sleep, as though my senses are coming back after being shut down. I have no clue as to how I got there--I may as well have fallen from the sky with a thud. As soon as I fully recognize and accept where I am--regardless of how or why I'm there--I jump to my feet and bolt like a scared rabbit, like I have to run with all my might so that I can make it home before . . . I don't know what. But running like the wind seems like a good plan, and I dig my Keds into the dirt and make each and every thrust of my legs and feet count, just in case whatever it is I might be running from changes its mind. Just to be sure I get away. To be sure that I make good my escape.

It takes me a few minutes of nonstop, good hard running. I pass some house frames to my left; the low golden sun streams through the spaces between their yellow East Texas pine skeletons, and a few late working carpenters and laborers pause to watch me with amusement as I run a blazing 50 yard dash past their construction site. Presently, in the distance ahead I can see my friends Charlie and Cathy Horton. They're killing time at the edge of the field at the tail end of Barbara Street with the sun behind their backs. Almost there. Finally I am. I run up to them, winded, scared, exhilarated.

"Goll-lee, Wade," Charlie says. "We been trying to find you for the lon-n-n-gest time. Where you been?" His Texas drawl stretches out his lazy syllables like wet salt water taffy.

"I don't know, I . . . guess I was pretty far away--out past the new houses," I say, pointing back to where I'd just run from.

"Must a been," he chuckles at me absent mindedly, kicking at the dirt as his hands dig in the pockets of his dungarees.

"He sure knows how to play hide 'n' seek, don't he?" Cathy giggles, wiping her dirty hands on her overalls.

"Well, I gotta go. Gotta get home!" I yell as I tear down the trail, my baggy shorts flapping against my legs. When I finally see our little red brick and white clapboard house, I feel my heart coming up into my throat. I'm really back, but is anyone at home? I rush around the side yard and head behind the house, streaking past the hedge of gardenia bushes and their pungent, welcoming scent. I make a beeline for the patio, my little blue Keds tearing through the rich green carpet of grass. Mom and Dad are just setting up their canvas lawn chairs, each with a beer can in their hands. I run up and hug each one of them around their legs, and they laugh at my high spirited antics as I dance and jog around the yard. For them, I'm coming in at the right time, just before dark. For me, in a tightly locked away part of me deep down inside, there's a running boy who knows he's barely escaped from a very strange place a long, long way from home. He's learned the hardest way possible, even though it wouldn't fully come to light for many years to come, that there really is no place like home.

Author's Note

Force Fields: Alien Visitations To A Planet Living In The Dark is a non-fiction, true life story. Due to the controversial nature of much of its material, I've changed the names of my family members and my relatives out of respect for their privacy. (The letter written to me by my brother, "Aaron," is quoted verbatim. I still have the original and the envelope it was mailed in.)

In the interest of simplifying the narrative, Angela DesJardin is a composite person; both of her real life counterparts, whom I knew in Miami from 1986 through my departure for Santa Fe in 1992, contributed to her persona. All of the dialog spoken by Angela is directly quoted from the two of them. Both of Ms. DesJardin's real life, flesh and blood co-creators are alive and well and flourishing at the time of this first printing in 2014. It has been my privilege to have known them and to have spent quality time with each of them. In all my life, they remain among my most treasured and unforgettable larger-than-life characters. In short, they are deeply loved.

Edmond M. Bazerghi, Psy.D., and Ellen Lyon, MSW, both appear in this book with their real names and real life identities intact. I owe them both an enormous amount of gratitude for this. In the realm of clinical therapy, each of them represents the highest level of therapeutic skills and ethics available anywhere.

Dr. John E. Salvaggio was the real name of my allergy specialist in New Orleans, Louisiana in 1969. All scenes with Dr. Salvaggio, and all of his quotes about my physical condition, my blood and my immune system, are accurate and true to life.

Dr. Jared R. Dixon, due to the extremely conservative mental health programs he is currently affiliated with in Miami, requested his true identity be withheld. Our session was real and is accurately reported, and the hypnosis transcript was transcribed verbatim from the tape recorded session.

Kimberly Lee, my gifted psychic friend, appears as herself, and all scenes and dialog with her are real and accurate. The remaining real-life Miami psychic readers portrayed here have had their names changed; but otherwise, all of the scenes with them at the psychic (900) phone line, including all dialog and situations of all parties portrayed, are real, true to life, and accurately depicted.

My heart-of-gold Coconut Grove landlord and owner of the bomb shelter, Kenneth Brandt, is also portrayed with his name changed, out of respect for his privacy. Everyone should be so lucky as to have at least one landlord like him. He was a wonderful man, and a godsend to me. His acts of kindness will never be forgotten.

Admiral Walt Shawnessy (Miami) and Billy Albrecht (Santa Fe), are both cover names for my Naval military friend and my ex-Air Force officer/former lover, respectively. Both men are portrayed exactly as they were, and all dialog, scenes and disclosures shared here are accurate and true to life.

Mark and Bob Dandridge, due to their positions in the Houston business and social communities, requested their identities be withheld. Both men are accurately portrayed. Bob Dandridge's UFO sighting in Monaco, and our discussion about it, have been reported faithfully, and to his satisfaction.

Lars Bloom, the successful Santa Fe artist, and Danny Dawson, his delivery truck driver, are both real, but their names have been changed out of respect for their privacy. Mr. Dawson's close up UFO encounter in New Mexico happened exactly as reported.

My Santa Fe hypnotherapy client, Amanda Black, and the psychologist who networked with me on her case, Dr. Landis, are both disguised identities, for obvious reasons. Ms. Black, her husband, and Dr. Landis have all been faithfully and accurately depicted, as well as our hypnosis sessions and discussions.

ACKNOWLEDGEMENTS

I would like to acknowledge all of the following people who directly or indirectly contributed to the evolution and driving force for the completion of this book: **Dr. Edmond M. Bazerghi**, my extraordinarily gifted, wise and kind hearted Austin psychologist and hypnotherapist who graciously agreed to appear as himself, thank you sir. **Ellen Lyon, MSW**, my Miami psychotherapist and hypnotherapist who adroitly finessed the emergence of my repressed memories--your peerless skills were a huge gift to me, as were your innate sensibilities about how best to facilitate and safely navigate our subconscious journeys; many thanks, Ms. Lyon for also appearing as yourself. To my dear friend and gifted seer **Kimberly Lee**, many thanks for your encouragement, moral support and crystal clear, accurate visions--your extraordinary gift has been a blessing to me in so many ways, as have been your warm heart, and our laughter and fun times together. For those who wish to experience Kimberly's psychic and God-given intuitive abilities, she may be contacted at: **kimberlytheseer@outlook.com.** To my long term good friend and top tier attorney, **Philip E. Glatzer, Esq.**, thanks always for your steel trap legal mind, your big hearted, good guy persona and your countless impromptu legal guidance counselings over the years; your many kind and selfless acts are never forgotten, Mr. G., you're the best. Kudos and many thanks also to **Frank Natoli, Esq.** who was relentlessly generous to a total stranger with sharing facts regarding Copyright Law and Fair Use--greatly appreciated. To my new professional contact **Linda Moulton Howe**--you have been exceptionally warm and generous with your time, expertise, global knowledge base, and the sharing of your out of this world, cutting edge material--it's been my honor and privilege to have networked with you, my high esteem for you before knowing you was well founded, you are a fearless, relentlessly inquisitive journalistic explorer non-pareil. Nuclear physicist, author and lecturer, **Stanton Friedman**, you are bar-none one of our most gifted and scientifically grounded UFO researchers and investigators, with the bulldog tenacity to dig up the facts in the depths of national archives and track down key witnesses far and wide. Your generosity regarding your time, both in person and in phone interviews, as well as Fair Use of quotes, were greatly appreciated. **Richard M. Dolan:** I am in awe of your encyclopedic UFOs and the National Security State books, I love reading them, they belong in every serious UFO library (and I'm looking forward to the final volume of the trilogy)--many thanks for your expertise and scholarly historian's approach to the phenomenon, as well as your generosity with quotes and paraphrasings for this book. **Michael Lindemann**, who kindly granted me much quoted material (and email and phone interview time) from your UFOs and the Alien Presence: Six Viewpoints book--still one of the best UFO interview books in the field--I thank you for all of your assistance and insight. **Leslie Kean**, my sincerest thanks go to you for your exceptionally fast, friendly and efficient

ACKNOWLEDGEMENTS

turn-around time for granting me Fair Use of some quotes and paraphrasings of your material. Your incredible, standard bearer of a book, UFOs: Generals, Pilots and Government Officials Go on the Record is a wonder in the genre of investigative journalism, I loved it. It is due for multiple more readings by me. **Whitley Strieber**, many thanks for being a fascinating subject, an insightful observer of life and the UFO phenomenon, and an amazingly gifted wordsmith; I've loved so many of your books, Hybrids is among my recent favorites; my sincerest gratitude goes to you, Mr. Strieber, for allowing me to share in this book a classic interview with you that still fully resonates today. **Travis Walton**, thanks for being a great interview subject, delivering many insightful and spot-on comments about UFOs and the larger reality of the world we live in. Your lucid and no-nonsense way of seeing things is greatly appreciated, all the best to you and yours. **Ryan Fugate**, you are one of my favorite contemporary artists, and deserve all the notice and acclaim you're getting for your beautiful oil and acrylic canvases (**Ryanfugatefineart.com**). Thanks so much for creating a hypnotic book cover image, and for being willing to work with me on all of my concepts, exceeding by far my expectations. To **Daphne Stratos Parker**, my wise and beautiful pal (and fellow book-a-holic) who has always been one of the brightest rays of sunshine in my life, I cherish my memories of the wonderful times in Miami, and of what a loyal ally and confidante you were when my heavy material surfaced. Your friendship has been my good fortune. **Dr. Steven M. Greer**, all my admiration goes out to you for your relentless pursuit of the truth, your groundbreaking Disclosure Project 2001, and likewise for your incomparable book DISCLOSURE: Military And Government Witnesses Reveal The Greatest Secrets In Modern History, bringing to light via numerous sources the truth about reverse engineered alien vehicles, as well as the fiercely suppressed free-energy technology that could transform the planet if it weren't trapped in the hands of terminally greedy men. **John Greenewald, Jr.**, creator and overseer of one of the most valuable hidden-information web sites out there (theBlackVault.com), kudos to you for bringing it to life and keeping it up and running. Your courtesy and assistance in our communiqués has been greatly appreciated. To the late **Budd Hopkins**, always you have my admiration for the kind hearted approach you extended to a diverse group of people who shared their at times traumatic UFO experiences with you. The insightful wisdom of your outlook was revealed in depth to me in your responses to my questions in our interview. To the prolific author **Jim Marrs**, many thanks to you, sir, for graciously granting me Fair Use of quotes and material from your superb UFO/alien book, Alien Agenda: The Untold Story Of The Extraterrestrials Among Us. It's one of my favorites, and is certainly destined for multiple more readings (two so far). To **Len Kasten**, thank you also, sir, for granting me Fair Use for quotes and paraphrasings from your marvelous book The Secret History Of Extraterrestrials: Advanced Technology and the Coming New Race; there is an abundance of fascinating and thought provoking information in those pages; it's likewise due for more enthusiastic readings. To my favor-

ACKNOWLEDGEMENTS

ite psychology professor, **Wendy K. Silverman, Ph.D.**, thanks always for selecting me to be one of your therapists-in-training at the FIU CAPP program in the Fall of 1990, you were a superb mentor. Your rigorous, diligent devotion to the scientific method has always been an inspiration to me--and your bubbly, friendly and fun personality was icing on the cake. Congratulations on your new post at Yale University--lucky them! To my good pal **Bill Zavatsky**: You will always be one of my favorite UT professors, who unceremoniously told his eager students on Day 1 of his Writing class, January 1979, that they should get up and leave if they didn't "own it" that they're completely out of their minds and certifiably mad for wanting to be writers (none of us left); and for sagely dispensing an invaluable maxim, near quote: "Everything that is ever going to be written, has already been written; the job for you as a writer is to express it in a way that's never been said before. Show it in a new light." Amen, Professor Z--and many thanks for the books of your wonderful poetry, and for kindly allowing me to quote a few lines from one of my favorites. And finally, merci beaucoups a/k/a "murky buckets" for all of the email volleys regarding grammar, usage and syntax of some tricky passages, and numerous insider tips about the publishing world. You're still my top dog English professor! (Was this an encomium?) To **Cristi Peterson**, thanks for helping transform my Word chapters and pour them into book pages. To hire Cristi for book formatting assistance, contact her at: **cristidesignsit@gmail.com**. To the superlative Reference Staff at downtown Austin's **Faulk Central Library**, many thanks for your countless and invaluable episodes of research assistance; the friendly, knowledgeable and efficient help from all of you has been a true blessing to me. Huge thanks beyond measure go out to my amazingly skilled and patient book printing rep **John Hudson** and his extremely savvy production team over at **OneTouchPoint Ginny's** in Austin, TX, and for all of their skill sets which came into play to craft the final, physical reality of the First Printing of **Force Fields**; and kudos and thanks to their top notch supervisor and relentlessly optimistic realist **Terry Sherrell** for overseeing the whole process. And last but not least, a huge acknowledgement goes to **My parents**, together on the Other Side: I send out all of my love and thanks to both of you for making the mistakes you made so that I could learn from them, and for being so kind and loving and generous to me while you were here; thanks also, Mom, for always encouraging my writing, proof reading parts of this book, and handling with unexpected aplomb all of my UFO stuff, including the beyond-bizarre material of my regression tapes. One of your comments after listening to them all still lingers in memory, "I'll never look at things the same way again." You and dad are prayed for and reflected upon daily. I wish you both great joy and peace and love in the Light with all of "The Posse," wherever y'all may be out there. Happy trails to all of you--until our next ride--see you on the Other Side. PS: Are there any serious free weight iron gyms, kickboxing dojos & lap pools over there?

APPENDIX

Contents

1) Research Methods: Dr. Wendy K. Silverman
   FIU (Florida International University): Fall 1990
   B.A. Thesis paper: Abstract: submitted by:
   Stephen W. Vernon

   The Presence Of Specific Personality Factors And Their Effect On A Person's UFO Belief Systems

2) Belief Systems: 15 Question UFO-Questionnaire

3) Test Results: 16PF scores correlated to
   15 Question UFO Questionnaire

4) *Chimayo Cattle Mutilation:* Abstract-Document
   (Santa Fe New Mexican archives)

5) *Santa Fe Space Science Center:* Abstract-Document (Santa Fe New Mexican archives)

6) *Sketch #1:* Hypnotic Regression Image: Left handed sketch

7) *Sketch #2:* Hypnotic Regression Image: Left handed sketch

The Presence Of Specific Personality Factors

And Their Effect On A Person's UFO Belief Systems

Stephen W. Vernon
Florida International University
December, 1990

The Presence Of Specific Personality Factors
And Their Effect On A Person's UFO Belief Systems

UFOs (unidentified flying objects) have been around for a
very long time. The legend of flying discs has existed
throughout recorded history, and ancient reports of strange craft
in the skies, often with mysterious occupants, can be found in
the Bible and in myriad cultural records scattered across the
planet. Dr. Jacques Vallee, an astrophysicist and computer
scientist, has spent over thirty-five years systematically and
scientifically studying these phenomena, and has on file more
than three hundred UFO sightings prior to the twentieth century
(Vallee, 1965).

In modern times, Kenneth Arnold's 1947 sighting of nine
shiny discs flying in formation near Mt. Rainier in Washington
state, which he described as skimming along at high speed "like a
saucer would if you skipped it across water," ushered in the era
of the flying saucer. Since then, many thousands of incidents
have been reported worldwide--clustered in periodic "UFO waves"--
ranging from distant visual sightings to close encounters with
spacecraft and even communication with/abduction by aliens (Blum,
1974; Hopkins, 1987; Kinder, 1987; Strieber, 1987).

In analyzing the UFO phenomenon it becomes clear that
psychology offers an ideal discipline through which to sieve the
incoming information, since in many cases it seems undeniable
that something physical is acting upon the psyches of otherwise

rational people. Possibly the first renowned psychologist to study the UFO enigma was Carl Jung, who wrote an article and a book about it. Most appropriate here is his comment that, "something is seen, but one doesn't know what" (Jung, 1959). This peculiarly elusive "almost but not quite there" quality is characteristic of all UFO reports, regardless of whether coming from a police officer, a co-ed, an engineer or a rural housewife. But this surreal, otherworldly air surrounding sightings shouldn't lead one to write off all UFOers; as Troy Zimmer's study (Zimmer, 1984) concluded, "UFO sighters were just as likely as nonsighters to major in the physical/biological sciences and to be high academic achievers. Hence, it is evident that those UFO critics who have labeled sighters as social marginals, maladjusted, malevolent, or simple minded have been incorrect."

In recent years, clinical psychologists have grown increasingly intimate with UFOers, especially with regards to proper treatment and care of patients who purportedly have had traumatic close encounters with spacecraft and/or aliens who have taken them onboard their craft (Fuller, 1966; Sprinkle, 1989). Over the past twelve years, Aphrodite Clamar, a psychologist in private practice in New York, has used hypnotherapy with nine abductees as a means to access their repressed encounters (Clamar, 1989). Ultimately her patients underwent an extensive battery of psychological tests, including projective tests, intelligence tests and the Minnesota Multiphasic Personality Inventory (MMPI). The resulting psychological report stated in

part, "To summarize, while this is a heterogenous group in terms
of overt personality style, it can be said that most of its
members share being rather unusual and very interesting. They
also share brighter than average intelligence and a certain
richness of inner life that can operate either favorably in terms
of creativity or disadvantageously to the extent that it can be
overwhelming. Shared underlying emotional factors include a
degree of identity disturbance, some deficits in the
interpersonal sphere, and generally mild paranoid phenomena
(hypersensitivity, wariness, etc.)." From this, Clamar
concluded, "Rather than ignoring the UFO phenomenon and those
individuals who report these experiences, it is time for
psychology to acknowledge the intrapsychic needs of these
individuals and to provide enlightened psychotherapeutic
services." (Clamar, 1989)

Another clinical psychologist, June Parnell, conducted
testing on a larger scale (Parnell, 1989); she analyzed data from
Sprinkle's study (1989) of 225 UFO experiencers who completed the
MMPI and Sixteen Personality Factors Test (16PF). A two-way
ANOVA involving levels of sighting and communication experience
showed that subjects claiming communication had a significantly
greater tendency to endorse unusual feelings, thoughts and
attitudes (significance level .003); to be suspicious or
distrustful (significance level .01); to be creative,
imaginative, or possibly have schizoid tendencies (significance
level .034). The results further stated, "the 16PF Test profile

shows persons who, let us say, march to the beat of a different drummer, tending toward being reserved, more intelligent, assertive, experimenting, liberal, freethinking, self-sufficient, resourceful, and preferring their own decisions (Factors A, B, E, Q_1 and Q_2)."

It is here that I find the relationship of specific personality factors and their influences upon UFOers' belief systems to be a salient one. Specifically, does the UFOer possess this unique personality profile as a result of an intense and bizarre real life encounter, or did said individual possess this profile to begin with, thus making him/her more likely to subscribe to the alternative reality of the UFO belief systems?

The possible significance here was brought out in Little's (1984) study: "Price-Williams (5) has pointed out that an important element not often addressed by ufologists is the psychology and epistemology underlying UFO interpretations. He suggests that more attention center upon the beliefs people hold about UFOs."

To take this belief system concept a final step: In the recent book Dimensions (Vallee, 1988), Dr. Vallee proposes that the human race is being subjected to a Skinnerian schedule of reinforcement, and he states, "The best schedule of reinforcement is one that combines periodicity with unpredictability. Learning is then slow but continuous. It leads to the highest level of adaptation. And it is irreversible. It is interesting to observe that the pattern of UFO waves has the same structure as a schedule of reinforcement...I believe that the UFO phenomenon is one of the ways through which an alien form of intelligence of incredible complexity is communicating with us *symbolically*."

I submit that this symbolic communication, whatever its true source, is having a subtle yet profound effect upon the belief systems of thousands--even millions--of people worldwide. On a large scale, it results in the formation of UFO interest groups and cults (Vallee, 1979). On the smaller, grassroots level, this sophisticated shaping of belief systems deserves careful scrutiny, if for no other reason than to explore the concept that Skinnerian conditioning may be at work here--to what end, no one knows.

Regardless of whether or not Dr. Vallee's hypothesis is correct, I concur with Price-Williams' assertion; therefore the purpose of this study then is to explore the relationship between specific personality factors and UFO belief systems of a general population, and possibly to replicate some of Parnell's 16PF correlations.

Please respond to the following statements by choosing the numbered response that most closely matches your true feelings, as follows:

[1]	[2]	[3]	[4]	[5]	[6]	[7]
strongly disagree	disagree	slightly disagree	neither agree nor disagree	slightly agree	agree	strongly agree

1) I am certain that I have seen a real UFO (unidentified flying object).

2) If aliens really were visiting us they would have announced their presence to the world by now.

3) The U.S. government is hiding a lot of information about UFOs from the general public.

4) Crossing the vast distances between stars is too difficult for any alien civilization.

5) Some UFOs are actually spaceships piloted by extra-terrestrials.

6) The Air Force's *Project Bluebook* was right when it concluded that studying UFOs is a waste of time.

7) If an alien landed his ship in my backyard and offered me a trip to his planet, I'd be tempted to go.

8) All UFO "alien abduction" reports (i.e., people claiming to have been forced aboard a flying saucer) are made up by the abductees.

9) Aliens from other star systems are visiting Earth.

10) If I ever saw what I thought was a UFO I probably wouldn't tell many people about it.

11) Out of my own interest I have read at least one book about UFOs in the last two years.

12) Earth is the only planet in the cosmos with intelligent life.

13) The government has a crashed flying saucer stored in some guarded hangar somewhere.

14) Our galaxy is full of civilizations far more advanced than ours.

15) All UFO reports are the result of people's misperceptions of everyday things like airplanes, meteors or satellites.

The UFO questionnaire created by the author in the Fall of 1990, for use in correlating answers on a 7 pt. Likert scale with the results of the 16PF Personality Test. A Test-Retest of content validity for this questionnaire resulted in a Pearson's r = .93.

Regardless of whether or not Dr. Vallee's hypothesis is correct, I concur with Price-Williams' assertion; therefore the purpose of this study then is to explore the relationship between specific personality factors and UFO belief systems of a general population, and possibly to replicate some of Parnell's 16PF correlations. Specifically, it is my hypothesis that subjects who score high on the UFO attitude survey developed for this study (indicating strong beliefs that UFOs/UFO-related phenomena are a reality) will also score significantly high on the Concrete vs. Abstract factor, the Practical vs. Imaginative factor and Conservative vs. Experimenting factor of the 16PF.

A Pearson product-moment correlation revealed near-zero correlations between the UFO attitude survey scores and the 3 16PF targeted scales. Of the 20 subjects whose UFO attitude survey scores fell above the mean, 13 (26%) were male and 7 (14%) were female. This ratio approaches Zimmer's findings (1985) that males tended to be more prone to UFO beliefs and/or have sighting reports by a factor of 2 to 1 over females. The mean score, averaged across all subjects, $x=+7.5$, standard deviation = 14.8. In all, 40% of the subject's UFO belief system scores fell above the mean; 14% of the scores fell at a point > 1 s.d. (>22.32) above the mean. The present data do not support the hypothesis proposed, that there would be a significantly positive correlation between high UFO attitude survey scores and high scores on the scales of the 16PF. The present data do support Zimmer's (1985) finding that males tend to have stronger positive UFO beliefs/belief systems or reported more UFO experiences by a factor of 2 to 1 over females.

The most convincing argument here would be that there is simply no relationship between strengths (greater positively) of a person's UFO belief systems and their scores on the scales of the 16PF. Furthermore, the fact that all 225 of Parnell's (1988) subjects had had UFO encounters prior to her study may have contributed to the significantly high scales (whereas in the present study, only 2 (1%) reported certainty of a UFO encounter).

Due to limitations of the current study, this still leaves unanswered the question posed before: do UFOers possess these unique psychological profiles posited by Parnell as a result of their close encounters, or were these prior and possibly pre-disposing characteristics for reporting said experiences? This would be an interesting avenue to pursue in future research; the remaining unanswered questions herein tend to support Price-Williams assertions that more attention should center on the beliefs people hold about UFOs, perhaps to try to ascertain how many UFOs really are flying in the skies, and how many are in people's heads.

CHIMAYO RANCHERS REPORT CATTLE MUTILATION

The Santa Fe New Mexican - Santa Fe, N.M.
Author: Ray Rivera with photo by Clyde Mueller
Date: Sep 11, 1996
Start Page: B.1
Section: SANTA FE/EL NORTE
Text Word Count: 490

Document Text

(Copyright 1996 Santa Fe New Mexican)

PENASCO An eerie trend that seems to resurface every few years since the 1960s apparently has popped up again.

A mutilated bull its genitals, tongue, intestines and heart cut out was discovered Monday in a high-mountain pasture here known as Llano de la Yegua. It was the third reported cattle mutilation in the last two months in Northern New Mexico. The last reported mutilations were in a cluster in 1993-94.

"You talk about aliens and UFOs and don't think much of it, then you see a mutilated animal and it's ... weird," said Chimayo rancher Carlos Trujillo.

Trujillo and the bull's owner, Wilfred Romero, also of Chimayo, graze their 60 head of cattle together through the summer in the high pastures above Penasco. They viewed the bull after Penasco resident Albert Ortiz discovered it Monday about 4 p.m. A Taos County Sheriff's deputy and a state livestock inspector also investigated the scene.

"We got up there around 7 p.m. and it was strange," Trujillo said. "There were no tracks, no blood, nothing."

Trujillo returned to the pasture Tuesday to dispose of the 2-year- old bull an 1,800-pound, pure-bred Santa Gertrudis Trujillo said was worth about $2,000.

The bull's heart was removed through a hole, approximately 6 inches in diameter, cut in its chest. The anus, testicles, intestines and tongue also were removed with incisions. Half of the bull's left ear was removed. There was no blood on any of the wounds, the carcass, nor on the grass around the animal. And there were no tire tracks or unusual markings around the carcass.

"Whatever did this has got more power than you or I can think of," Trujillo said. "An animal of this size and weight would have to be tranquilized."

Trujillo said the livestock inspector did not ask to perform an autopsy on the animal.

State livestock inspectors could not be reached Tuesday.

The livestock board recorded about 30 mutilation incidents in 1993- 94 in Northern New Mexico and Colorado. An Eagle Nest rancher alone reported 14 mutilations. The occurrences resembled a rash of mutilations in the 1960s and '70s in several states.

A 1980 investigation by a former FBI agent concluded that most of the mutilations in the '60s and '70s were caused by predators and scavengers.

There were no reports of mutilations in 1995 and the first half of 1996 until two mutilated cattle were discovered in Questa and Arroyo Hondo near Taos in July. These latest mutilations, along with the 1993-94 incidents, are still being investigated by the state livestock board. The board has concluded that the 1993-94 incidents were not caused by predators or scavengers.

In a 1994 report to U.S. Sen. Pete Domenici, the livestock board said the investigation into the mutilations revealed "possible involvement of clandestine Satanic groups."

The livestock board also said at the time that the State Police, the state Veterinary Diagnostic Lab and the Albuquerque Police Department were assisting in the investigation.

Abstract (Document Summary)

SPACE SCIENCE CENTER GIVES WAY TO UFOS

The Santa Fe New Mexican - Santa Fe, N.M.
Author: Zack Van Eyck
Date: Aug 27, 1994
Start Page: A.1
Section: MAIN
Text Word Count: 666

Document Text

His Space Science Center is now closed, but Felip Cabeza de Vaca is still open to the idea that humans are not the only intelligent life in the universe.

Cabeza de Vaca, whose nonprofit St. Francis Drive museum served the public for nearly 15 years, said Friday he will focus his future work solely on UFOs.

Eventually, he hopes to open a new facility in Santa Fe dedicated entirely to collecting and spreading information about unidentified aircraft and related phenomena.

"It's the most suppressed field on this planet," the 54-year-old Santa Fe native said of UFO studies. "I intend to continue my research in this field to liberate this knowledge which will serve a cross-section of all inhabitants of the many universes."

He also has tried to make Santa Fe a better place by getting elected into office, but has yet to succeed. He has run for City Council 11 times, including last fall, and has made one run at mayor.

He has never accepted a dime in contributions. Cabeza de Vaca, who grows garlic and onions but makes his living as a plumber, said he will likely run for office again.

His Space Science Center, which closed Aug. 4, housed films, video tapes, books, catalogs and other information from the National Aeronautics and Space Administration.

It also contained a vast collection of materials on UFOs and extraterrestrials, which Cabeza de Vaca now keeps in storage. The center was open to anyone, at just about any time, free of charge.

Suzanne Martinez, a retired special education teacher for Santa Fe Public Schools, said many teachers took their classes to the Space Science Center each year.

"We were interested basically in the science end of it," she said. "It's going to be missed."

When he first started the museum, Cabeza de Vaca said, there was tremendous interest in NASA and the products of its scientific research. In the last six or seven years, however, about 90 percent of the people who visited the center expressed interest primarily in UFOs, he said.

That has been the experience of the founders of the International UFO Museum and Research Center in Roswell, where more than 33,000 have visited since it opened in October 1992.

UFO phenomena "is what is happening the most and what (the public) knows the least about," Cabeza de Vaca said. "When I started the Space Science Center, you couldn't find a picture of the moon here. Now there's a lot of (space science) material all over Santa Fe, so I don't see a need for it."

He sent an entire room full of materials back to NASA in June just about everything except a classified film that he says shows footage of UFOs taken by astronauts on NASA missions.

Cabeza de Vaca closed the center later for reasons he is not comfortable talking about, although he hinted that the cost of rent might have been a factor. He is now busy remodeling his home. Soon, he said, he will begin looking for an affordable location for his new venture.

Cabeza de Vaca is convinced that beings from another planet are not only visiting earth but may be planning to stay. "The government, whoever that is," is aware of the visitations but is engaged in a massive cover-up to conceal what it knows, he said.

"These people are coming from a dying star," he said. "It's what we're going to be faced with later on when that happens in this solar system."

He believes the aliens, who he calls "grays," are responsible for a rash of cattle mutilations across the globe, including

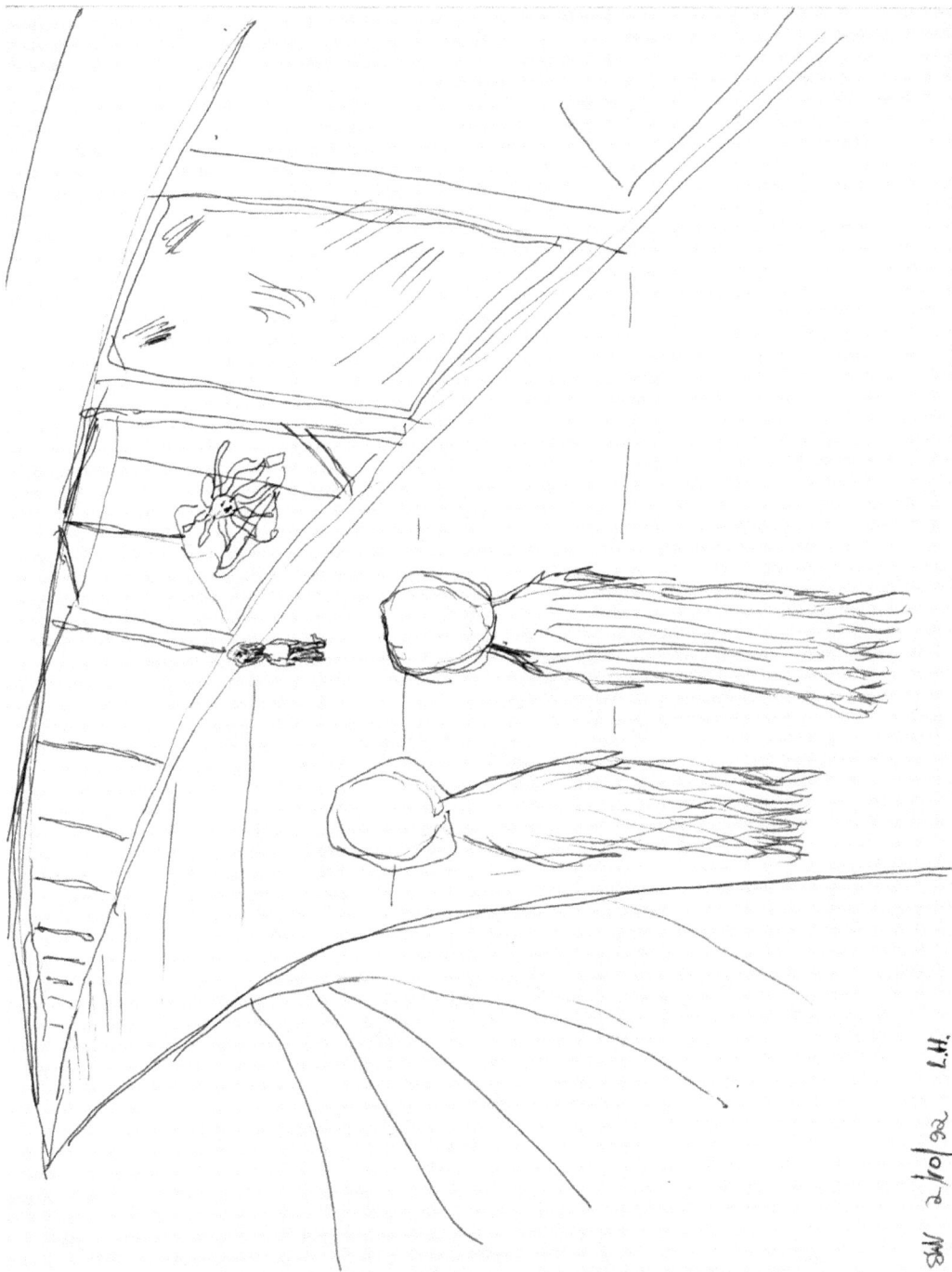

Sny #7 2/01/00

BIBLIOGRAPHY

I. BEGINNINGS

1. Reader's Digest, Outer Space Ghost Story: May 1966; Pp. 72 - 73.
2. Fuller, John G. Incident At Exeter. New York: MJF Books (1966).
3. Booth, Billy. UFO Casebook(online).
   http://ufos.about.com/od/bestufocasefiles/p/arnold.htm
4. Project Sign: Wikipedia:
   http://en.wikipedia.org/wiki/Project_Sign
5. Sturrock, Peter A. The UFO Enigma: A New Review Of The Physical Evidence. New York: Warner Books (1999); Pp. 12-13.
6. Ibid. : p. 13.
7. UFO Evidence: UFO Photographs: 1950's: August 31, 1951: Lubbock Lights:
   http://www.ufoevidence.org/photographs/section/1950s/Photo45.htm
8. Dolan, Richard M. UFOs and the National Security State: Chronology of a Cover-up 1941-1973. Charlottesville: Hampton Roads Publishing Company, Inc. (2002); Pp. 90-91.
9. Keyhoe, Maj. Donald E. Flying Saucers From Outer Space. New York: Henry Holt and Company(1953); Chapter 4.
10. Ibid.; Pp. 62-64.
11. Ibid.; Pp. 64-66.
12. Ibid.; Pp. 66.
13. Ibid.; Pp. 66-67.
14. Ibid.: Pp. 68-70.
15. Ibid.: Pp. 73-89.
16. UFO: Unidentified Flying Objects: The True Story Of Flying Saucers. MGM/Metro-Goldwyn-Mayer Studios, Inc. (1956); IMDB:
    http://www.imdb.com/media/rm3093870592/tt0131627?ref_=tt_ov_i
17. Keyhoe, Maj. Donald E. Flying Saucers From Outer Space. New York: Henry Holt and Company (1953); P. 1.
18. Newsweek, Michigan: Pi in the Sky: April 4, 1966; Pp. 22-23.
19. Time, Michigan: Fatuus Season: April 1, 1966; p. 25B.
20. Newsweek, Michigan: Pi in the Sky: April 4, 1966; p. 22.
21. Blum, Ralph and Judy. Beyond Earth: Man's Contact With UFOs. New York: Bantam Books (1974); p. 157.
22. Ibid.
23. Sturrock, Peter A. The UFO Enigma: A New Review Of The Physical Evidence. New York: Warner Books (1999); p. 16.

BIBLIOGRAPHY

24. Time, Michigan: Fatuus Season: April 1, 1966; p. 25B.
25. Newsweek, Michigan: Pi in the Sky: April 4, 1966; p. 22.
26. Blum, Ralph and Judy. Beyond Earth: Man's Contact With UFOs. New York: Bantam Books (1974); p. 157.
27. Newsweek, Michigan: Pi in the Sky: April 4, 1966; p. 22.
28. U.S. News & World Report: March Of The News: Flying Saucers-Illusions Or Reality?: April 4, 1966
29. U.S. News & World Report: UFOs: They're Back In New Sizes, Shapes, Colors: August 22, 1966.

II. INTERLUDE

1. Blum, Ralph and Judy. Beyond Earth: Man's Contact With UFOs. New York: Bantam Books (1974); Pp. 9-11.
2. James Harder; Wikipedia: http://en.wikipedia.org/wiki/James_Harder
3. The Pascagoula, Mississippi Abduction (Hickson/Parker); from The UFO Casebook; located online at: http://www.ufocasebook.com/Pascagoula.html
4. Blum, Ralph and Judy. Beyond Earth: Man's Contact With UFOs. New York: Bantam Books (1974); Pp. 30-36.
5. Ibid.; Pp. 35-36.
6. Blum, Ralph and Judy. Beyond Earth: Man's Contact With UFOs. New York: Bantam Books (1974); p. 24.
7. Ibid.; Pp. 24-25.
8. Ibid.; Pp. 16-17.
9. Ibid; p. 17.
10. Pascagoula Abduction; Wikipedia; located online at: http://en.wikipedia.org/wiki/Pascagoula_Abduction
11. Blum, Ralph and Judy. Beyond Earth: Man's Contact With UFOs. New York: Bantam Books (1974); p. 196.
12. Ibid.; p. 200.
13. Ibid.; Pp.201-205.
14. Ibid.; p. 201.
15. Ibid.; p. 203.
16. Ibid.; p. 209.
17. Ibid.; p. 205.
18. The Pascagoula, Mississippi Abduction (Hickson/Parker); from The UFO Casebook; http://www.ufocasebook.com/Pascagoula.html
19. Blum, Ralph and Judy. Beyond Earth: Man's Contact With UFOs. New York: Bantam Books (1974). p. 109.
20. Ibid.

BIBLIOGRAPHY

III. CONTACT: BEHIND THE GRAY WALL

1. Vallee, Jacques. DIMENSIONS: A Casebook of Alien Contact. New York: Ballantine Books (1988); p. 103.
2. Ibid.; p. 105.
3. Vallee, Jacques. REVELATIONS: Alien Contact and Human Deception. New York: Ballantine Books (1991); Pp. 276-277.
4. Ibid.; p. 268.
5. Ibid.; p. 278.
6. Ibid.; p. 256-257.
7. Elders, Lee J. & Welch, Thomas K. UFO. . . Contact From The Pleiades, Volume I; Phoenix: Genesis III Publishing (1980).
8. Stevens, Lt. Col. Wendelle (Ret.). UFO Contact From The Pleiades: A Preliminary Investigation Report; Tucson; Stevens (1978); p. 35.
9. Kinder, Gary. Light Years: An Investigation Into The Extraterrestrial Experiences Of Eduard Meier. New York: The Atlantic Monthly Press (1987); p. 186.
10. Ibid.; p. 187.
11. "tachyon": Encyclopedia Britannica. 2006. Encyclopedia Britannica Premium Service. 12 Sept. 2006: http://www.britannica.com/eb/article-9070867
12. Blum, Ralph. The Book Of Runes. New York: Oracle Books/St. Martin's Press (1982); Pp. 111-112.

IV: THE VIEW FROM THE MOUNTAINTOP

1. Blum, Ralph and Judy. Beyond Earth: Man's Contact With UFOs. New York: Bantam Books (1974); p. 155.
2. Lindemann, Michael. UFOs And The Alien Presence: Six Viewpoints. Santa Barbara: The 2020 Group (1991); Pp. 26-28.
3. Fawcett, Lawrence & Greenwood, Barry J. Clear Intent: The Government Coverup of the UFO Experience. Englewood Cliffs, NJ: Prentice Hall, Inc. (1984); Pp. XIV-XV.
4. Lindemann, Michael. UFOs And The Alien Presence: Six Viewpoints. Santa Barbara: The 2020 Group (1991); Pp. 61-63.5. Ibid.; Pp. 64-65.
6. Ibid.; Pp. 67-68.
7. Howe, Linda Moulton. Strange Harvests 1993: An investigation of strange aerial lights and unusual animal deaths. Albuquerque: LMH Productions (1994).
8. Ibid.
9. Ibid.

BIBLIOGRAPHY

10. Ibid.
11. Ibid.
12. Ibid.
13. Ibid.
14. Rivera, Ray. Chimayo Ranchers Report Cattle Mutilation. The Santa Fe New Mexican. Santa Fe/El Norte; September 11, 1996.
15. Walton, Travis. Fire In The Sky: The Walton Experience. Arizona: Skyfire Productions (2010).
16. Weiner, Tim. Blank Check: The Pentagon's Black Budget. New York: Warner Books (1990).
17. Friedman, Stanton T. & Berliner, Don. Crash At Corona. New York: Paragon House (1992); Pp. 99-100.
18. Ibid.; p.101.
19. Friedman, Stanton T. TOP SECRET/MAJIC. New York: Marlowe & Company (1996).
20. Ibid.; Pp. 20-21.
21. Ibid.; Appendix A, p. 225.
22. Ibid.; Pp. 42-55.
23. Ibid.; Pp. 38-39.
24. Ibid.; Pp. 167-184.
25. Lindemann, Michael. UFOs And The Alien Presence: Six Viewpoints. Santa Barbara: The 2020 Group (1991); Pp. 80-81.
26. Ibid.; p.26.

V. INTEGRATION: ASSEMBLING THE PUZZLE PIECES

1. Marrs, Jim. Alien Agenda: The Untold Story Of The Extraterrestrials Among Us. London: HarperCollins Publishers (1997); Pp. 300-303.
2. Ibid.; Pp. 305-308.
3. Ibid.; Pp. 315-316.
4. Ibid.; p. 313.
5. Ibid.; Pp. 322-323.
6. Ibid.; p. 19.
7. Ibid.
8. Sci-Fi Channel. Out Of The Blue (2002); narrated by Peter Coyote.
9. Marrs, Jim. Alien Agenda: The Untold Story Of The Extraterrestrials Among Us. London: HarperCollins Publishers (1997); p. 20.
10. Ibid.
11. Ibid.; p. 24.
12. Ibid.
13. Ibid.

BIBLIOGRAPHY

14. Ibid. p. 25.
15. Sci-Fi Channel. Out Of The Blue (2002); narrated by Peter Coyote.
16. Marrs, Jim. Alien Agenda: The Untold Story Of The Extraterrestrials Among Us. London: HarperCollins Publishers (1997); Pp. 340-342.
17. Ibid.; Pp. 342-344.
18. Blum, Ralph and Judy. Beyond Earth: Man's Contact With UFOs. New York: Bantam Books (1974); p. 117.
19. Sci-Fi Channel. Out Of The Blue (2002); narrated by Peter Coyote.
20. Price, Richard. ARIZONA UFO UPDATE; USA TODAY, 6/18/97. Located online at: http://www.nhne.com/newsbriefs/nhnenb67.html.
21. Davenport, Peter B. 2nd Anniversary of 'Phoenix Lights Incident.' ufoevidence.org; March 13, 1999. Located online at: http://www.ufoevidence.org/documents/doc1266.htm.
22. Sci-Fi Channel. Out Of The Blue (2002); narrated by Peter Coyote.
23. Dolan, Richard M. UFOs and the National Security State. Charlottesville: Hampton Roads Publishing Company, Inc. (2002); p. xi.
24. Ibid.; Pp. 42-43.
25. Sci-Fi Channel. Out Of The Blue (2002); narrated by Peter Coyote.
26. Ibid.
27. Strieber, Whitley. Majestic. New York: The Berkley Publishing Group (1990); Pp. 6-7, 250-252, 301.
28. Dolan, Richard M. UFOs and the National Security State. Charlottesville: Hampton Roads Publishing Company, Inc. (2002); Pp. 391-393.
29. Ibid.; Pp. 390-391.
30. Kean, Leslie. UFOs: Generals, Pilots and Government Officials Go on the Record. New York: Three Rivers Press, Inc. (2010). Pp.104-106.
31. Ibid.; p. 107.
32. Dolan, Richard M. UFOs and the National Security State. Charlottesville: Hampton Roads Publishing Company, Inc. (2002); p. 313.
33. Ibid.; p. 329-330.
34. TIME, August 4, 1967; p. 33. A Fresh Look At Flying Saucers.
35. Kean, Leslie. UFOs: Generals, Pilots and Government Officials Go on the Record. New York: Three Rivers Press, Inc. (2010). Pp. 145-146.
36. Ibid.; Pp. 65-72.
37. Dolan, Richard M. & Zabel, Bryce. A.D. After Disclosure: When The Government Finally Reveals The Truth About Alien Contact. New Jersey: Career Press/New Page Books (2012). Pp. 188-189.

BIBLIOGRAPHY

38. Fawcett, Lawrence & Greenwood, Barry J. Clear Intent: The Government Coverup Of The UFO Experience. New York: Prentice Hall Press (1984). Pp. 65-66.
39. Ibid.; Pp. 66-68.
40. Greer, Steven M. M.D. DISCLOSURE: Military And Government Witnesses Reveal The Greatest Secrets In Modern History. Crozet, Virginia: Crossing Point Publications (2001); Pp. 274-277.
41. Wilber, Ked (Ed.) The Holographic Paradigm And Other Paradoxes: Exploring The Leading Edge Of Science. Boulder: Shambhala Publications (1982); Pp. 44-57.
42. McTaggart, Lynne. The Field: The Quest For The Secret Force Of The Universe. New York: HarperCollins (2008).
43. Greer, Steven M. M.D. DISCLOSURE: Military And Government Witnesses Reveal The Greatest Secrets In Modern History. Crozet, Virginia: Crossing Point Publications (2001); Pp. 505-506.
44. Dolan, Richard M. UFOs and the National Security State, Volume Two: The Cover-Up Exposed. New York: Keyhole Publishing Company (2009);Pp. 458-460.
45. Greer, Steven M. M.D. DISCLOSURE: Military And Government Witnesses Reveal The Greatest Secrets In Modern History. Crozet, Virginia: Crossing Point Publications (2001); Pp. 357-364.
46. Kasten, Len. The Secret History Of Extraterrestrials: Advanced Technology And The Coming New Race. Rochester, Vermont: Bear & Company (2010); Pp. 1-5.
47. Harbinson, W.A. Genesis. New York: Dell (1980). Author's Note & Sources; Pp. 587-605.
48. Cook, Nick. The Hunt For Zero Point: Inside The Classified World of Antigravity Technology. New York: Broadway Books (2001); Pp. 205-213.
49. Kasten, Len. The Secret History Of Extraterrestrials: Advanced Technology And The Coming New Race. Rochester, Vermont: Bear & Company (2010); Pp. 161-167.
50. Greer, Steven M. M.D. DISCLOSURE: Military And Government Witnesses Reveal The Greatest Secrets In Modern History. Crozet, Virginia: Crossing Point Publications (2001); Pp. 419-421.
51. The Guardian (theguardian.com): Gary McKinnon timeline: events leading up to extradition decision; located at: http://www.theguardian.com/world/2012/oct/16/gary-mckinnon-time line-extradition
52. Project Camelot: Kerry Cassidy & Gary McKinnon (2006): Transcript http://projectcamelot.org/lang/en/gary_mckinnon_interview_transcript_ en.html
53. Kasten, Len. The Secret History Of Extraterrestrials: Advanced

BIBLIOGRAPHY

Technology And The Coming New Race. Rochester, Vermont: Bear & Company (2010); Pp. 76-78.

54. Ibid.; p. 10.

55. De Chardin, Pierre Teilhard. Human Energy. New York: Harcourt Brace Jovanovich, Inc. (1962); p. 34.

56. Kasten, Len. The Secret History Of Extraterrestrials: Advanced Technology And The Coming New Race. Rochester, Vermont: Bear & Company (2010); Pp. 147-148.

57. MacLaine, Shirley. Out On A Limb. New York: Bantam Books (1983). Pp. 306-307; 319-320; 328-331; 342.

58. Ibid.; Pp. 330-331.

59. Mack, John E. M.D. Passport To The Cosmos: Human Transformation and Alien Encounters. Commemorative Edition. Guildford, UK: White Crow Books (1999, 2011). p.7.

60. Ibid.; Pp. 215-216.

61. Ibid.; p. 13.

62. Howe, Linda Moulton. Glimpses Of Other Realities, Volume II: High Strangeness. Cheyenne, Wyoming: Pioneer Printing (1998); Pp. 2, 10, 14-21, 58, 63.

63. Ibid.; p. 131.

64. Ibid.; p. 136.

65. Ibid.; p. 141.

66. Dolan, Richard M. UFOs And The National Security State, Volume Two: The Cover-Up Exposed. New York: Keyhole Publishing Company (2009); p. 551.

67. Howe, Linda Moulton. Glimpses Of Other Realities, Volume II: High Strangeness. Cheyenne, Wyoming: Pioneer Printing (1998); Pp. 1, 133.

68. Strieber, Whitley. Hybrids. New York: Walker & Collier/Tor Book (2011); Pp. 102-103.

69. Harbinson, W.A. Genesis. New York: Dell (1980). p. 178.

70. Dolan, Richard M. UFOs and the National Security State. Charlottesville: Hampton Roads Publishing Company, Inc. (2002); p. 60.

71. Kaku, Michio. quoted in: How The Universe Works. S1E1: Big Bang. Discovery Channel (2010).

APPENDIX

Research Methods B.A. Thesis: FIU (Florida International University)
Fall 1990
The Presence Of Specific Personality Factors And Their Effect On A Person's
UFO Belief Systems (abstract presented by Stephen W. Vernon)

BIBLIOGRAPHY

1. Vallee, Jacques. Anatomy Of A Phenomenon: Unidentified Objects In Space. NTC Contemporary Publishing (1965).
2. Blum, Ralph and Judy. Beyond Earth: Man's Contact With UFOs. New York: Bantam Books (1974).
3. Hopkins, Budd. Intruders: The Incredible Visitations At Copley Woods. New York: Bantam Books (1987).
4. Kinder, Gary. Light Years: An Investigation Into The Extraterrestrial Experiences Of Eduard Meier. New York: The Atlantic Monthly Press (1987).
5. Strieber, Whitley. Communion. New York: Morrow/Beech Tree Books (1987).
6. Zimmer, Troy A. Social Psychological Correlates of Possible UFO Sightings. The Journal Of Social Psychology, Vol. 123, Second Half, August 1984.
7. Fuller, John G. The Interrupted Journey. New York: Dell (1966).
8. Sprinkle, R. Leo. Psychological Services for Persons Who Claim UFO Experiences. Psychotherapy in Private Practice, Vol. 6(3) 1988.
9. Parnell, J.O. Personality Characteristics on the MMPI, 16PF, and ACL of persons who claim UFO experiences. Dissertation Abstracts International, Vol. 47, No. 7.
10. Little, Gregory L. Educational Level And Primary Beliefs About Unidentified Flying Objects Held By Recognized Ufologists. Psychological Reports (1984); 54, 907-910.
11. Vallee, Jacques. Dimensions: A Casebook Of Alien Contact. New York: Ballantine (1988).
12. Vallee, Jacques. Messengers Of Deception: UFO Contacts and Cults (1979).

Wade Vernon grew up in Texas, Louisiana and Florida, and earned his B.A. in Psychology at F.I.U. in Miami, FL in 1991. That same year he also attained certification as a hypnotherapist at the Omni Hypnosis Training Center in Ft. Lauderdale. Time regression hypnosis is his specialty, as well as creating custom hypnosis cd's for behavior modification.

The son of a scientist, Wade Vernon learned early on that he shared his father's passion for research, and focused on UFOs, having had 2 sightings in the Houston area growing up. **Force Fields** is the culmination of decades of his investigations. Currently, Wade is a hypnotherapist in Austin, TX, where he lives. This is his first published book.

FR

RYAN FUGATE
FINE ART

Ryanfugatefineart.com

www.ingramcontent.com/pod-product-compliance
Lightning Source LLC
Chambersburg PA
CBHW080242030426

42334CB00023BA/2669